ULTRASOUND IN MEDICINE

Ultrasound in Medicine, The Third Mayneord–Phillips Summer School
St Edmund Hall, Oxford, June 1997

Ultrasound in Medicine, The Third Mayneord–Phillips Summer School
St Edmund Hall, Oxford, June 1997

1 Victor Humphrey
2 Elizabeth Moore
3 Barry Ward
4 Matthew Reilly
5 Kat Dixon
6 Jim Greenleaf
7 Osiyah Papayi

8 Jeff Bamber
9 Nadine Pay
10 Jim Williams
11 Anders Olsson
12 Ben Khoo
13 Malcolm Sperrin
14 Elvin Nix

15 Darko Grošev
16 Sandy Mather
17 Tim Spencer
18 Mike Halliwell
19 Phil Burford
20 Meg Warner
21 Tony Whittingham

22 John Truscott
23 Francis Duck
24 Andrzej Jastrzebski
25 Frank Rakebrandt
26 Panagiotis Tsiganos
27 Gareth Price
28 Kit Hill

Medical Science Series

ULTRASOUND IN MEDICINE

Edited by

Francis A Duck

Royal United Hospital, Bath and University of Bath, UK

Andrew C Baker

Christian Michelsen Research AS, Bergen, Norway
formerly University of Bath

Hazel C Starritt

Royal United Hospital, Bath, UK

Based on Invited Lectures presented at the
Third Mayneord–Phillips Summer School 1997
sponsored by
Institute of Physics and Engineering in Medicine
British Institute of Radiology
Institute of Physics
British Medical Ultrasound Society

CRC Press
Taylor & Francis Group
Boca Raton London New York

CRC Press is an imprint of the
Taylor & Francis Group, an **informa** business

CRC Press
Taylor & Francis Group
6000 Broken Sound Parkway NW, Suite 300
Boca Raton, FL 33487-2742

First issued in paperback 2019

© 1998 by Taylor & Francis Group, LLC
CRC Press is an imprint of Taylor & Francis Group, an Informa business

No claim to original U.S. Government works

ISBN-13: 978-0-7503-0593-8 (hbk)
ISBN-13: 978-0-367-40071-2 (pbk)

British Library Cataloguing-in-Publication Data

A catalogue record for this book is available from the British Library.

Library of Congress Cataloging-in-Publication Data are available

Visit the Taylor & Francis Web site at
http://www.taylorandfrancis.com

and the CRC Press Web site at
http://www.crcpress.com

The Medical Science Series is the official book series of the International Federation for Medical and Biological Engineering (IFMBE) and the International Organization for Medical Physics (IOMP).

IFMBE

The IFMBE was established in 1959 to provide medical and biological engineering with an international presence. The Federation has a long history of encouraging and promoting international cooperation and collaboration in the use of technology for improving the health and life quality of man.

The IFMBE is an organization that is mostly an affiliation of national societies. Transnational organizations can also obtain membership. At present there are 42 national members, and one transnational member with a total membership in excess of 15 000. An observer category is provided to give personal status to groups or organizations considering formal affiliation.

Objectives

- To reflect the interests and initiatives of the affiliated organizations.
- To generate and disseminate information of interest to the medical and biological engineering community and international organizations.
- To provide an international forum for the exchange of ideas and concepts.
- To encourage and foster research and application of medical and biological engineering knowledge and techniques in support of life quality and cost-effective health care.
- To stimulate international cooperation and collaboration on medical and biological engineering matters.
- To encourage educational programmes which develop scientific and technical expertise in medical and biological engineering.

Activities

The IFMBE has published the journal *Medical and Biological Engineering and Computing* for over 34 years. A new journal *Cellular Engineering* was established in 1996 in order to stimulate this emerging field in biomedical engineering. In *IFMBE News* members are kept informed of the developments in the Federation. *Clinical Engineering Update* is a publication of our division of Clinical Engineering. The Federation also has a division for Technology Assessment in Health Care.

Every three years, the IFMBE holds a World Congress on Medical Physics and Biomedical Engineering, organized in cooperation with the IOMP and the IUPESM. In addition, annual, milestone, regional conferences are organized in different regions of the world, such as the Asia Pacific, Baltic, Mediterranean, African and South American regions.

The administrative council of the IFMBE meets once or twice a year and is the steering body for the IFMBE. The council is subject to the rulings of the General Assembly which meets every three years.

For further information on the activities of the IFMBE, please contact Jos A E Spaan, Professor of Medical Physics, Academic Medical Centre, University of Amsterdam, PO Box 22660, Meibergdreef 9, 1105 AZ, Amsterdam, The Netherlands. Tel: 31 (0) 20 566 5200. Fax: 31 (0) 20 691 7233. Email: IFMBE@amc.uva.nl. WWW: http://vub.vub.ac.be/~ifmbe.

IOMP

The IOMP was founded in 1963. The membership includes 64 national societies, two international organizations and 12 000 individuals. Membership of IOMP consists of individual members of the Adhering National Organizations. Two other forms of membership are available, namely Affiliated Regional Organization and Corporate Members. The IOMP is administered by a Council, which consists of delegates from each of the Adhering National Organization; regular meetings of Council are held every three years at the International Conference on Medical Physics (ICMP). The Officers of the Council are the President, the Vice-President and the Secretary-General. IOMP committees include: developing countries, education and training; nominating; and publications.

Objectives

• To organize international coopération in medical physics in all its aspects, especially in developing countries.
• To encourage and advise on the formation of national organizations of medical physics in those countries which lack such organizations.

Activities

Official publications of the IOMP are *Physiological Measurement, Physics in Medicine and Biology* and the *Medical Science Series*, all published by Institute of Physics Publishing. The IOMP publishes a bulletin *Medical Physics World* twice a year.

Two Council meetings and one General Assembly are held every three years at the ICMP. The most recent ICMPs were held in Kyoto, Japan (1991), Rio de Janeiro, Brazil (1994) and Nice, France (1997). The next conference is scheduled for Chicago, USA (2000). These conferences are normally held in collaboration with the IFMBE to form the World Congress on Medical Physics and Biomedical Engineering. The IOMP also sponsors occasional international conferences, workshops and courses.

For further information contact: Hans Svensson, PhD, DSc, Professor, Radiation Physics Department, University Hospital, 90185 Umeå, Sweden. Tel: (46) 90 785 3891. Fax: (46) 90 785 1588. Email: Hans.Svensson@radfys.umu.se.

CONTENTS

CONTRIBUTING AUTHORS

Dr Andrew C Baker
Christian Michelsen Research AS
Fantoftvegen 38, Postboks 6031
5020 Bergen
Norway
(Formerly Department of Physics,
University of Bath, UK)
Andrew.Baker@cmr.no

Dr Jeffrey C Bamber
Joint Department of Physics
The Royal Marsden NHS Trust
Downs Road
Sutton
Surrey SM2 5PT
UK
jeff@icr.ac.uk

Professor David O Cosgrove
Department of Radiology
Hammersmith Hospital
Du Cane Road
London W12 0NN
UK
dcosgrov@rpms.ac.uk

Dr Francis A Duck
Medical Physics Department
Royal United Hospital
Combe Park
Bath BA1 3NG
UK
f.duck@bath.ac.uk

Professor James F Greenleaf
Biodynamics Research Unit
Department of Physiology and
    Biophysics
Mayo Foundation
Rochester MN 55905
USA
jfg@mayo.edu

Dr Michael Halliwell
Medical Physics and
    Bioengineering
Bristol General Hospital
Guinea Street
Bristol BS1 6SY
UK
mike.halliwell@bris.ac.uk

Dr Jeffrey W Hand
Radiological Sciences Unit
Department of Imaging
Hammersmith Hospital
Du Cane Road
London W12 0NN
UK
jhand@rpms.ac.uk

Professor Christopher R Hill
Stoney Bridge House
Castle Hill
Axminster
Devon EX13 5RL
UK
(Formerly Physics Department,
Royal Marsden Hospital, UK)

Dr Victor F Humphrey
Department of Physics
University of Bath
Claverton Down
Bath BA2 7AY
UK
v.f.humphrey@bath.ac.uk

Dr Timothy G Leighton
Fluid Dynamics and Acoustics
    Group
Institute of Sound and Vibration
    Research
University of Southampton
Highfield
Southampton SO17 1BJ
UK
tgl@isvr.soton.ac.uk

Dr Gareth J Price
Department of Chemistry
University of Bath
Claverton Down
Bath BA2 7AY
UK
g.j.price@bath.ac.uk

Dr Hazel C Starritt
Medical Physics Department
Royal United Hospital
Combe Park
Bath BA1 3NG
UK
h.c.a.starritt@bath.ac.uk

Dr Thomas L Szabo
Hewlett Packard
Imaging Systems Division
3000 Minuteman Road
Andover, MA 1810-1099
USA
tom.szabo@
hp.andover.om3.om.hp.com

Dr Gail ter Haar
Joint Department of Physics
The Royal Marsden NHS Trust
Downs Road
Sutton
Surrey SM2 5PT
UK
gail@icr.ac.uk

Dr John G Truscott
Centre for Bone and Body
    Composition Research
Institute of Physical Science
Department of Clinical Medicine
Wellcome Wing
Leeds LS1 3EX
UK
j.g.truscott@leeds.ac.uk

Professor Peter N T Wells
Medical Physics and
    Bioengineering
Bristol General Hospital
Guinea Street
Bristol BS1 6SY
UK
peter.wells@bris.ac.uk

Dr Tony Whittingham
Regional Medical Physics
    Department
Newcastle General Hospital
Westgate Road
Newcastle-Upon-Tyne
NE4 6BE
UK

GLOSSARY

The following summary includes substantially all the defined symbols and acronyms which have been used throughout the book. Wherever possible the same symbol has been used uniquely for a particular quantity. On the few occasions when this has proved to be impossible, local use is identified in the text. Conversely when, rarely, it has been necessary to use the same symbol in different chapters for different quantities, this is noted in the list. On occasions subscripts have been used in the text to define the material (w for water, t for tissue and so on). This level of detail is excluded from the list of symbols. The list also includes several symbols which have been used arbitrarily as constants within particular equations.

a, b	constants in equation for bubble behaviour
A	backscattered signal amplitude
	radiating area
$A(f)$	amplitude spectrum
ACF	autocorrelation function
A_L	acoustic loss factor
B/A	nonlinearity parameter
B	constant relating temperature rise to time
BUA	broadband ultrasonic attenuation
c	acoustic wave velocity
c_{cc}	mean pulse velocity
c^D	elastic stiffness constant (clamped)
c_g	acoustic group velocity
c_p	acoustic phase velocity
C_0	capacitance (clamped)
d	beam diameter
	piezoelectric element thickness
d_3	-3 dB focal zone width
D	radiation force (drag) coefficient
	effective beam diameter
	dielectric displacement
$\mathrm{DIFF}(z)$	diffraction correction factor
E_0	energy density
$\langle E \rangle$	time-averaged energy density

E_p	pulse energy
E_L	electric loss factor
f	acoustic frequency
f_D	Doppler shifted frequency
f_{opt}	optimal frequency for hyperthemia
f_r	resonant frequency (bubble)
F	radiation force
FEM	finite element modelling
FUS	focused ultrasound surgery
g	gravitational acceleration
$g(x, y)$	amplitude beam profile
G	geometric factor for acoustic streaming
	amplitude focusing gain
$G(t)$	magnetic field gradient
h	height of liquid column
	depth of spherical transducer surface
	a piezo constant
$h(x, y, z)$	point spread function
H	forward propagation operator
H^{*t}	backward propagation operator
i	$\sqrt{-1}$
I	electrical current
	acoustic intensity
$I(z)$	axial intensity distribution
I, I_{ta}	time-averaged intensity
I_{pa}	pulse-averaged intensity
I_{sp}	spatial peak intensity
I_{sppa}	spatial peak pulse-averaged intensity
I_{spta}	spatial peak time-averaged intensity
$I_{.3}$	'de-rated' intensity
I_0	source intensity
I_R	intensity at the centre of curvature of a spherical source
k	wave number
	thermal conductivity
k_{eff}	effective thermal conductivity
k_T	electromechanical coupling coefficient
K	constant relating to thermal equilibrium
$\langle K \rangle$	time-averaged kinetic energy density
l_d	discontinuity length
l_v	length of vessel
l_3	-3 dB focal zone length
L	perfusion length
m	mass
MI	mechanical index

MR	magnetic resonance
MRE	magnetic resonance elastography
ODS	Output Display Standard
p	acoustic pressure
p	complex pressure
p_0	acoustic pressure amplitude at source or for plane wave
p_c	acoustic pressure at peak compression
p_r	acoustic pressure at peak rarefaction
p_f	acoustic pressure at the focus
p_n	acoustic pressure amplitude at harmonic n
p_{opt}	least peak rarefaction pressure causing inertial cavitation
$p()$	probability density
prf	pulse repetition frequency
P	excess pressure
P^E	excess pressure (Eulerian co-ordinates)
P^L	excess pressure (Lagrangian co-ordinates)
P_i	generalised property value of tissue i
P_{Lan}	Langevin radiation pressure
P_{Ray}	Rayleigh radiation pressure
PII	pulse intensity integral
PPSI	pulse pressure squared integral
PSF	point spread function
PVDF	polyvinylidene fluoride
PZT	lead zirconate titanate
Q	heat flux
	Q factor at resonance
Q_{BF}	heat term to account for blood flow
Q_e	electrical Q
r	vessel radius
r_a	radius of a circular source
r_z	radial distance at depth z
$r(t)$	postion vector of the moving spin
R	reflection coefficient
	radius of curvature
R_A	radiation resistance
R_{opt}	critical bubble radius for inertial cavitation
R_0	bubble radius
RF	radio frequency
s	specific heat capacity
$s(\Delta t)$	cross-correlation function
S	area
	speckle cell size
S, S_{ij}	radiation stress (tensor)
S_i	generalised signal value associated with tissue i

SAR	specific absorption rate
SNR	signal-to-noise ratio
t	time
t_p	pulse period
T	temperature
	strain
$T(x, y, z)$	tissue backscatter impulse response
TI	thermal index
TIB	bone-at-focus thermal index
TIC	cranial thermal index
TIS	soft tissue thermal index
TOA	time-of-arrival
TOF	time-of-flight
u	particle velocity
u_0	particle velocity amplitude (sine wave)
$\boldsymbol{u}$	complex particle velocity
USAE	ultrasound stimulated acoustic emission
v	streaming velocity
	wave velocity of piezoelectric material
$\boldsymbol{v}$	vector velocity of source or observer
V	volume
	voltage
$\langle V \rangle$	time-averaged potential energy density
w	perfusion volume flow rate
W, W_A	acoustic power
W_E	electrical power
$W_{.3}$	'de-rated' acoustic power
W_1	acoustic power from a 1 cm length of array
W_V	absorbed power per unit volume
x, y	dimensions orthogonal to the beam axis
X	reactance
X_A	radiation reactance
X_{eq}	vessel thermal equilibrium length
X_6	-6 dB beam width
z	dimension parallel with acoustic axis
z_f	focal distance
Z, Z_A	acoustic impedance
$Z(f)$	Fourier transform
$Z(f)^*$	complex conjugate Fourier transform
Z_T	electrical impedance
α	amplitude attenuation coefficient
α_a	amplitude absorption coefficient
α_{a0}	amplitude absorption coefficient at 1 MHz
α_s	amplitude scattering coefficient

β	nonlinearity parameter
δ_{ij}	Kronecker delta
ε	acoustic Mach number
ε^S	dielectric constant (clamped)
γ	gyromagnetic ratio
Γ	Gol'dberg number
κ, κ_0	adiabatic bulk compressibility
λ	wavelength
Λ	constant associated with thermal contact between vessel and tissue
μ	intensity attenuation coefficient
μ_a	intensity absorption coefficient
μ_s	intensity scattering coefficient
μ_{bs}	backscattering coefficient
μ_{ds}	differential scattering coefficient
ν	kinematic viscosity
φ	angle of refraction
	phase offset
$\varphi(f)$	phase spectrum
$\Phi(\tau)$	transverse magnetisation phase
η	shear viscosity
ρ, ρ_0	equilibrium density
σ	standard variation
	shock parameter
σ_m	nonlinear propagation parameter
$\sigma(\theta, \varphi)$	differential scattering cross-section
τ	time shift
ξ_0	peak displacement of spin
ψ	beam profile phase function
Ψ	tissue orientation angle
ω	angular frequency
Ω	solid angle

INTRODUCTION

Francis A Duck

The weather of the week of 7 June 1997 was almost perfect in Oxford. The venue for the Third Mayneord–Phillips Summer School 'Ultrasound in Medicine' had been chosen to be St Edmund Hall, Oxford only on precedent: the two earlier Medical Physics Summer Schools had been successfully held there. In the event, the week turned out to be a very special occasion for the forty or so lecturers and students who attended; unique and memorable (frontispiece). Apart from the loss of Barry's glasses during the punting expedition, the memories of the week remain very positive, a week of learning and companionship, and new and renewed friendships.

This book is one outcome of that Summer School. All the lecturers who contributed to the School have prepared chapters, each based around the topic of their own lecture. In a number of cases the chapter has been limited to the material contained within the lecture in Oxford, while other authors have extended their material to include details more easily presented in a printed form. More details of the book's content and structure are described below. Initially, however, a short background to the Mayneord–Phillips Memorial Trust will be given, since without its establishment, this Summer School, and this book, would not have been created.

The Mayneord–Phillips Memorial Trust was established in 1994, the first Trustees representing the three founding bodies: the British Institute of Radiology, the Institute of Physics and the Institute of Physical Sciences in Medicine (now the Institute of Physics and Engineering in Medicine). One of the original Trustees, Professor Kit Hill, is a welcome contributor to this book. The Trust deed identified one objective as 'the furthering, for the benefit of the public, the knowledge and understanding of all aspects and all applications of medical physics and kindred sciences ... by the organisation of educational meetings to be called the Mayneord–Phillips Summer Schools'. In addition the Trustees should 'arrange for the publication either in full or in part of any such Schools'. The Trustees decided that the Third School should take the topic 'Ultrasound in Medicine' and to use the School and subsequent publication to explore a broad-ranging study of medical ultrasound, including ultrasound propagation, interaction with tissue, and an exploration of a number of contemporary innovations in the application of ultrasound in medicine. Given this background, it is clear that the content of both School and publication was and is rather narrower than the title might imply. The focus is specifically on the science and technology, the physics and the engineering, rather than on the clinical applications. This is not to say that clinical applications are absent, since it is the nature of applying physics to medicine that the link between scientific and engineering development and clinical application must be firmly made. Nevertheless the emphasis always remains thus: to draw from the basic sciences those aspects which relate most closely to the challenge of applying this science to a particular clinical need: and to review against the clinical need how successful technological innovation has been in using natural science to improve medical diagnosis and treatment.

W V Mayneord and C E S Phillips were two outstanding pioneers of the applications of physics to medicine. In their nature as pioneers they both had a strong concern to help and encourage younger colleagues to develop their own interests and expertise. The following paragraphs briefly summarise their lives and contributions to medical physics. Further details may be found elsewhere [1, 2].

Major Phillips has been described as the first British medical physicist. Born in 1871, his early experiments with discharge tubes led to his description of the 'Phillips' phenomenon', the rotation of a luminous ring in an electrical discharge tube within a static magnetic field. In 1897, he published a complete bibliography of X-ray literature, probably the last occasion when this was a possibility. His work on radiation standards during the first decade of the twentieth century led him to be commissioned to prepare three radium standards for the Roentgen Society. He became the physicist to the X-Ray Committee of the War Office during the 1914–18 war. He worked with the radiologist Robert Knox as honorary physicist at the Cancer Hospital (now the Royal Marsden Hospital) up to his retirement in 1927, during which time he helped to develop the scientific basis of radiotherapy, handling radioactive materials, and radiation protection.

Val Mayneord was 22 years old when he gained his first job as a medical physicist at St Bartholomew's Hospital in 1924. Very soon after moving to the Cancer Hospital on Phillips' retirement, Mayneord started making major contributions to radiation dosimetry. Unlike Phillips, who published little, Mayneord was a prolific writer. His first book [3], published before his 30th birthday, remains one of the clearest early discussions of the scientific realities of medical radiation therapy and protection. His year's secondment to Canada by the UK government after the war only fired his enthusiasm on his return for the application of physics to a wide range of medical problems. Perhaps even more important than his own personal scientific achievements was his building up of a department of physics applied to medicine at the

Royal Marsden Hospital, which earned itself an international reputation for work not only in radiotherapy, but also in nuclear medicine, diagnostic ultrasound and several other areas of medical physics. He achieved this by understanding that good medicine must be based on good science, and that good physical scientists need a strong and stimulating environment in which to thrive. This is as true now as it was then.

Ultrasound has been a late starter in its application to medicine. Even during the period of vigorous growth in applying physics to medical problems which Val Mayneord and his contemporaries experienced following the war, ultrasound still had a somewhat secondary place to the innovations in nuclear medicine, diagnostic radiology and radiotherapy. Interestingly, acoustics, the study of the science of sound waves, and the knowledge of piezoelectricity both substantially predate the discovery of X-rays and of radioactivity during the last decade of the nineteenth century. Perhaps it was the astonishing breadth of Lord Rayleigh's book 'The Theory of Sound' [4], first published in 1877, which discouraged others from attempting any deeper study. Maybe the dramatic overturn of classical theories of physics at the turn of the century caused acoustics to become a poor relation in physics. Or maybe it required an appropriate practical objective to draw together acoustic science and transducer technology towards the exploitation which characterises medical ultrasound at the end of the twentieth century.

The Curies rediscovered piezoelectric phenomena in crystals following Becquerel's work earlier in the nineteenth century [5], which was itself based on work by Hauy in the late 1700s. The Curies' work seems to have generated interest mostly among scientists rather than practical people (both Roentgen and Kelvin showed active interest in the phenomenon). Nevertheless, the only practical outcome of their observations of the reverse piezoelectric effect of quartz [6] seems to be the 'Quartz piézo-électrique' [7] which was used so effectively by Marie and Pierre Curie in their careful experimental studies of the radioactivity of radium. It was left to Langevin, who had previously been a student of the Curies, to exploit the strong piezoelectric properties of quartz as a resonant electro-acoustic transducer for underwater echo-location for depth sounding and submarine detection [8]. While acoustic depth sounding had been suggested in the early part of the nineteenth century, it was Langevin's work during the First World War which established two key elements for its success. Firstly, he recognised the compromise required in the selection of the optimum frequency, balancing resolution against penetration. This led him to identify the potential advantage of using frequencies above the audible limit (about 20 kHz) for underwater echo-location. Shortly after the *Titanic* disaster, Lewis Fry Richardson had taken out a patent for the use of 100 kHz sound for the same purpose [9], but the device seems never to have been implemented. Richardson is better known for his book on the mathematical prediction of weather. Langevin realised however that to make such a device work, greater sensitivity was required. His second innovation was to exploit the electronic methods already available for radio communication to develop resonant modes of transmission and reception, so substantially enhancing the output power, and the detection sensitivity of his quartz transducers. (As a footnote, Langevin also describes clearly the measurement of acoustic power using a radiation force method.) It lies outside the scope of this introduction to trace further the development from this early use of ultrasonic echo detection to the extreme sophistication of modern medical diagnostic systems. Much of the early medical work, during the 1940s and 1950s, has been well described elsewhere [10, 11].

In the remaining paragraphs a brief overview will be given of the chapters contained in the book, and of the way in which they relate to one another.

Ultrasound is rapidly becoming the imaging method of choice for much of diagnostic medicine, and in some specialist areas it has all but replaced other diagnostic methods. It has been estimated that a quarter of all medical imaging studies world-wide is now an ultrasound study [12]. This is supported by UK Department of Health data which suggest that in NHS Trusts in England alone, over 4 million ultrasound imaging studies were carried out in one year, of which about 1.7 million were obstetric scans (1996/97 figures). Even this huge number excludes the studies carried out in primary care (GP practices) and in the private sector. It represents three scans for each live birth. This astonishing growth has been encouraged in part because ultrasound is perceived, it must be said with good reason, to be a diagnostic method with no risk overhead. The assertion that medical ultrasound is safe has become a truism, and many of the developments in diagnostic ultrasound, especially those using Doppler methods and described by Peter Wells in Chapter 6, were made possible because of this view. This was true in spite of the considerable increases in intensity and acoustic power required for some applications, and these are reviewed by Tony Whittingham (Chapter 7). Undoubtedly many of the advances have come about because of the development of techniques in array technology which are described by Tom Szabo in Chapter 5. These have allowed parallel operation in Doppler and pulse-echo modes so as to fully exploit the ability of ultrasound to image both structure (through pulse-echo scanning) and blood perfusion (through Doppler). Miniaturisation of arrays now gives access to deep structures by array insertion into the rectum, oesophagus and vagina, and even through the vascular tree as far as the cardiac arteries. The development of very high frequency miniature probes is an exciting development area which, sadly, is one activity not covered here. Vascular ultrasound is becoming further enhanced by developments in echo-enhancing contrast materials. David Cosgrove introduces this rapidly growing clinical area in Chapter 12.

By comparison, therapeutic and surgical uses of ultrasound, some of which predated the diagnostic methods [13], have developed in parallel but have failed thus far to exert the impact on medicine that was promised from the early work. Perhaps this has been in part because insufficient emphasis has been placed on the proper scientific development of these methods, including a full recognition of the care needed in acoustic design, and in establishing a more complete understanding of the interaction between ultrasonic waves and tissue. From recent successes in hyperthermia and focused surgery, discussed fully by Jeff Hand in Chapter 8, and Gail ter Haar in Chapter 9, and also the use of a variety of ultrasonic approaches to lithotripsy described by Michael Halliwell in Chapter 10, it may be confidently predicted that ultrasound will indeed find a valuable place in the surgical and therapeutic armoury of the future, perhaps comparable with that achieved already by diagnostic ultrasound.

However, strong and successful applications only develop from a strong understanding of the basic science. It is surprising how often insufficient attention is placed in standard medical ultrasound texts on the difficulties in describing fully the propagation of diagnostic ultrasound pulses through tissue. Even the description of pulse propagation in a focused near-field under linear conditions in a loss-less fluid poses some difficulties, and these are described carefully in Chapter 1 by Victor Humphrey. The reality of finite-amplitude effects and nonlinear acoustic propagation is now known to be not an esoteric side issue but central to all meaningful discussions of medical ultrasound. Andy Baker introduces some aspects of this difficult topic in Chapter 2, while Francis Duck's subsequent chapter describes two practical nonlinear phenomena, acoustic pressure and acoustic streaming. Chapter 4 by Jeff Bamber gives a complete overview of a subject which is central to all discussions of ultrasound in medicine, the acoustic properties of tissue. Attenuation, absorption, scattering, sound speed, nonlinear parameter, and their frequency dependencies are all described.

An understanding of biophysical processes is important in the interpretation of tissue/ultrasound interactions, both for an understanding of ultrasound dosimetry in therapy and surgery, and in safety discussions. Thermal processes have already been noted (Chapter 8) as has streaming (Chapter 3). The third process is acoustic cavitation, which is described by Tim Leighton in Chapter 11. Often acoustic cavitation seems to be a topic on the boundaries of interest in medical applications, but the use of contrast materials (Chapter 12) has brought a new interest in the topic, in addition to its importance in ultrasound surgery and in safety. Cavitation is however central to the use of ultrasound in chemical processing, and Gareth Price describes how this scientific cousin may instruct and illuminate applications in medicine, for example in drug delivery (Chapter 13).

The purpose of the Summer School was not only to review topics which are part of present clinical practice, but also to allow an exploration of some research topics where new developments are active. It is rarely possible to separate fully state-of-the-art applications from new developments, but the last three chapters in the book each represent in one way or another new steps forward. Linking the exciting capabilities for imaging presented by magnetic resonance imaging, Jim Greenleaf and his colleagues describe methods of great originality for studying the mechanical properties of tissue, which have fascinating potential in diagnostic medicine. Kit Hill returns to a fundamental issue in diagnostic ultrasound imaging, signal-to-noise ratio. John Truscott reviews critically some of the methods currently being used to investigate bone using ultrasound, and suggests alternative methods which may have the potential of improved precision.

All the chapters in this book have been prepared with a view to bridging the gap between the tutorial texts widely available for sonographer and medical training, and books of acoustics which contain few links between theoretical acoustics and the applications of ultrasound to medicine. Some material which is very well addressed in standard medical ultrasound texts has been deliberately omitted; for example the description of basic pulse-echo and imaging methods, quality assurance tests using phantoms, artefact generation and avoidance and so on. Readers are directed towards one of many texts which now include this material. The present book is offered as a unique collection of chapters containing well-referenced material of direct relevance to any student wishing to explore medical ultrasound at depth. Where space limited the scope of any chapter, ample referencing will allow the serious student to discover a much wider base of knowledge. It is hoped that these pages, which have been prepared with the spirit of Mayneord and Phillips in mind, will serve to illuminate and instruct any who wish to learn at greater depth of the science and technology in the application of ultrasound to medicine.

ACKNOWLEDGMENTS

I would like to add my heartfelt thanks to all the lecturers and authors who were cajoled into taking part in this enterprise, more or less willingly. I also acknowledge the enormous support given by Andy Baker and Hazel Starritt, co-organisers of the Summer School, and co-editors of this book, whose warm and steady support was essential in the success of both projects. And finally we would like to dedicate the book to all those who attended the Summer School who, not knowing what they were letting themselves in for, enjoyed it anyway.

REFERENCES

[1]   Hill C R and Webb S 1993 *The Mayneord–Phillips Summer Schools: Background to the Schools and Short Profiles of the Two Pioneering Physicists* (Sutton and London: Institute of Cancer Research and Royal Marsden Hospital)

[2]   Spiers F W 1991 William Valentine Mayneord *Biographical Memoirs of the Royal Society* **37** 341–64

[3]   Mayneord W V 1929 *The Physics of X-Ray Therapy* (London: Churchill)

[4]   Rayleigh, Baron: Strutt J W 1877 *The Theory of Sound* (London: Macmillan)

[5]   Becquerel A-C 1823 Expériences sur le développement de l'électricité par la pression; lois de ce développement *Annales de Chimie et de Physique* **22** 5–34

[6]   Curie J and Curie P 1881 Contractions et dilatations produites par des tensions électriques dans les cristaux hémiedres a faces inclinées *Compte Rendu Acad. Sci. Paris* **93** 1137–40

[7]   Curie J and Curie P 1893 Quartz piézo-électrique *Phil. Mag. (5th Ser.)* **36** 340–2

[8]   Langevin P 1924 The employment of ultra-sonic waves for echo sounding *Hydrographic Rev.* **II** No 1, Nov 1924, 57–91

[9]   Richardson L F 1912 Apparatus for warning a ship at sea of its nearness to large objects wholly or partly under water *UK patent* 11125

[10] White D N 1976 Historical survey *Ultrasound in Medical Diagnosis* ed D White (Kingston: Ultramedison) pp 1–36

[11] Levi S 1997 The history of ultrasound in gynaecology 1950–1980 *Ultrasound Med. Biol.* **23** 481–552

[12] WFUMB 1997 *WFUMB News* **4** (2); *Ultrasound Med. Biol.* **23** following p 974

[13] Bergmann L 1938 *Ultrasonics and Their Scientific and Technical Applications* (New York: Wiley)

PART 1

THE PHYSICS OF MEDICAL ULTRASOUND

CHAPTER 1

ULTRASONIC FIELDS: STRUCTURE AND PREDICTION

Victor F Humphrey and Francis A Duck

INTRODUCTION

Pulsed ultrasound beams such as those used daily during medical examinations have acoustic structures of considerable complexity. This is true even when considering their propagation simply through an idealised acoustically uniform medium with no loss. Propagation through the acoustic inhomogeneities of body tissues results in further alterations in the acoustic field, both from scattering at small scale and from large scale interface effects. The purpose of this chapter is to discuss the factors which control the acoustic beams used for medical applications, and to describe these beams and the methods for their prediction. The propagation models used will be limited to those where the beam is assumed to be of sufficiently small amplitude that linear assumptions may be made about the acoustic wave propagation. While this is an invalid assumption for very many medical ultrasound beams in practice, it allows instructive analyses to be developed. Some of the beam characteristics which arise due to finite-amplitude effects are described in Chapter 2. The second broad limitation in this chapter is that consideration is limited to a loss-less liquid medium which is acoustically homogeneous. As will be seen, the use of these two assumptions allows the propagation of the ultrasound wave to be described in terms of only two acoustic quantities, the wave number and the acoustic impedance, together with information about the source geometry. Consideration of the source of the ultrasound wave, the transducer, is given in Chapter 5.

For the majority of diagnostic applications of ultrasound the range of frequencies used is quite narrow, 2–10 MHz, where the particular frequency is selected in order to achieve the best compromise between spatial resolution and depth of penetration. Higher frequencies are only used in specialised applications such as ophthalmology, skin imaging and intravascular investigations, reaching experimentally as high as 100 MHz or more. Frequencies below 2 MHz are used in Doppler systems for fetal heart monitoring, and the lower part of the ultrasonic spectrum is also used for therapeutic and surgical applications such as lithotripsy (Chapter 10) or hyperthermia and focused ultrasound surgery (Chapters 8 and 9). Sonochemistry (Chapter 13) and some bio-effects studies use ultrasonic frequencies below 100 kHz; this chapter will not consider the particular issues associated with beams using such long wavelengths.

Having established the simplifying assumptions, several complicating factors of particular relevance in medical applications of ultrasound are introduced. It is recognised that it is pulsed rather than continuous wave ultrasound beams which are of most interest. In order to achieve good spatial resolution, the ultrasound pulses are less than 1 mm in length. The velocity of sound through soft tissues, c_t, lies in the range approximately 1450–1600 m s^{-1} (see Chapter 4), so the range of acoustic wavelengths, λ (= c_t/f), is about 0.15 to 0.75 mm. The pulses themselves are thus commonly less than 1 μs long and consist of very few acoustic cycles: they are all end and no middle. This fact results in significant differences between the beam profiles in such pulsed beams and those at a single frequency from the same source. A second important practical consideration is that medical ultrasound practice usually couples the transducer directly to the tissue to be investigated. This means that signals are returned from the 'near field' of the transducer, in a region which may be strongly influenced by the size and shape of the transducer itself. The analysis of the near field is therefore of significance.

Symmetry is of considerable importance in the structure of acoustic fields. The analysis of beams with circular symmetry has been well developed in the literature, and this will be described below. However the majority of modern scanners do not use circular sources of ultrasound, linear, curvilinear and 'phased' arrays being almost universally used. For this reason, the analysis of rectangular sources is very important. The final significant factor is that all practical medical ultrasound transducers vary in both amplitude and phase over their aperture: that is they are apodised and they are focused.

Recognising these complexities, this chapter will commence with a simple description of the beam from a circular plane piston source of ultrasound, and proceed to describe the way in which each of the more complex features which are relevant to the description of the structure of medical ultrasound beams have been addressed.

1.1. CIRCULAR PLANE SOURCES

It is common first to consider the acoustic field generated by a plane circular 'piston' source which is vibrating with a sinusoidal motion only in a direction

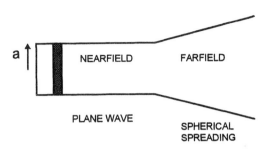

Figure 1.1. *A simplified axial section of the acoustic field from a plane circular single-frequency source whose diameter is significantly greater than the acoustic wavelength in the propagating medium.*

perpendicular to its surface. The analysis of this field is included in many texts (Wells 1977, Kinsler *et al* 1982). This field has particular characteristics which arise from the very specific temporal and spatial symmetries of the source. In practice, the beams from many physiotherapy transducers approximate in structure to the following description.

The simplest view of the field of a circular transducer would consider the field to be a plane wave of the same diameter as the transducer in the near field (the Fresnel region) and then to be an expanding spherical wave in the far field (Fraunhofer region). This is shown in figure 1.1.

The transition occurs at the Rayleigh distance z_R, where

$$z_R = \frac{\pi a^2}{\lambda} = \frac{ka^2}{2} \tag{1.1}$$

where a is the transducer radius, λ is the wavelength and k ($= 2\pi/\lambda$) is the wave number.

This model is conceptually simple, and useful. It is possible, for example, to calculate the approximate Rayleigh distance for a typical 1 MHz physiotherapy transducer with $a = 12.5$ mm in water to be about 33 cm, demonstrating that for such a transducer, all treatment occurs in the near field. Nevertheless this simple model does not allow for diffraction and interference and further development is required.

Consider a planar, circular, transducer mounted in a rigid baffle (surface) and radiating into a fluid. In order to calculate the field due to the transducer assume that each small element dS of the transducer surface vibrates continuously with the same velocity u normal to the surface (the x, y plane) where

$$u = u_0\, e^{i\omega t}. \tag{1.2}$$

Then each element dS gives rise to a spherical wave contributing an elemental pressure contribution dp at a range r' of

$$dp(r', t) = i\frac{\rho_0 c k}{2\pi r'} u_0\, e^{i(\omega t - kr')}\, dS \tag{1.3}$$

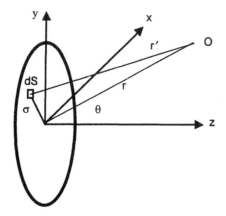

Figure 1.2.   *The geometry for the calculation of the pressure field $p(r, \theta, t)$ at an observer point O, due to a plane circular piston source.*

where ρ_0 is the density of the fluid. The resultant field of the transducer can be evaluated by adding up all of the contributions due to the small elements. In the limit this summation becomes an integral and the resultant pressure field $p(r, \theta, t)$ is given by

$$p(r, \theta, t) = i\frac{\rho_0 c k}{2\pi} u_0 \int_{\text{Surface}} \frac{e^{i(\omega t - kr')}}{r'} \, dS. \tag{1.4}$$

The surface integral is bounded by the condition $\sigma \leqslant a$, where σ is the radial position of the surface element dS (see figure 1.2). The expression in equation (1.4) is often known as the Rayleigh integral.

   In general it is only possible to find simple closed form solutions to this integral for special situations, that is along the symmetry axis of the transducer and in the far field. Otherwise alternative numerical strategies are needed. For example, Stepanishen and Benjamin (1982) and Williams and Maynard (1982) have developed methods for the prediction of acoustic fields using a spatial Fourier transform approach, recognising that the far field beam pattern is the Fourier transform of the aperture function. This approach can be of considerable value in investigating the far field of non-circular sources, such as the rectangular sources discussed in section 1.5. In principle such methods can be fast and accurate, provided that the spatial grid is fine enough to prevent spatial aliasing. For the near field the time-domain numeric methods used by Zemenek (1971) have been very effective in evaluating the diffraction integral.

1.1.1.   Pressure variation on the axis

The axial pressure variation may be derived from equation (1.4). Geometrical considerations give:

$$r' = \sqrt{r^2 + \sigma^2} \tag{1.5}$$

and

$$dS = 2\pi\sigma \, d\sigma \tag{1.6}$$

where $d\sigma$ is the incremental width of a surface annulus of radius σ. Substitution in equation (1.4) gives

$$p(r, 0, t) = \text{const} \int_0^a \frac{\exp(-ik\sqrt{r^2 + \sigma^2})}{\sqrt{r^2 + \sigma^2}} 2\pi\sigma \, d\sigma. \tag{1.7}$$

However

$$\frac{d}{d\sigma}\left(\frac{\exp(-ik\sqrt{r^2 + \sigma^2})}{ik}\right) = \frac{\exp(-ik\sqrt{r^2 + \sigma^2})}{ik}\frac{1}{2}\frac{(-ik)2\sigma}{\sqrt{r^2 + \sigma^2}} \tag{1.8}$$

that is, the integrand is a perfect differential. Substituting and evaluating at $\sigma = a$ and $\sigma = 0$ gives the axial complex pressure $p(r, 0, t)$:

$$p(r, 0, t) = \rho_0 c u_0 \exp(i\omega t)\left[\exp(-ikr) - \exp(-ik\sqrt{r^2 + a^2})\right]. \tag{1.9}$$

The pressure amplitude is the magnitude (i.e. the real component) of $p(r, 0, t)$. Expressed in rectangular coordinates, replacing r by z, the distance along the beam axis perpendicular to the source, the axial variation of the pressure amplitude $p(z)$ is

$$p(z) = 2\rho_0 c u_0 \left|\sin\left\{\frac{kz}{2}\left[\sqrt{1 + \left(\frac{a}{z}\right)^2} - 1\right]\right\}\right|. \tag{1.10}$$

This variation is shown in figure 1.3, from which it may be seen that the axial pressure variation in the near field is characterised by a series of unequally spaced pressure maxima, with value $2\rho_0 c u_0$, separated by localised field nulls where the pressure is zero. Since the pressure amplitude at the source, p_0, is $\rho_0 c u_0$, the near field pressure maxima have an amplitude $2p_0$. In those regions where the sine is negative the phase of the pressure wave is reversed.

The positions of the maxima and minima may be calculated from the conditions giving the sine function in equation (1.10) values of ± 1 (maxima) and 0 (minima). That is,

$$\frac{kz}{2}\left[\sqrt{1 + \left(\frac{a}{z}\right)^2} - 1\right] = \frac{m\pi}{2} \tag{1.11}$$

where the positions of the maxima arise when m is odd ($m = 1, 3, 5, \ldots$) and the positions of the nulls are when m is even ($m = 2, 4, 6, \ldots$). The most

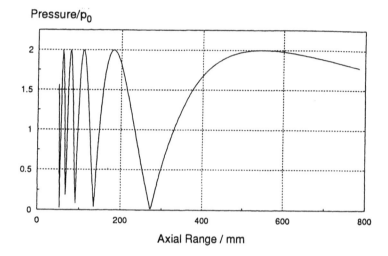

Figure 1.3. *The calculated variation of acoustic pressure on the axis of a plane circular transducer of 38 mm diameter at 2.25 MHz. The pressure is normalised to p_0, the pressure at the source.*

distant maximum from the source is commonly referred to as the 'last axial maximum'; its position, z_{lam}, may be calculated approximately by setting $m = 1$ in equation (1.11), and assuming that $a \ll z$, when $z_{lam} = a^2/\lambda$. Similarly the approximate position of the last axial null is $a^2/2\lambda$, half the distance to the last axial maximum.

It has become conventional to consider that the distance z_{lam} marks the boundary between the near and far fields for plane circular sources. As will be shown below, this definition becomes less meaningful when considering sources which do not have radial symmetry, for example rectangular sources from array transducers, and for focused beams where the position of the 'last axial maximum' is controlled as much by the focusing geometry as by the diffraction pattern from the source. Comparison with equation (1.1) shows that the Rayleigh distance $z_R = \pi z_{lam}$; for the example cited of a 1 MHz physiotherapy source, $a = 12.5$ mm, the last axial maximum is approximately 11 cm from the transducer face.

Referring again to equation (1.10), it is possible to approximate the axial behaviour in the far field by considering conditions when $z/a \gg ka$, when equation (1.10) reduces to

$$p(z) = \frac{\rho_0 c u_0 k a^2}{2z} = p_0 \frac{S}{\lambda z} \qquad (1.12)$$

where S is the area of the source. Equation (1.12) shows that the axial pressure in the far field reduces with $1/(\text{distance})$.

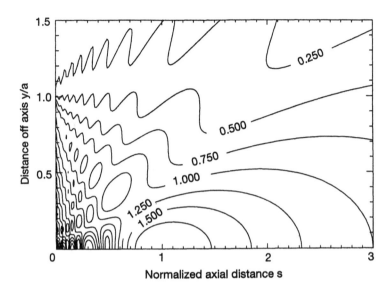

Figure 1.4. *Contour plot of the normalised acoustic pressure (p/p_0) for a circular transducer of radius $a = 5\lambda$. The axial distance is normalised to a^2/λ.*

1.1.2. Pressure variation off the axis

At positions off the axis of symmetry the ultrasound pressure field has considerable complexity. A calculated example of the normalised pressure field amplitude p/p_0 for a circular aperture of radius $a = 5\lambda$ is shown in figure 1.4, using a numerical approach (see Zemanek 1971). The complete pressure field may be thought of as being formed by rotating this radial section around the z axis, and consists of rings of higher acoustic pressure whose number and radial frequency increase as the source is approached. It may also be seen that the -6 dB beam width at z_{lam} (a_2/λ) is only about 0.4 that at the aperture, demonstrating the so-called 'self-focusing' of a plane source.

The alternative representation of the near field pattern shown in figure 1.5 emphasises an alternative approach to the analysis of the ultrasonic near field, originally used by Schoch (1941). He showed that the field could be considered as a convolution between two parts, one a plane wave propagating normally from the source, and the other a wave from its boundary. The interference between these two waves may be clearly seen in the field pattern shown in figure 1.5. This view of the field as being composed of a plane wave and an 'edge wave' is particularly useful when considering the characteristics of pulsed ultrasonic beams (see below).

In the far field the off-axis acoustic pressure p_θ may be expressed in terms of its directivity function D_θ:

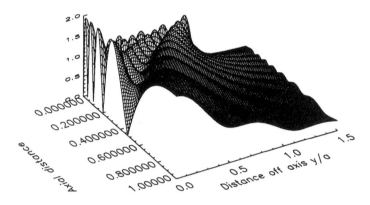

Figure 1.5. *Spatial distribution of the normalised acoustic pressure* (p/p_0)
for a circular transducer of radius $a = 5\lambda$. *The axial distance is normalised*
to a^2/λ.

$$p_\theta = D_\theta \frac{i\rho_0 c u_0 ka}{2} e^{i(\omega t - kr)} \tag{1.13}$$

where

$$D_\theta = \frac{2J_1(ka \sin\theta)}{ka \sin\theta} \tag{1.14}$$

and J_1 is Bessel's function of the first kind. $J_1(ka \sin\theta) = 0$ when
$ka \sin\theta = 3.83, 7.02, 10.17, 13.32$ etc. That is the field is formed of a
central lobe, and side lobes. The boundary of the central lobe occurs where
$\theta = \sin^{-1}(0.61\lambda/a)$.

1.2. PULSED FIELDS

The importance of the differences between the single-frequency and
broad-band pulsed beams cannot be overemphasised. Several authors have
reviewed the structure of pulsed fields (Friedlander 1958, Harris 1981a,
Wells 1977, Duck 1980, Krautkramer and Krautkramer 1990). While the far
field main lobe is not much affected, the spatial variations in the pressure
near field become smaller, and side lobes may diminish in amplitude and
merge. In addition the pressure pulse waveform, and its spectrum, vary
with position, including the potential for pulse splitting. These alterations
become important once the pulse length reduces to less than six cycles of
oscillation (Krautkramer and Krautkramer 1990). They are thus important for
all medical pulse-echo ultrasound beams, and also for many pulsed Doppler
applications when shorter pulse lengths are used (Duck and Martin 1992).
   One approach to the analysis of a pulsed field is to consider it as a
summation of the component fields of all the spectral components comprising

the pulse spectrum (Papadakis and Fowler 1971). For any source radius, the positions of near field maxima and minima depend on the wave number k (see equation (1.11)), and hence on the frequency. Summation of all the spectral components will thus result in a smearing of the local spatial variations in acoustic pressure. This approach has been valuable in its ability to introduce attenuative loss into the calculation of the pulsed acoustic field, with its associated frequency dependence. However, it is necessary to generate a sufficient sampling of the frequency-dependent fields to avoid errors, and generally this method may only be expected to give good approximations rather than exact solutions to the prediction of the pulsed acoustic field.

A widely used alternative method has developed from the analysis of the temporal impulse response of a source. This allows the pressure $p(r, t)$ to be calculated from the convolution between a source function and the pressure impulse response function $h(r, t)$:

$$p(r, t) = \rho_0 u_0(t) * h(r, t) \tag{1.15}$$

where $*$ indicates a temporal convolution. Since $h(r, t)$ is a function only of the source shape, and $u_0(t)$ is a function of the source vibration only, equation (1.15) gives a powerful general approach to the analysis of a variety of source geometries, in addition to the circular piston source which has been considered so far. Following Stepanishen's original publications for the circular piston (Stepanishen 1971, 1974, Beaver 1974), expressions for $h(r, t)$ for a number of other source geometries have been published: including those for rectangular sources (Lockwood and Willette 1973), shallow bowl (focused) sources (Penttinen and Luukkala 1976a), and sources with a variety of apodising functions (Harris 1981b).

An example of a calculation of the pressure wavefront in the near field of a plane piston source, excited using a single sinusoidal cycle, is shown in figure 1.6 ($a = 8$ mm, $f = 4$ MHz, $z = 40$ mm). The time scale is exaggerated in order to emphasise the pulse structure across the beam. The pulse waveform on the axis consists of two components separated in time. The first is that from the plane wave propagated from the source. The second occurs from the constructive interference of the edge wave, and has been termed a 'replica pulse'. It arrives on the axis at a time delay

$$\Delta t = \frac{\sqrt{z^2 + a^2} - z}{c}. \tag{1.16}$$

Its phase is inverted, and its amplitude is the same as that of the plane wave. Off the axis there are two replica pulses with lower amplitudes, because of incomplete constructive interference between the edge wave components.

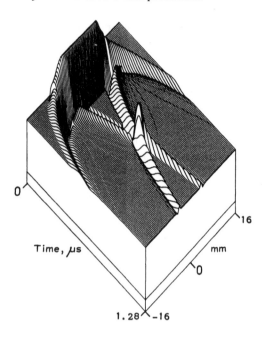

Figure 1.6. *Calculated pulse pressure profile for a circular piston source, radius 8 mm, vibrating with one cycle at 4 MHz. $c = 1500\ m\ s^{-1}$; $z = 40\ mm$ and $a^2/\lambda = 170\ mm$.*

In the region outside the projection of the source area, the plane wave component is absent, and only the two edge wave components exist. In this example it is only at the lateral boundaries of the beam that there is overlap between the edge wave and the plane wave. As the wave propagates, so the delay becomes smaller between the plane wave component and its replica pulse (see equation (1.16)). Interference between the two components can only occur at distances where the time delay Δt has become less than the pulse length, and only beyond this distance does on-axis variation in pulse pressure amplitude appear. Experimental observations of edge waves and replica pulses have been demonstrated by, for example, Weight and Hayman (1978).

Other methods appropriate to the calculation of both pulsed and single-frequency beams are finite element/boundary element methods, and finite difference methods. Finite difference methods have found most effective application in the prediction of beams within which finite amplitude effects are important, giving rise to nonlinear acoustic behaviour. Because of the importance of these effects in medical ultrasound, they are dealt with separately in Chapter 2.

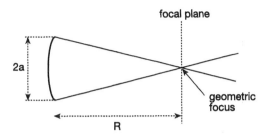

Figure 1.7. *Geometry of a spherical focusing source.*

1.3. FOCUSED FIELDS

For many medical applications of ultrasound the need arises to reduce the beam width, or to increase the local pressure amplitude, or both, from the values which are easily achieved using plane transducers. This requires the beams to be focused, so achieving improved spatial resolution for imaging, or intensities of sufficient magnitude to destroy tissue (see Chapter 9). Useful analyses have been published by O'Neil (1949), Kossoff (1979) and Lucas and Muir (1982).

The simplest means for focusing a beam is by the use of a spherical cap, or bowl transducer (see figure 1.7).

The amplitude gain G for a beam from a source with radius of curvature R can be calculated approximately by

$$G = \frac{z_R}{R} = \frac{\pi a^2}{\lambda R}. \tag{1.17}$$

It is common to consider three types of focused beam, categorised by their degree of focusing. These are:

weak focus: $0 < G \leqslant 2$
medium focus: $2 < G \leqslant 2\pi$
strong focus: $G > 2\pi$.

Diagnostic beams are generally of medium focus, while those used for hyperthermia and therapy are commonly of strong focus (see Chapters 8 and 9). While it is common in acoustics to specify focusing in terms of the amplitude gain of a beam, it is sometimes useful to specify the intensity or power gain, especially when energy delivery is required, as with hyperthermia. Intensity gain may therefore be quoted, which is G^2.

For a single-frequency, continuous-wave, focused source there are axial pressure variations similar to those in the beam from a plane source. The main differences are that the maxima vary in magnitude, the spacing between the maxima and minima are different, and for strongly focused fields,

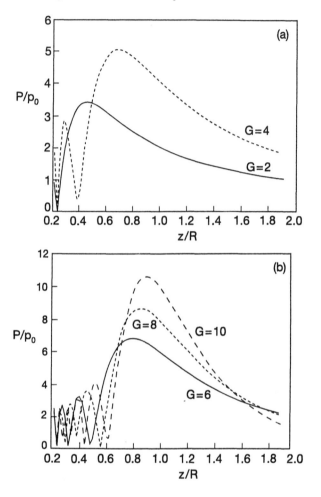

Figure 1.8. *Axial variation of normalised acoustic pressure (p/p_0) in spherically focused fields with different focal gains (a) $G = 2$ and 4; (b) $G = 6$, 8, and 10. The distance is normalised to the geometric focal length, R.*

on-axis minima may occur beyond the focus. Examples of axial profiles with amplitude focal gains of 2, 4, 6, 8, and 10 are shown in figure 1.8(*a*) and (*b*). The axes are normalised to the source pressure p_0 and to the radius of curvature R. The axial variation is given approximately by

$$p_z \cong p_0 G \frac{R}{z} \frac{\sin X}{X} \tag{1.18}$$

where

$$X = \frac{G}{2}\left(\frac{R}{z} - 1\right). \tag{1.19}$$

At the geometric focus (in figure 1.8 where $z/R = 1$), the pressure $p_{R0} = Gp_0$. However, the maximum axial pressure always exceeds this value, and the maximum (i.e. the acoustic focus) is always reached at a position closer to the source than the geometric focus. Furthermore, this separation between the geometric and acoustic foci decreases as the focusing gain increases. While a gain of 10 is associated with an approximate 10% shift in position, for a gain of 4 (used in some diagnostic systems) the acoustic focus may occur only at about $0.7R$. It has become conventional in pulsed diagnostic beams to give the focal length in terms of the acoustic focus, where the beam intensity reaches a maximum (see Chapter 7). Inspection of figure 1.8a,b also demonstrates that the length of the focal zone decreases as the gain and p_{R0}/p_0 increases. The radial pressure variation in the focal plane $p_R(y)$ is given approximately by

$$p_R(y) \cong \frac{2J_1(kya/R)}{kya/R} \tag{1.20}$$

where y is the off-axis distance. $p_R(y)$ reduces to -3 dB of its axial value when $y = 1.62R/ka$. In other words the -3 dB beam width at the focus, d_3, is

$$d_3 = \frac{3.24R}{ka}. \tag{1.21}$$

Pulsed focused beams differ from single-frequency beams in a manner comparable with the differences for plane sources. Impulse response functions have been developed for spherical bowl sources (Penttinen and Luukkala 1976a), for focusing lenses (Penttinen and Luukkala 1976b) and for conical radiators (Patterson and Foster 1982). Figure 1.9 demonstrates a computed pulse profile for a circular bowl source, radius of curvature 80 mm, under conditions otherwise similar to those used for figure 1.6. The edge wave component is still visible, but the plane wave is now a spherical wave converging towards the geometric focus.

1.4. SOURCE AMPLITUDE WEIGHTING

The theoretical development in all the preceding sections has assumed that movement of all elements dS over the source has been of equal amplitude. Focusing can be considered simply as the alteration of the relative phase of the movement of the elements. However most real transducers are not true 'piston' sources, that is, there is some variation in the source amplitude over the source area. This may come about deliberately, as happens with some arrays for which weighting, or apodisation, is applied across the array. It may occur because of the physical mounting of the transducer, for example if the edges are less free to move than the centre of a piezoelectric element,

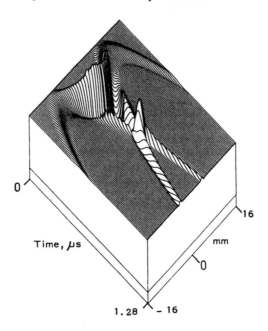

Figure 1.9. *Calculated pulse pressure profile for a circular spherical bowl source, radius 8 mm, vibrating with one cycle at 4 MHz. $c = 1500$ m s^{-1}; $z = 40$ mm. Compare with figure 1.6.*

or, when using a lens, from the transmission loss which varies from the lens axis to its edge. Smaller scale variations across the source area may also occur. These occur because of the construction of an array and may also characterise the behaviour of a physiotherapy transducer driven at its third harmonic.

Each of these examples indicates that fields from real transducers may well differ from the description derived from the formal theory set out in this chapter. In practice it is quite common to apodise an array used in a diagnostic scanner, for example using a Gaussian amplitude weighting function (Du and Breazeale 1985). Figure 1.10 shows the outcome of Gaussian apodisation for a plane single-frequency circular source with $a = 5\lambda$. The effect of apodisation is to reduce the pressure variations in the near field, and to reduce the side lobe level in the far field. A more complete description of the effects of radial weighting on pulsed fields has been given by Harris (1981b).

For a plane circular source the effect of apodisation caused by edge tethering is to alter slightly the position of the last axial maximum. For this reason it is common to consider the source as having an effective radius a_{eff} such that the measured last axial maximum lies at $(a_{eff})^2/\lambda$.

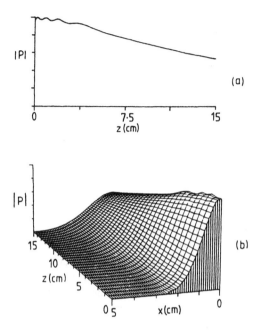

Figure 1.10. *Calculated single-frequency variation in peak pressure amplitude, for a Gaussian source apodisation with a = 5λ: (a) axial pressure variation; (b) off-axis pressure variation.*

1.5. RECTANGULAR SOURCES

The radial symmetry associated with both plane and spherically focused circular sources has a particular effect on the beams produced, particularly along the beam axis. This is true whether single-frequency or impulse behaviour is considered, and also for practical applications using broadband pulses. Any alteration from the circular symmetry of the source serves to alter the geometric conditions and hence the beam. Of particular practical interest for medical applications is the behaviour of rectangular sources of ultrasound, because of their common use in the arrays used for diagnostic imaging and associated Doppler applications. The design and fabrication of such arrays is described in greater detail in Chapter 5. Here we will be concerned only with the theoretical considerations of the acoustic beams generated by such rectangular sources of ultrasound.

While circularly symmetric beams may be analysed in terms of only two (rectangular) coordinates (y, z), any other shape of source, including rectangular, requires three (x, y, z) (figure 1.11). In simple terms the beam formation, both in terms of the length of the near field and the divergence in the far field, is controlled separately by the two orthogonal dimensions of the source, $2a$ and $2b$. If $a \ll b$ then there will be a shorter region in

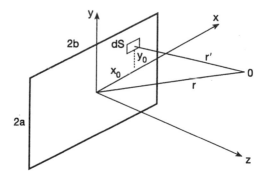

Figure 1.11. *Coordinate system for a rectangular source.*

which there is off-axis amplitude modulation in the y direction than in the x direction, and the beam will diverge more strongly beyond this region. If $a = b$, the source will behave somewhat like a circular source with radius a, and there will be a maximum on the axis at $z \cong a^2/\lambda$, although the near field amplitude modulation will not be so pronounced.

Figure 1.11 shows an observation point O at (x, y, z) and a source element $dS = dx_0\,dy_0$ at (x_0, y_0). Geometric considerations give

$$(r')^2 = z^2 + (x - x_0)^2 + (y - y_0)^2$$

and using a binomial expansion r' can be approximated by

$$r' \approx z\left(1 + \frac{(x - x_0)^2}{2z^2} + \frac{(y - y_0)^2}{2z^2}\right).$$

Further, substituting in the equation for the Rayleigh integral (equation (1.4)) and replacing the $1/r'$ term with $1/r$ we have

$$p = i\rho_0 c u_0 \frac{k}{2\pi} \frac{\exp(i(\omega t - kz))}{r} \int_{-b}^{b} \exp(-ik(x - x_0)^2/2z)\,dx_0$$

$$\times \int_{-a}^{a} \exp(-ik(y - y_0)^2/2z)\,dy_0. \tag{1.22}$$

Substituting $g = x_0\sqrt{(2/z\lambda)}$; $g_0 = b\sqrt{(2/z\lambda)}$; $h = y_0\sqrt{(2/z\lambda)}$; $h_0 = a\sqrt{(2/z\lambda)}$; and defining the aspect ratio of the rectangle $N = a/b$ so $h_0 = b/N\sqrt{(2/z\lambda)}$; we have on axis

$$p = \frac{i\rho_0 c u_0}{2}\exp(i(\omega t - kz)) \int_{-g_0}^{g_0} \exp(-i\pi g^2/2)\,dg \int_{-h_0}^{h_0} \exp(i\pi h^2/2)\,dh. \tag{1.23}$$

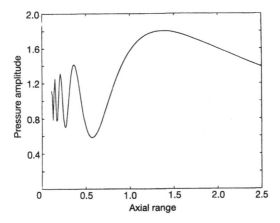

Figure 1.12. *Axial variation in normalised acoustic pressure amplitude* (p/p_0) *for a square transducer. Compare with figure 1.3 for a circular source.*

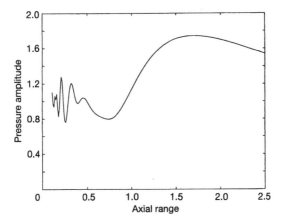

Figure 1.13. *Axial variation in normalised acoustic pressure* (p/p_0) *for a rectangular source with aspect ratio 1:2.*

The two integrals on the right hand side of equation (1.23) give an amplitude dependency on $b^2/z\lambda$ and $a^2/z\lambda$. The relative locations of the acoustic features in the beam will depend on the aspect ratio N.

The calculated axial pressure variation is shown for a square transducer in figure 1.12 and for a rectangular aperture with aspect ratio $N = 1:2$ in figure 1.13. Pressure maxima and minima variations are reduced for the square transducer, and largely absent for the rectangular transducer. In addition the maximum pressure amplitude on axis is less than $2p_0$, which is reached theoretically on the axis in the near field of a plane circular source.

Figure 1.14 shows the pulse pressure profile at 40 mm from a square

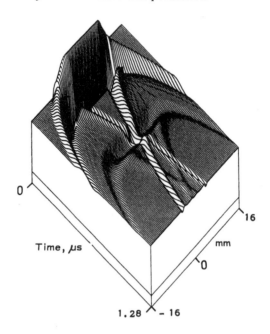

Figure 1.14. *Calculated pulse pressure profile for a square piston source, side 2a = 16 mm, vibrating with one cycle at 4 MHz. c = 1500 m s⁻¹; z = 40 mm. Compare with figure 1.6.*

source, $a = 8$ mm, operating for a single cycle at 4 MHz, calculated using the impulse response function derived by Lockwood and Willette (1973). The conditions are the same as those used to calculate the profile shown in figure 1.6 for a circular source, with which comparison may be made. This shows the presence of further replica pulses being caused by the edge waves from each of the four edges of the source.

Medical ultrasonic scanners not only use rectangular transducers, but apply astigmatic (cylindrical) focusing. The analysis of these focused pulsed fields must be carried out using numerical methods such as the finite difference methods which were mentioned above.

1.6.   CONCLUSION

This chapter has shown that it is possible to describe the acoustic pressure field structure from a variety of source geometries, using appropriate approximations. The advent of powerful computers has enabled numerical methods to be applied to previously intractable analyses. The particular symmetry associated with the acoustic field from an ideal circular piston source vibrating with a single frequency is not shared by other more practical ultrasound beams used for medical applications. Apodisation, pulsing and the use of rectangular sources all serve to smooth out the pressure variations, resulting in beams with low amplitude modulation in the near field, and low acoustic side lobe amplitudes. The full analysis and prediction of the field generated by a pulsed rectangular source with astigmatic focusing and apodisation, propagating nonlinearly through an inhomogeneous absorbing and scattering medium, remains a challenge.

REFERENCES

Beaver W L 1974 Sonic near fields of a pulsed piston radiator *J. Acoust. Soc. Am.* **56** 1043–8

Du G and Breazeale M A 1985 The ultrasonic field of a Gaussian transducer *J. Acoust. Soc. Am.* **78** 2083–6

Duck F A 1980 The pulsed ultrasonic field *Physical Aspects of Medical Imaging* ed B M Moores *et al* (London: Wiley)

Duck F A and Martin K 1992 Exposure values for medical devices, in *Ultrasonic Exposimetry* eds M Ziskin and P Lewin (Boca Raton: CRC)

Friedlander F 1958 *Sound Pulses* (Cambridge: Cambridge University Press)

Harris G R 1981a Review of transient field theory for a baffled planar piston *J. Acoust. Soc. Am.* **70** 10–20

—— 1981b Transient field of a baffled planar piston having an arbitrary vibration amplitude distribution *J. Acoust. Soc. Am.* **70** 186–204

Kinsler L E, Frey P, Coppens A B and Sanders J V 1982 *Fundamentals of Acoustics* 3rd edition (New York: Wiley)

Kossoff G 1979 Analysis of focusing action of spherically curved transducers *Ultrasound Med. Biol.* **5** 359–65

Krautkramer J and Krautkramer H 1990 *Ultrasonic Testing of Materials* 4th edition (Berlin: Springer) pp 87–92

Lockwood J C and Willette J G 1973 High-speed method for computing the exact solution for the pressure variations in the nearfield of a baffled piston *J. Acoust. Soc. Am.* **53** 735–41

Lucas B G and Muir T G 1982 The field of a focusing source *J. Acoust. Soc. Am.* **72** 1289–96

O'Neil H T 1949 Theory of focusing radiators *J. Acoust. Soc. Am.* **21** 516–26

Papadakis E P and Fowler K A 1971 Broad-band transducers: radiation field and selected applications *J. Acoust. Soc. Am.* **50** 729–45

Patterson M S and Foster S F 1982 Acoustic fields of conical radiators *IEEE Trans. Sonics Ultrasonics Freq. Contr.* **SU-29** 83–92

Penttinen A and Luukkala M 1976a The impulse response and pressure nearfield of a curved ultrasonic radiator *J. Phys. D: Appl. Phys.* **9** 1547–57

—— 1976b Sound pressure near the focal area of an ultrasonic lens *J. Phys. D: Appl. Phys.* **9** 1927–36

Schoch A 1941 Betrachtungen uber das Schallfeld einer Kolbenmembran *Akust. Z.* **6** 318–26

Stepanishen P R 1971 Transient radiation from pistons in an infinite planar baffle *J. Acoust. Soc. Am.* **49** 1629–38

—— 1974 Acoustic transients in the far field of a baffled circular piston using the impulse response approach *J. Sound Vibr.* **32** 295–310

Stepanishen P R and Benjamin K C 1982 Forward and backward projection of acoustic fields using FFT methods *J. Acoust. Soc. Am.* **71** 803–12

Weight J P and Hayman A J 1978 Observations of the propagation of very short ultrasonic pulses and their reflection by small targets *J. Acoust. Soc. Am.* **63** 396–404

Wells P N T 1977 *Biomedical Ultrasonics* (London: Academic)

Williams E G and Maynard J D 1982 Numerical evaluation of the Rayleigh integral for planar radiators using the FFT *J. Acoust. Soc. Am.* **72** 2020–30

Zemanek J 1971 Beam behaviour within the near field of a vibrating piston *J. Acoust. Soc. Am.* **49** 181–91

CHAPTER 2

NONLINEAR EFFECTS IN ULTRASOUND PROPAGATION

Andrew C Baker

INTRODUCTION

In a fluid, ultrasound propagates as longitudinal waves of alternate compressions and rarefactions. To a first approximation the wave travels at a constant speed (c) and so its shape remains unchanged as it propagates. This level of approximation corresponds to the simplest possible form of wave equation and is widely applicable to many acoustic systems (e.g. normal sound levels in air and most sonar systems in water). The methods of linear systems theory are appropriate to the solutions of problems in these fields and great use is made of methods such as superposition and linear scaling of solutions. The introduction of a frequency-dependent absorption causes no great difficulties either since the system is linear. The linear wave equation depends on two main assumptions: firstly that the particle velocity (u) of the wave is infinitesimal (or at least small compared to c) and secondly that the pressure–density relationship of the fluid is linear.

If the acoustic amplitude is sufficiently high then assumptions of linearity are no longer valid and will introduce significant errors. The resulting wave has compressional phases that travel at a speed ($c + \beta u_0$), which is faster than the speed of the rarefactions ($c - \beta u_0$); β is a parameter characterising the nonlinearity of the medium (the nonlinearity parameter is often expressed as $B/A = 2(\beta - 1)$: measurement methods and typical values are given in Chapter 4). Note that the finite particle velocity and the nonlinearity of medium both produce the same effect. Thus we get distortion that will cause a waveform that is initially sinusoidal to become more like a sawtooth (figure 2.1). The amount of distortion will increase with distance propagated and shock-like waveforms are commonly encountered, with an abrupt increase from peak negative pressure to peak positive pressure as the wave passes any point. In terms of frequency content, the waveform

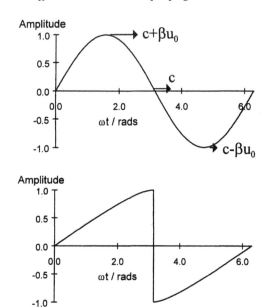

Figure 2.1. *Initial waveform (top, $\sigma = 0$) and distorted waveform (bottom, $\sigma = \pi/2$).*

distortion is equivalent to harmonic generation at integer multiples of the original frequency. Thus energy is pumped to higher frequencies where the absorption losses will be higher causing, among other things, increased intensity loss which can lead to enhanced heating and streaming effects. In a real beam of ultrasound there will also be diffraction effects which interact with nonlinearity and absorption to further complicate matters. A general history of nonlinearity in fluids can be found in the book by Beyer (1984).

2.1. NONLINEAR PROPAGATION IN MEDICAL ULTRASOUND

The use of ultrasound in medicine only become prevalent during the 1960s and although nonlinear acoustic effects have been known about since the eighteenth century, it was 1980 when the first papers highlighted the importance of nonlinear effects in medical ultrasound. Muir and Carstensen (1980) and Carstensen *et al* (1980) discussed the potential for shock formation, enhanced absorption due to harmonic generation and beam broadening, caused by the transfer of energy from the main lobe of the fundamental beam to higher harmonics. One of the measures of the strength of nonlinearity they used was the plane wave shock parameter $\sigma = \beta \varepsilon k z$, $\varepsilon \ (= u_0/c)$ is the acoustic Mach number where u_0 is the particle velocity amplitude at the source, $k \ (= 2\pi f/c)$ is the wavenumber and z is the distance that the wave has travelled. A value of $\sigma = 1$ indicates that a shock is just starting to form (i.e. a vertical discontinuity is just appearing in the pressure waveform). At this point the distance travelled (z) is often denoted by the plane wave shock distance $l_d = 1/\beta \varepsilon k$. In the case of a plane wave, u_0 can be determined from the acoustic pressure amplitude, p_0, using the plane wave impedance relation $u_0 = p_0/\rho_0 c$ where ρ_0 is the density of the medium. When $\sigma = \pi/2$ the wave is fully shocked, with a discontinuity from the peak positive pressure to the peak negative pressure (figure 2.1). Further distortion leads to reductions of the peak positive and negative pressures as more of the wave moves into the shocked region. Note that σ is proportional to both the acoustic pressure (p_0) and the distance travelled (z) hence it is possible to determine experimentally whether nonlinear propagation is occurring in a system by noting whether waveform distortion decreases as either drive pressure and/or observation distance is decreased. In water at 20°C, $\beta = 3.5$, hence if we take ultrasonic parameters that are typical of current imaging systems (e.g. $f = 3.5$ MHz, $p_0 = 1$ MPa; see Chapter 7) we will have a plane wave shock distance $l_d = 43$ mm (assuming $\rho = 1000$ kg m^{-3} and $c = 1486$ m s^{-1}). We would therefore expect to observe nonlinear waveform distortion relatively easily at distances greater than this. It should be noted that in clinical systems focusing and diffraction will also affect the shock formation distance so l_d should only be used as a rough estimate of nonlinear effects in clinical beams.

The situation is similar in human tissues where β values are typically in the range 4 to 6 (Duck 1990) with the higher values due to fatty tissues. However, attenuation losses are higher in tissue which tends to counteract nonlinear distortion. This is indicated by the Gol'dberg number $\Gamma = 1/l_d\alpha = \beta \varepsilon k/\alpha$ where α is the linear attenuation coefficient (in neper m^{-1}). The Gol'dberg number accounts for the fact that absorption counteracts nonlinear generation so it is the ratio of the two which is important. Methods of measuring the nonlinearity parameter and the possibility of mapping it to form *in vivo* images have been reviewed by Bjørnø (1986).

Figure 2.2 represents the nonlinear propagation of a 3.5 MHz, 500 kPa plane wave in water (i.e. the Gol'dberg number $\Gamma = 38$). We can see that at zero range only the fundamental is present. The harmonics build up with distance and eventually settle in almost constant ratio to the fundamental. Energy is lost from the fundamental and is pumped into the harmonics. In the case of a linear wave we would expect no harmonics and the fundamental to remain almost constant; linear absorption would only account for a loss of a few per cent over this sort of distance in water. A range of 85 mm corresponds to $\sigma = 1$ for this wave and it can be seen that there is appreciable second and third harmonic content. Only the first five harmonics are plotted here although many more will be present especially at longer ranges. At a range of 133 mm we have $\sigma = \pi/2$ and appreciable energy has been lost from the fundamental to the higher harmonics.

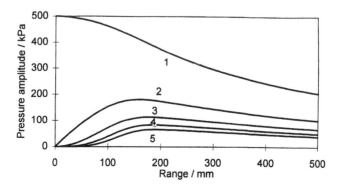

Figure 2.2. *Fundamental and second to fifth harmonics for a nonlinear plane wave in water ($f_0 = 3.5$ MHz, $P_0 = 500$ kPa, $\Gamma = 38$).*

σ is a useful quantity when trying to estimate the significance of nonlinearity in a given situation but a plane wave represents a rather idealised case. In an attempt to include the effects of focusing, Bacon (1984) proposed a nonlinear propogation parameter (σ_m) which takes account of amplitude focal gain G:

$$\sigma_m = \frac{2\pi f \beta p_f z_f \ln(G + \sqrt{G^2 - 1})}{\rho_0 c^3 \sqrt{G^2 - 1}} \tag{2.1}$$

where the focal pressure p_f is defined as $(p_c + p_r)/2$ (p_c and p_r are compression and rarefaction pressure magnitudes at the focal distance z_f), and f is the acoustic frequency. This equation can be used also to calculate σ at the focus up to $\sigma = \pi/2$. Above this value σ_m no longer depends linearly on source amplitude, and ultimately reaches a saturation value of 2π.

The inclusion of diffraction and focusing in the nonlinear problem causes phase shifts in waveform so that instead of resembling a sawtooth, a nonlinearly distorted ultrasonic pulse looks more like the measurements shown in figure 2.3. The waveform shown is a relatively low amplitude 2.25 MHz pulse generated by a heavily damped, shock-excited transducer. The diffractive phase shifts cause the top–bottom asymmetry of the distorted pulse.

Diagnostic ultrasound tends to operate over shorter distances than 600 mm but with correspondingly higher drive levels and focusing gain, hence the distorted waveform shape is typical of the distorted pulses observed from clinical systems (Duck and Starritt 1984). Even the high absorption of tissue is not sufficient to suppress nonlinearity hence similar waveform distortion and harmonic generation have been observed in biological tissues (Starritt *et al* 1985, 1986). An extensive survey of the output of diagnostic systems (Duck *et al* 1985) showed that almost all the systems surveyed were likely to be subject to nonlinear propagation. A more recent survey (Henderson

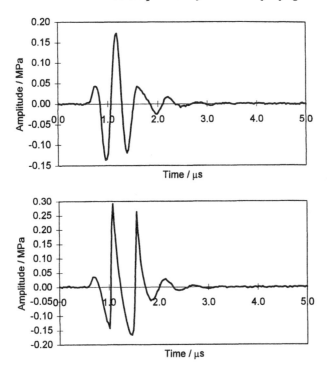

Figure 2.3. *Initial pulse (top) and nonlinear distortion of pulse (bottom) after propagating 600 mm in water.*

et al 1995) indicated that the acoustic output levels of new diagnostic systems had increased considerably and thus nonlinear effects are now even more significant. Other medical ultrasound systems such as lithotripters and hyperthermia systems will also be subject to nonlinear propagation since although they usually have lower fundamental frequencies they also have high acoustic drive levels.

2.2.   CONSEQUENCES OF NONLINEAR PROPAGATION

2.2.1.   *Experimental measurements*

The most obvious practical consequence of nonlinear propagation is that an increased measurement bandwidth is necessary. This is true both for acoustic field measurements with hydrophones (see Chapter 7) and for pulse-echo imaging of harmonic backscatter as is used clinically, for example in harmonic imaging of contrast materials (see Chapter 12). Figure 2.4 shows the frequency content of the pulses in figure 2.3. The initial pulse has its energy concentrated around the centre frequency of the transducer (2.25 MHz

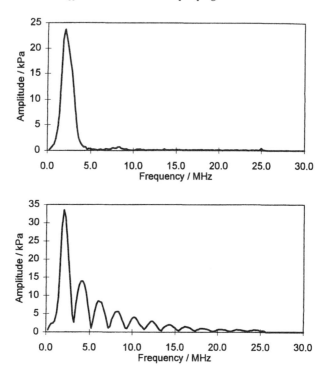

Figure 2.4.   *Initial spectrum (top) and spectrum of distorted pulse (bottom).*

in this case). The growth of distortion leads to peaks at multiples of the centre frequency up to about 25 MHz where the hydrophone bandwidth starts to limit the width of the spectrum. The hydrophone used in this case was a GEC-Marconi bilaminar PVDF (polyvinylidene fluoride) membrane device which has a smooth response up to its main resonance at about 20 MHz. Other variants of this device have a higher resonant frequency providing even greater bandwidth (Bacon 1982). Until the membrane hydrophone was developed it was difficult to observe the distorted waveforms produced by medical ultrasound systems. In 1980 it was noted that 'Although microprobes with a flat response to 10 MHz have been reported, they are difficult to construct and are not commercially available' (Carstensen *et al* 1980). The choice of hydrophone is an important one when dealing with such distorted waveforms and few devices can currently approach the GEC-Marconi membrane in terms of the width and smoothness of their operational frequency range. In recent years though there have been some improvements in needle probe hydrophone design, and these devices usually have a significant price advantage over the GEC-Marconi membrane hydrophone. Nevertheless it is recognised that the frequency response may be quite variable, particularly at lower frequencies, and needle hydrophones need to

be used critically if high fidelity measurements are required (Preston 1991).
Several factors are important when handling distorted waveforms of this
type including:

2.2.1.1. System bandwidth. The loss of high frequency components due
to limited hydrophone bandwidth (or limited bandwidth in any part of the
signal processing system) is particularly noticeable in the observed peak
positive pressure which can appear to be significantly reduced. The choice
of digitising frequency must also be appropriate to the system bandwidth. It
is common to use digitising frequencies of 100 MHz or higher for the capture
of typical medical imaging pulses which have their centre frequencies in the
low MHz range. Care must also be taken to avoid aliasing in digitisers.

2.2.1.2. Hydrophone calibration and frequency response. The presence of
nonlinearity means that measurements cannot be scaled from one drive level
to another. In a linear system, for example, the directivity plot of an acoustic
beam does not change with drive level; under nonlinear conditions it will
change with drive level. It is therefore useful to know the acoustic pressure
at which measurements were made. The hydrophone calibration is also
important in removing waveform artefacts due to hydrophone resonances.
The frequency response of a hydrophone is usually determined by one or
more of its electro-mechanical resonances which may not be apparent in
undistorted waveforms. A distorted waveform however can have sufficient
energy at higher frequencies to excite the hydrophone resonances hence
the observed electrical signal is not a true representation of the pressure
waveform being measured. It is hardly ever justifiable to use a single
frequency calibration over the bandwidths required.

2.2.1.3. Hydrophone size. The harmonic beam patterns are narrower than
the fundamental beam (figure 2.5). Thus a hydrophone chosen to be small
enough to avoid spatial averaging problems at the fundamental frequency
may well be rather large in comparison with the higher harmonic beams.
This will lead to under-estimates of the peak values (Smith 1989, Zeqiri and
Bond 1992, Baker *et al* 1996).

2.2.1.4. Hydrophone alignment. The narrower beamwidths of the harmonics make hydrophone alignment more critical. The harmonic amplitude
beamwidths vary as $1/\sqrt{n}$, where n is the harmonic number (Reilly and
Parker 1989). The higher harmonics, however, can provide a useful guide
to alignment since the peak positive pressure is sensitive to their presence.

2.2.1.5. Hydrophone linearity. The acoustic pressures are sufficiently
high that the hydrophone itself can generate harmonic components. The
magnitude of these components will only depend on the amplitude of

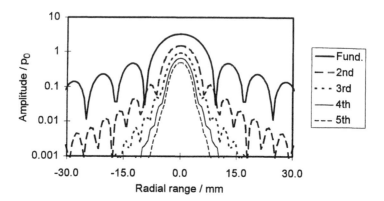

Figure 2.5. *Calculated harmonic beam patterns in focal plane of focused transducer ($f_0 = 2.25$ MHz).*

acoustic field being measured whereas nonlinearity due to propagation also accumulates with distance. It is thus possible to distinguish between these two sources of nonlinearity by moving the hydrophone close to the ultrasonic source where the propagation nonlinearity should be negligible. Care is still needed as the ultrasonic source will often transmit low levels of harmonic directly. The effect of hydrophone nonlinearity and direct transmission of harmonics is not usually serious since the levels are small in comparison with the harmonic levels generated by nonlinear propagation (Preston *et al* 1983).

2.2.1.6. Choice of propagating medium. Laboratory measurements are invariably made in water, but this can create difficulties. The absorption of ultrasound in water is low (relative to tissue) which allows a greater degree of nonlinear distortion to occur and hence increased signal bandwidth. It is not a simple matter to translate water-based measurements to *in vivo* values. The 'derating' procedure which is commonly used in standards (AIUM/NEMA 1992) is based on assumptions of linear propagation. Christopher and Carstensen (1996) conclude that applying the linear derating factor to strongly shocked measurements in water can lead to significant underestimates of the pressure field in tissue.

The effect of nonlinear propagation can also be observed in measurements of ultrasonic properties such as absorption coefficient which become dependent on the drive level and measurement geometry (Zeqiri 1992, Wu 1996). Both of these studies concluded that it is advisable to minimise the transmitter–receiver separation and to keep the plane wave shock parameter $\sigma \leqslant 0.1$ in order to avoid significant nonlinear errors.

Characteristics for hydrophones and guidance for making measurements in this frequency range can also be found in the relevant international standards

(IEC 1987, 1991, 1993) and Preston (1991). Other practical aspects of the use of hydrophones for exposure measurement are given in Chapter 7, section 7.7.

2.2.2. Theoretical predictions

Theoretical models for ultrasound propagation are useful in the design and analysis of ultrasound systems, especially since *in vivo* measurements are not easy to carry out. The main difficulty in modelling is the presence of nonlinearity which rules out most of the methods that are applicable to linear systems. The cumulative nature of the distortion with distance and its interaction with diffraction and absorption mean that it is not normally possible to calculate the amplitude of a single field point at a distance from the source without calculating the full field in the region between the field point and the source. Thus straightforward analytical solutions can only be found for relatively simple geometries and in general it is necessary to use computationally intensive numerical methods. In addition to calculating the acoustic field, there is also a requirement to be able to predict effects such as heating, streaming and cavitation since these are potential sources of bio-effects and will depend on the acoustic field.

A number of approaches to predicting the ultrasonic fields of medical ultrasound systems have been tried but the method that has probably received most attention to date is a finite-difference solution to an approximate nonlinear wave equation. The wave equation is known as the Khokhlov–Zabolotskaya–Kuznetsov (or KZK) equation and it accounts for nonlinearity, absorption and diffraction (Kuznetsov 1971). The most important assumption in the KZK equation, in this context, is that the acoustic energy propagates in a fairly narrow beam; this is known as the parabolic approximation or the paraxial approximation. The parabolic approximation is valid for acoustic sources which are many wavelengths across and for field points that are not too close to the source or too far off axis. For circular sources of ultrasound we need $(kr_a)^2 \gg 1$ where r_a is the source radius and the minimum axial distance is $r_a(kr_a/2)^{1/3}$ (Naze Tjøtta and Tjøtta 1980). In practice, for most weakly focused diagnostic beams, these conditions do not usually pose serious difficulties. A finite-difference solution for the KZK equation was described by Aanonsen, Barkve, Naze Tjøtta and Tjøtta of the University of Bergen, Norway (Aanonsen *et al* 1984). Naze Tjøtta and Tjøtta have been responsible for developing much of the mathematical background in the field of nonlinear sound beams; the resulting numerical solutions and computer programs are now widely known as the Bergen code. The approach used in the Bergen code is to substitute a Fourier series for the time waveform into the KZK equation and solve the resulting set of coupled differential equations using finite difference methods.

The Bergen code has been applied to ultrasonic sources similar to those found in medical systems and has proved to be a reliable model of the beam

Pressure amplitude / MPa

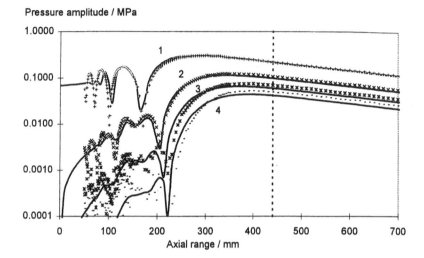

Figure 2.6. *Comparison of the KZK equation (——) with measurements of fundamental and second, third and fourth harmonics (+, ×, *, −) for a focused ultrasound source in water (f_0 = 2.25 MHz, p_0 = 68 kPa, a = 19 mm). The vertical dashed line indicates the position of the focal plane.*

behaviour in water. Plane circular sources of continuous-wave ultrasound have been studied (Baker *et al* 1988, TenCate 1993, Nachef *et al* 1995) as well as focused sources as shown in figure 2.6 (Baker 1992, Averkiou and Hamilton 1995).

Figure 2.6 shows that close to the source there are no harmonic components, only the fundamental. The harmonics build up with axial range with various maxima and minima mirroring those in the fundamental until the final axial maxima settle at roughly constant levels (approximately $1/n$) relative to the fundamental. The KZK solution does not show the expected nearfield oscillations at very short ranges, this is a consequence of the step size used. Smaller steps would have shown more detail at short ranges; instead we see the average value of the solution in that region.

The Bergen code may also be initialised with a pulse spectrum hence pulsed fields similar to diagnostic systems have been examined (Baker and Humphrey 1992, Baker 1991). Rectangular geometries have also been modelled (Berntsen 1990, Baker *et al* 1995). The Bergen code has recently been applied to ultrasound systems with rectangular arrays (Cahill and Baker 1997a, b) and it was found that nonlinearity can interact with diffraction to cause the region of peak intensity loss to move from the acoustic axis (at low drive levels) to off-axis locations at high drive levels (figure 2.7). This shift is contrary to the predictions of linear theories. The first part of figure 2.7 shows the linear (low drive level) pressure field of a square aperture as would be

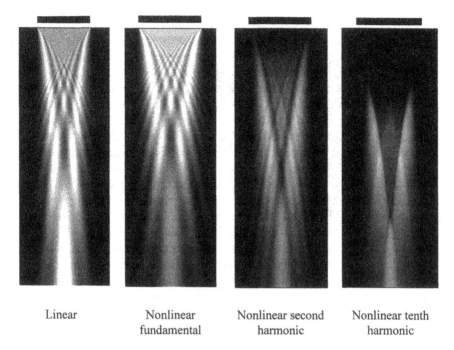

| Linear | Nonlinear fundamental | Nonlinear second harmonic | Nonlinear tenth harmonic |

Figure 2.7. *Nonlinear generation along the diagonal of a plane square aperture (side length = 20 mm). The beam propagates down the page for a distance of 200 mm (f_0 = 2.25 MHz).*

measured across its diagonal; diffraction effects are strongest on the diagonal due to the interaction of edge diffraction from the two sides. It can be seen that in the linear case the peak amplitude occurs on the acoustic axis at the bottom of the plot. The second plot corresponds to the fundamental when the source pressure is increased to 1 MPa. The region of peak fundamental amplitude now occurs off axis and much closer to the source. The second harmonic is strongest where the fundamental is strongest so it too has its peak values off axis nearer the source. The second harmonic also shows twice as many fringes across the beam when compared to the fundamental; this effect can be seen in figure 2.5. The tenth harmonic (i.e. 22.5 MHz) has a sharply defined edge to the off-axis region and again exhibits a maximum amplitude off axis. This has consequences for the prediction of potential bio-effects since the intensity loss from the beam determines the heat source distribution for thermal effects and the driving force for streaming.

Computational requirements can become an issue for nonlinear modelling. The continuous wave circular case at moderate drive levels can be run on a personal computer (e.g. a Pentium-based PC) in a matter of minutes. The Bergen code, however, works in the frequency domain and if the drive level is increased, more harmonics are needed in the solution which requires more memory and more CPU time. The inclusion of pulsed waveforms requires more frequency components in the initial spectrum which again requires more memory and more CPU time. The rectangular code is another order of magnitude bigger in its requirements for memory and CPU time. The results of Cahill and Baker (1997b) required about 500 MB physical memory and took of the order of 40 hours CPU per run on a DEC Alpha 8400 computer. Some savings in computer effort were made by introducing an artificially high absorption factor for the highest harmonics which can reduce the total number of harmonics required in the solution. Fourier transform methods also enabled considerable CPU time savings by calculating the nonlinear interactions in the time domain. Spatial resolution sometimes has to be traded for savings in run time: the more closely spaced the grid points in the finite-difference scheme, the more memory and CPU time that is required.

Apart from the Bergen code a number of other methods of solution are also feasible. Christopher and Parker (1991) have demonstrated a nonlinear model which operates in the frequency domain but uses a spatial Fourier transform method to calculate diffraction effects. It is claimed that it has no restrictions on the positions of field points and its use for lithotripter modelling has been demonstrated (Christopher 1994). Currently the main drawback of the Christopher and Parker model appears to be its lack of a single mathematical basis. The use of Fourier transform methods for dealing with diffraction is well established but the inclusion of nonlinearity does not appear to have been fully justified mathematically.

Others have developed nonlinear solutions which operate entirely in the time domain (Cleveland *et al* 1996, Lee and Hamilton 1995). Time domain models have the advantage that nonlinear distortion and pulsed fields can be accounted for more easily; however, attenuation becomes a convolution operator instead of a simple multiplying factor in the frequency domain.

All nonlinear modelling methods require the properties of the medium to be known over a wide frequency range; this poses serious restrictions for all biological tissues and fluids. The only medium that could be said to be well characterised is water. Currently the most comprehensive catalogue of acoustic properties of biological media is that of Duck (1990) but even commonly used ultrasound paths such as urine and amniotic fluid are represented by only a few data points. Further uncertainty arises since many of those measurements were made at a time when the importance of nonlinear propagation in making absorption measurements was not widely appreciated.

2.2.3. *Clinical systems*

In practice nonlinearity can have a number of both useful and unwanted side-effects for the ultrasound system. All clinical ultrasound systems (diagnostic and therapeutic) are liable to suffer from enhanced absorption due to

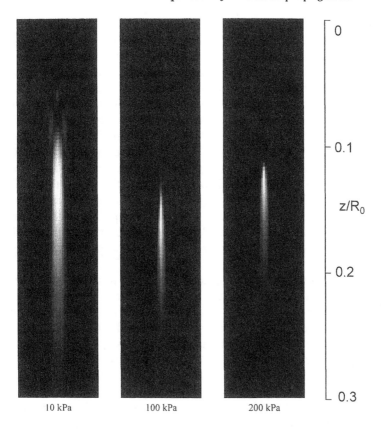

Figure 2.8. *The effect of nonlinearity on the pattern of intensity loss for a focused circular source. The beam propagates down the page ($f_0 = 2.25$ MHz, $a = 19$ mm, gain $= 6$).*

nonlinearity which can lead to acoustic saturation. In a system that has become saturated an increase in acoustic drive level does not lead to an increase in the acoustic pressure at some distant field point. The extra energy is converted to harmonic frequencies and becomes absorbed. The same effect can lead to increased sidelobe levels as the main lobe of the beam will saturate before the sidelobes which have lower acoustic amplitudes. The energy pumped into the harmonics will have a greater contribution to bio-effect mechanisms such as heating (Bacon and Carstensen 1990) and streaming (Starritt *et al* 1989). An example of nonlinear propagation is observed in ultrasonic surgery (Watkin *et al* 1996) where the lesions are generated closer to the ultrasound source as the drive level is increased. An explanation of this can be seen in figure 2.8 which shows the spatial distribution in a focused field of the rate of intensity loss $\partial I / \partial z \propto \sum_{n=1}^{m} |p_n|^2 n^2$, where p_n is the amplitude of the nth harmonic component and *m* is the total number of harmonics generated. This quantity can be considered as the source term for thermal calculations (or the driving force for streaming). Whereas the measurement of all the harmonic components over an entire pressure field would be very time consuming, a numerical model can generate these data relatively easily. In figure 2.8 the source is 38 mm in diameter and is positioned at the top of each plot, the width of the plots is equal to the source diameter and the propagating medium is water. At the lowest drive level (10 kPa) the intensity loss follows the same pattern as the linear intensity plot. At 100 kPa the region of intensity loss is much narrower and peaks at a greater axial range. Note that each plot is normalised to its maximum intensity loss and the total intensity loss in the second plot is much higher than the first and its spatial distribution has changed. At the highest drive level (200 kPa) the position of peak loss has moved even closer to the source. Corresponding observations of the shift of peak streaming velocity (Starritt *et al* 1989) can be explained by the same mechanism (Baker 1997).

There are however several potential benefits many of which can be achieved without further increase in ultrasonic drive levels. Most current imaging systems use the same transducer elements to transmit and receive and the bandwidth is not sufficient to capture the harmonic frequencies present in the returning echo. An imaging system with a wideband receiving array would have the advantage of higher signal to noise ratio and better resolution due to the inclusion of the harmonics (Ward *et al* 1996, 1997). Such systems are now becoming widely available commercially as so-called 'harmonic imaging' modes on clinical scanners. Evidence of improved signal-to-clutter ratios has been reported during their application. Much interest has arisen recently from the use of ultrasonic contrast media including second harmonic imaging, and its benefits are being demonstrated clinically. Nonlinear image enhancement has also been demonstrated in the field of acoustic microscopy. Calibration techniques have also been based on nonlinear propagation where the increased signal bandwidth is used to calibrate a hydrophone at several frequencies in a single set of measurements (Smith and Bacon 1990).

REFERENCES

Aanonsen S I, Barkve T, Naze Tjøtta J and Tjøtta S 1984 Distortion and harmonic generation in the nearfield of a finite amplitude sound beam *J. Acoust. Soc. Am.* **75** 749–68

AIUM/NEMA 1992 *Standard for Real-time Display of Thermal and Mechanical Acoustic Output Indices on Diagnostic Ultrasound* (Rockville, MD: AIUM & NEMA)

Averkiou M A and Hamilton M F 1995 Measurements of harmonic generation in a focused finite-amplitude sound beam *J. Acoust. Soc. Am.* **98** 3439–42

Bacon D R 1982 Characteristics of a pvdf membrane hydrophone for use in the range 1–100 MHz *IEEE Trans. Sonics Ultrasonics* **SU-29** 18–25

Bacon D R 1984 Finite amplitude distortion of the pulsed fields used in diagnostic ultrasound *Ultrasound Med. Biol.* **10** 189–95

Bacon D R and Carstensen E L 1990 Increased heating by diagnostic ultrasound due to nonlinear propagation *J. Acoust. Soc. Am.* **88** 26–34

Baker A C 1991 Prediction of non-linear propagation in water due to diagnostic medical ultrasound equipment *Phys. Med. Biol.* **36** 1457–64

—— 1992 Nonlinear pressure fields due to focused circular apertures *J. Acoust. Soc. Am.* **91** 713–17

—— 1997 A numerical study of the effect of drive level on the intensity loss from an ultrasonic beam *Ultrasound Med. Biol.* **23** 1083–8

Baker A C and Humphrey V F 1992 Distortion and high frequency generation due to nonlinear propagation of short ultrasonic pulses from a plane circular piston *J. Acoust. Soc. Am.* **92** 1699–705

Baker A C, Anastasiadis K and Humphrey V F 1988 The nonlinear pressure field of a plane circular piston: theory and experiment *J. Acoust. Soc. Am.* **84** 1483–7

Baker A C, Berg A M, Sahin A and Naze Tjøtta J 1995 The nonlinear pressure field of plane, rectangular apertures: experimental and theoretical results *J. Acoust. Soc. Am.* **97** 3510–7

Baker A C, Ward B and Humphrey V F 1996 The effect of receiver size on nonlinear pressure field measurements *J. Acoust. Soc. Am.* **100** 2062–70

Berntsen J 1990 Numerical calculations of finite amplitude sound beams, in *Frontiers of Nonlinear Acoustics* eds M F Hamilton and D T Blackstock (London: Elsevier) pp 191–6

Beyer R T 1984 *Nonlinear Acoustics in Fluids* (New York: Van Nostrand Rheinhold)

Bjørnø L 1986 Characterisation of biological media by means of their non-linearity *Ultrasonics* **24** 254–9

Cahill M D and Baker A C 1997a Increased off-axis energy deposition due to diffraction and nonlinear propagation from rectangular sources *J. Acoust. Soc. Am.* **102** 199–203

—— 1997b Numerical simulation of the acoustic field of a phased-array medical ultrasound scanner *J. Acoust. Soc. Am.* **104** 1274–83

Carstensen E L, Law W K, McKay N D and Muir T G 1980 Demonstration of nonlinear acoustical effects at biomedical frequencies and intensities *Ultrasound Med. Biol.* **6** 359–68

Christopher P T and Parker K J 1991 New approaches to nonlinear diffractive field propagation *J. Acoust. Soc. Am.* **90** 488–99

Christopher T 1994 Modelling the Dornier HM3 lithotripter *J. Acoust. Soc. Am.* **95** 3088–95

Christopher T and Carstensen E L 1996 Finite amplitude distortion and its relationship to linear derating formulae for diagnostic ultrasound systems *Ultrasound Med. Biol.* **22** 1103–16

Cleveland R O, Hamilton M F and Blackstock D T 1996 Time-domain modeling of finite-amplitude sound in relaxing fluids *J. Acoust. Soc. Am.* **99** 3312–8

Duck F A 1990 *Physical Properties of Tissue: A Comprehensive Reference* (London: Academic Press)

Duck F A and Starritt H C 1984 Acoustic shock generation by ultrasonic imaging equipment *Br. J. Radiol.* **57** 231–40

Duck F A, Starritt H C, Aindow J D, Perkins M A and Hawkins A J 1985 The output of pulse-echo ultrasound equipment: a survey of powers, pressures and intensities *Br. J. Radiol.* **58** 989–1001

Henderson J, Willson K, Jago J R and Whittingham T A 1995 A survey of the acoustic outputs of diagnostic ultrasound equipment in current clinical use *Ultrasound Med. Biol.* **21** 699–705

IEC 1987 *Characteristics and Calibration of Hydrophones for Operation in the Frequency Range 0.5 MHz to 15 MHz* IEC 866 (Brussels: CENELEC)
—— 1991 *Measurement and Characterisation of Ultrasonic Fields Using Hydrophones in the Frequency Range 0.5 MHz to 15 MHz* IEC 1102 (Brussels: CENELEC)
—— 1993 *Ultrasonics—Fields—Measurement and Characterisation of Ultrasonic Fields Generated by Medical Ultrasound Equipment Using Hydrophones in the Frequency Range 0.5 MHz to 15 MHz* IEC 1220 (Brussels: CENELEC)
Kuznetsov V P 1971 Equations of nonlinear acoustics *Sov. Phys.–Acoust.* **16** 467–70
Lee Y-S and Hamilton M F 1995 Time-domain modeling of pulsed finite-amplitude sound beams *J. Acoust. Soc. Am.* **97** 906–17
Muir T G and Carstensen E L 1980 Prediction of nonlinear acoustic effects at biomedical frequencies and intensities *Ultrasound Med. Biol.* **6** 345–57
Nachef S, Cathignol D, Naze Tjøtta J, Berg A M and Tjøtta S 1995 Investigation of a high intensity sound beam from a plane transducer. Experimental and theoretical results *J. Acoust. Soc. Am.* **98** 2303–23
Naze Tjøtta J and Tjøtta S 1980 An analytical model for the nearfield of a baffled piston transducer *J. Acoust. Soc. Am.* **68** 334–9
Preston R C 1991 *Output Measurements for Medical Ultrasound* (London: Springer)
Preston R C, Bacon D R, Livett A J and Rajendran K 1983 PVDF membrane hydrophone performance properties and their relevance to the measurement of the acoustic output of medical ultrasound equipment *J. Phys. E: Sci. Instrum.* **16** 786–96
Reilly C R and Parker K J 1989 Finite-amplitude effects on ultrasound beam patterns in attenuating media *J. Acoust. Soc. Am.* **86** 2339–48
Smith R A 1989 Are hydrophones of 0.5 mm diameter small enough to characterize diagnostic ultrasound equipment? *Phys. Med. Biol.* **34** 1593–607
Smith R A and Bacon D R 1990 A multiple-frequency hydrophone calibration technique *J. Acoust. Soc. Am.* **87** 2231–43
Starritt H C, Duck F A, Hawkins A J and Humphrey V F 1986 The development of harmonic distortion in pulsed finite-amplitude ultrasound passing through liver *Phys. Med. Biol.* **31** 1401–9
Starritt H C, Duck F A and Humphrey V F 1989 An experimental investigation of streaming in pulsed diagnostic ultrasound beams *Ultrasound Med. Biol.* **15** 363–73
Starritt H C, Perkins M A, Duck F A and Humphrey V F 1985 Evidence for ultrasonic finite-amplitude distortion in muscle using medical equipment *J. Acoust. Soc. Am.* **77** 302–6
TenCate J A 1993 An experimental investigation of the nonlinear pressure field produced by a plane circular piston *J. Acoust. Soc. Am.* **94** 1084–9
Ward B, Baker A C and Humphrey V F 1996 Nonlinear propagation applied to the improvement of lateral resolution in medical ultrasound scanners *Proceedings of 1995 World Congress on Ultrasonics* 965–8
—— 1997 Nonlinear propagation applied to the improvement of resolution in diagnostic medical ultrasound equipment *J. Acoust. Soc. Am.* **10** 143–54
Watkin N A, ter Haar G and Rivens I 1996 The intensity dependence of the site of maximal energy deposition in focused ultrasound surgery *Ultrasound Med. Biol.* **22** 483–91
Wu J 1996 Effects of nonlinear interaction on measurements of frequency-dependent attenuation coefficients *J. Acoust. Soc. Am.* **99** 3380–4
Zeqiri B 1992 Errors in attenuation measurements due to nonlinear propagation effects *J. Acoust. Soc. Am.* **91** 2585–93
Zeqiri B and Bond A D 1992 The influence of waveform distortion on hydrophone spatial-averaging corrections: theory and measurement *J. Acoust. Soc. Am.* **92** 1809–21

CHAPTER 3

RADIATION PRESSURE AND ACOUSTIC STREAMING

Francis A Duck

INTRODUCTION

This chapter deals with two linked but separate phenomena, radiation force and acoustic streaming, both of which have some practical importance in medical applications of ultrasound. We will start with simple qualitative descriptions of these phenomena, before giving an outline of their theoretical basis, then discuss some reports in the literature of their appearance in a medical context, and speculate about the difficulties still remaining.

When a solid object is placed in a progressive ultrasound wave it experiences a small force which is directed along the beam in the direction of propagation. This is called the *radiation force*. If the solid target is larger than the beam, this force is proportional to the total acoustic power. As a result, radiation force has been used as the basis of well established methods for measuring acoustic power. When an ultrasound wave propagates in a fluid (liquid or gas), the fluid within the beam flows away from the transducer in the direction of propagation. This is called *acoustic streaming*. (In some older texts it is referred to as 'quartz wind'.)

These two phenomena both exist because of the inherent nonlinearity of acoustic wave propagation in real media. While the magnitude of the force on the target, or of the streaming velocity, also depend on a number of other factors (absorption and reflection coefficient, geometry of the beam and target/container and so on) streaming and radiation force would not exist if pressure and density were linearly related (Beyer 1997).

3.1. RADIATION PRESSURE

The link between radiation force and acoustic streaming lies in the concept of *radiation pressure* (Beyer 1950, Borgnis 1953, Beissner 1985). Note that

39

the radiation pressure does not vary with time in a continuous-wave beam, unlike the *acoustic pressure*. It is a steady 'dc' stress which arises from nonlinear propagation effects.

There are a number of confusions in the literature in the discussions of radiation pressure, and the details of the debate will not be entered into here. What is important is a broad understanding of the phenomena, and a pragmatic approach to the mathematics which allows the general concepts to be laid down and some useful results to be seen. The first source of confusion stems from the use of the same term 'radiation pressure' to mean two different things. The first has to do with the observed radiation force which is experienced by a wall or interface in a beam of sound, and is determined by the wave alone. This is the Langevin radiation pressure. The second is due to Lord Rayleigh, and has to do with the field and its constraints.

3.2.   LANGEVIN RADIATION PRESSURE, P_{Lan}

The *Langevin radiation pressure* is defined as the difference between the mean pressure at a reflecting or absorbing wall and the pressure immediately behind the wall. The derivation of this pressure has been dealt with by several authors (see, for example, Hueter and Bolt 1955). While the pressure derives from the inherent nonlinearity of acoustic wave propagation, the radiation pressure P_{Lan} may be expressed very simply.

For a plane wave impinging normally on a perfect absorber the radiation pressure on the surface is equal in value to the *time-averaged energy density* $\langle E_0 \rangle$ at the surface, or the total energy per unit volume,

$$P_{Lan} = \langle E_0 \rangle = p_0^2/2\rho_0 c^2 = I/c \tag{3.1}$$

where p_0 is the pressure amplitude of the wave, ρ_0 the fluid density, and c the sound speed. I is the intensity of the wave. In a real beam the surface of a target will experience a radiation pressure field which will extend over its surface with a profile identical to the 2D intensity profile of the beam.

If the dimensions of the target extend beyond the boundary of the beam, the radiation force F on the target is given by the integral of the radiation pressure over the target area, S. For an absorbing target the force is

$$F = \int (I/c) \, dS = W/c \tag{3.2}$$

where W is the total acoustic power. The radiation force on the target is therefore a measure of the total power in the beam, the constant of proportionality being the acoustic velocity in the fluid.

Table 3.1. *Examples of radiation pressure for plane waves with energy density E_0.*

Physical situation	Energy densities at interface	Coefficient D Radiation force $F = DE_0 S$
Perfect absorber normal to sound beam	Front: $E_1 = E_0$ Back: $E_2 = 0$	$D = 1$
Perfect reflector normal to sound beam	Front: $E_1 = 2E_0$ Back: $E_2 = 0$	$D = 2$
Perfect reflector at angle θ to the beam	Front: $E_1 = 2E_0$ Back: $E_2 = 0$	$D = 2\cos^2\theta$
Non-reflecting interface normal to beam: $c_1 \neq c_2$	Front: $E_1 = E_0$ Back: $E_2 = E_0(1 - D)$	$D = 1 - c_1/c_2$ for $c_1 < c_2$, force away from source for $c_1 > c_2$, force towards source
Absorbing medium, no interface, $I_z = I_0\, e^{-2\alpha z}$ (α = amplitude attenuation coefficient)	$(\mathrm{d}E/\mathrm{d}z)_z = (\mathrm{d}I_z/\mathrm{d}z)/c$	$D = 2\alpha$ $-(\mathrm{d}E/\mathrm{d}z) = E_z D$ Causes streaming in fluid medium

In practice the proportionality depends not only on the sound speed but also on the surface material and its reflection or absorption properties, and on the geometry of the surface. The coefficients for several situations are set out in table 3.1. A number of details about radiation pressure and force should be noted relating to table 3.1. Firstly a totally reflecting interface doubles the energy density at the surface in the liquid, so doubling the force on the target. For a totally absorbing target $D = 1$ in the direction of the beam, for all angles of incidence. However, if the beam is incident at an angle θ to the normal, the coefficient for the radiation force normal to the surface is $\cos\theta$. A radiation force balance with an absorbing target will still therefore be dependent on the angle of incidence (see Chapter 7 for further remarks on radiation force balances).

The radiation force exerted on a water/air interface is sufficient to raise the water surface until the height is balanced by the hydrostatic pressure in the column produced. This generates a so-called acoustic fountain. The height of the liquid column is given approximately by

$$h = \frac{2I}{\rho_0 c g} \tag{3.3}$$

where I is intensity, ρ_0 is the liquid density, c the speed of sound and g the acceleration due to gravity. It should be noted that the process generating this surface elevation is distinct from that generating streaming, which is discussed below.

Table 3.2. *Estimated radiation forces and pressures in a 3 MHz pulsed beam, I_{ta} 1 W cm^{-2} and I_{pa} 500 mW cm^{-2}, beam area 10 mm^2.*

Totally absorbing surface	
Time-average force	65 μN
Time-average radiation pressure	6.5 Pa
Pulse-average force	32.5 mN
Pulse-average radiation pressure	3.25 kPa

Tissue-like attenuating medium	
Time-average force gradient	0.2 μN mm^{-1}
Time-average radiation pressure gradient	0.02 Pa mm^{-1}
Pulse-average force gradient	0.1 mN mm^{-1}
Pulse-average radiation pressure gradient	10 Pa mm^{-1}

The radiation pressure on an interface between two media with different acoustic speeds, but zero transmission loss, causes a interesting result. When $c_1 > c_2$, the radiation force is backwards towards the transducer. This counter-intuitive result occurs because the force is towards the region of lower energy density. Particularly strong examples of this effect are shown by Beyer (1978) and more recently by Hertz (1993) for water/carbon tetrachloride, and water/aniline interfaces.

The final example in table 3.1 notes that it is possible to extend the simple analysis to include propagation through a uniform attenuating medium. Under these conditions the radiation pressure on an element of the medium depends on its attenuation coefficient. Since there is no interface between media, a radiation force per unit volume is calculated. This is equivalent to a radiation pressure gradient in the direction of the beam.

For measurements in water the radiation force on an absorbing target in a beam of 1 watt total acoustic power would be balanced by a mass of approximately 69 mg. Diagnostic beams have acoustic powers which are typically in the range 1–200 mW, and therefore radiation force balances of the type described in Chapter 7 must be very delicate to allow accurate measurement. Table 3.2 gives estimates of forces, pressures and pressure gradients in a typical diagnostic beam, for both time-average and pulse-average calculations.

3.3. RADIATION STRESS TENSOR

The simple analysis which leads to an equivalence between radiation pressure and scaled energy density has significant deficiencies, however. One example is the failure to explain why the force on a target alters with the angle between the beam and surface (Herrey 1955). The radiation pressure calculated in the simple way described above is a scalar quantity, and thus operates normally on a surface. It is unfortunate that the word 'pressure' has been used to describe what is strictly a stress tensor (Beyer 1978, Brillouin 1964).

The proper representation of the stress exerted on a medium supporting an acoustic wave is by means of the *radiation stress tensor* S_{ij}. For a plane wave propagating in a loss-less medium this is

$$S_{ij} = -\langle P \rangle \delta_{ij} - \rho \langle u_i u_j \rangle \tag{3.4}$$

where $\langle P \rangle$ is the *time-averaged excess pressure* and $\langle u_i u_j \rangle$ is the *time-averaged particle velocity product*. These quantities are second-order quantities, and are non-zero at the nonlinear level. δ_{ij} is the Kronecker delta. So for a sound wave travelling in the x direction the stress tensor is

$$\begin{vmatrix} -\langle P \rangle - \rho_0 \langle u^2 \rangle & 0 & 0 \\ 0 & -\langle P \rangle & 0 \\ 0 & 0 & -\langle P \rangle \end{vmatrix}.$$

The sign notation used reflects the convention that a tension is positive and a compression negative.

The stress tensor is seen to comprise two elements: the time-averaged excess pressure and the term $\langle \rho_0 u^2 \rangle$. The second term represents the time-averaged transport of the *momentum density* $\rho_0 u$.

3.3.1. The excess pressure

The problem with the use of the radiation stress tensor lies with the calculation of the time-averaged excess pressure $\langle P \rangle$. Expressions for the excess pressure depend on the frame of reference. This may be either *Eulerian*, that is evaluated at a point in space; or *Lagrangian*, that is evaluated at a particular particle. Generally these are different, although for the specific circumstance of a rigid reflecting wall, they are identical at its surface.

Beissner (1986) usefully summarises two expressions termed 'Langevin's first and second relations', which are good approximations to these two time-averaged pressures. Langevin's first relation gives the excess pressure in Lagrangian coordinates. It is

$$\langle P^{\mathrm{L}} \rangle = \langle V \rangle + \langle K \rangle + C \tag{3.5}$$

where $\langle V \rangle$ is the *time-averaged potential energy density* and $\langle K \rangle$ is the *time-averaged kinetic energy density*.

$$\langle V \rangle = \langle p^2 \rangle / 2\rho_0 c^2 \qquad (3.6a)$$

and

$$\langle K \rangle = \rho_0 \langle u^2 \rangle / 2. \qquad (3.6b)$$

C is a constant in space and time and depends on the system constraints. According to Lee and Wang (1993), C is zero for the Langevin radiation pressure, which depends only on the wave. Langevin's second relation gives the excess pressure in Eulerian coordinates:

$$\langle P^E \rangle = \langle V \rangle - \langle K \rangle + C. \qquad (3.7)$$

The radiation pressure experienced at a wall is, in general, Lagrangian; that is it is experienced by the particles. The radiation pressure in the radiation stress tensor is at a point in space and is therefore Eulerian. The 1D case clarifies the link between these quantities. The radiation stress is

$$S_{xx} = -\langle P^E \rangle - \rho_0 \langle u^2 \rangle = -(\langle V \rangle - \langle K \rangle + C) - \rho_0 \langle u^2 \rangle. \qquad (3.8)$$

Assuming particle motion is only in one direction, the time-averaged kinetic energy density $\langle K \rangle$ is $\rho_0 \langle u^2 \rangle / 2$, so

$$S_{xx} = -(\langle V \rangle + \langle K \rangle + C) = -\langle P^L \rangle. \qquad (3.9)$$

So in the 1D case the excess pressure expressed in Lagrangian coordinates is the negative of the radiation stress. This is not generally true for the 2 or 3D case.

3.4.   RAYLEIGH RADIATION PRESSURE, P_{Ray}

Rayleigh (1902) originally concerned himself more with a general field quantity rather than with the actual pressure experienced at a real interface. The Rayleigh radiation pressure P_{Ray} is defined as the difference between the average pressure at a surface moving with a particle $\langle P^L \rangle$ and the pressure existing in the absence of sound, with the fluid at rest. It is more of theoretical than practical interest. Calculations of Rayleigh radiation pressure depend on the field constraints in a way that the equivalent Langevin pressures do not, and generally require the constant C to be evaluated. For plane wave propagation in a uniform loss-less medium:

$$P_{Ray} = \frac{(1 + B/2A)\langle E \rangle}{2} \qquad (3.10)$$

where $\langle E \rangle$ is the time average of the energy density and B/A describes the material nonlinearity. It may be considered to be the pressure on an interface between two identical media (Lee and Wang 1993).

Expressed in tensor form, P_{Ray} is (Beyer 1978)

$$\begin{vmatrix} -[(1 + B/2A)\langle E \rangle]/2 & 0 & 0 \\ 0 & -[(B/2A - 1)\langle E \rangle]/2 & 0 \\ 0 & 0 & -[(B/2A - 1)\langle E \rangle]/2 \end{vmatrix}.$$

The *Rayleigh radiation pressure* can be considered to be a property of the field. For diffractive beams it varies throughout the three dimensional extent of the field and it follows therefore that three dimensional radiation pressure gradients can be generated.

Hertz (1993) has described an elegant method for mapping the variation of radiation pressure throughout an acoustic beam. Noting that the radiation pressure of 1 Pa corresponds to a sound intensity of 150 mW cm^{-2}, very sensitive local pressure measurement is required. Hertz developed a bubble technique: the radiation pressure is transferred to air through a bubble at the end of an air-filled tube placed at the point of measurement. Using a reference tube at a position of known pressure, outside the ultrasound beam, the differential pressure between the tubes was measured by a gas pressure transducer. The method is independent of the static pressure due to surface tension, and measures the difference between the Lagrangian pressure (exerted on the bubble) and the Eulerian pressure given by the reference tube outside the beam. The method can operate at radiation pressures less than 40 Pa, above which the calibration ceases to be linear. Figure 3.1 shows the results of measurements made in this way in the field of a weakly focused 1 MHz transducer of 16 mm diameter. The focal intensity is approximately 6 W cm^{-2}.

The presence of continuous radiation force on a discrete target may cause its displacement. While this target movement is noted and used in the measurement of total, time-averaged acoustic power in some radiation force balance designs, there is very little experimental evidence of strain in homogeneous targets, nor of intermittent movement of macroscopic targets during the period of a single pulse, both of which might be expected at high enough intensities. Slight displacement of agar gel was reported by Dyer and Nyborg (1960) during continuous wave exposure, but unfortunately this study was carried out at 25 kHz, and its extrapolation to higher frequencies is uncertain. Of greater interest perhaps is the report given in this volume by Greenleaf *et al* (Chapter 14) which shows acoustic emission from a target at the pulse repetition frequency. This observation would imply that the target position is being modulated in the presence of a pulsed acoustic beam, so demonstrating a local radiation force effect.

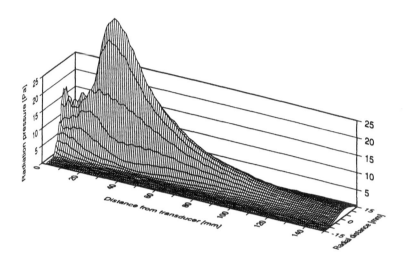

Figure 3.1. *Measured radiation pressure field from a weakly focused 1 MHz transducer. (From Hertz (1993) with permission.)*

3.5.   ACOUSTIC STREAMING

As shown above, acoustic waves are capable of generating radiation pressure gradients within the exposed medium supporting the wave. If the medium is a fluid, and therefore free to move, it will do so. The result bulk flow of fluid is called 'acoustic streaming', or in older texts, 'quartz wind'. An example of the visualisation of acoustic streaming is shown in figure 3.2. The stream was generated in water from the weakly focused 1 MHz beam associated with the radiation pressure field mapped in figure 3.1. It may be noticed that the stream commences predominantly in the region of the focus, and not at the source, as observed by others. A number of authors have looked at the theoretical problem of predicting streaming in plane progressive waves, either within a limiting tube or as a free stream, using a variety of simplifications (Eckart 1948, Markham 1952, Tjøtta 1959, Tjøtta and Naze Tjøtta 1993, Nyborg 1998). The difficulties stem from the origin of the force causing the stream, which we have seen to be a radiation stress tensor field with acoustically nonlinear origins, the range of geometries both for the beam and for the boundaries of the fluid, and in the degree of nonlinearity associated with the ultrasound propagation. However, the following equation summarises the situation adequately to describe many situations:

$$v = (2\alpha I / c v) d^2 G \qquad (3.11)$$

where v is the streaming velocity, I is the intensity, c is the acoustic speed, v is the kinematic viscosity $= \eta/\rho$, η is the shear viscosity, d is the beam diameter and G is a geometric factor.

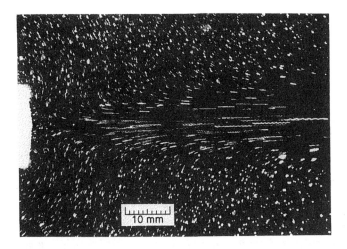

Figure 3.2. *Photograph of streaming motion induced in water by a weakly focused 1 MHz transducer. The radiation pressure field is shown in figure 3.1. Exposure time 1 s.*

Examination of equation (3.11) shows that the pressure gradient which drives the stream is that associated with the radiation pressure noted earlier, $2\alpha I/c$ or equivalently $2\alpha E_0$. Examples of the radial profiles for acoustic pressure gradient for a number of materials are shown in figure 3.3. These profiles were derived from high resolution measurements of pulse-average intensity in a 3 MHz beam propagating to a high degree of nonlinear distortion at its focus in water. The radiation pressure gradient was derived from the linear summation of the contributions from all the harmonic components using equation (3.11), and taking account of the frequency dependence of attenuation in each of the materials: water, amniotic fluid, blood and average soft tissue.

Any absorption mechanism can contribute to α (Nyborg 1953), including shear viscosity, bulk viscosity and relaxation. Excess absorption from strong nonlinearity and shock propagation also contributes to the radiation pressure gradient (Starritt *et al* 1989). The pressure gradient, and hence the streaming velocity, are frequency dependent because of the frequency dependence of α. In this way it differs from the radiation pressure on a totally reflecting or totally absorbing interface, which has no dependence on the frequency of the wave. The retarding forces which limit the maximum velocity are from two sources: firstly the viscous drag of the fluid itself, given by $1/\nu$ and secondly the geometry of the boundaries of the fluid space, represented by the factor G.

The equation in this form represents the steady state, in a continuous beam of uniform cross-section. A number of other considerations come into play when evaluating streaming in biological fluids under conditions relevant to

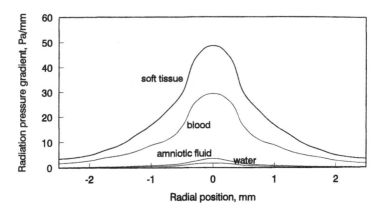

Figure 3.3. *Radial profiles of radiation pressure gradients calculated from measured pulse intensities at the focus of a shocked, weakly focused beam in water. Frequency 3 MHz; pulse-average intensity 118 W cm^{-2}.*

medical ultrasound. Several authors have considered the time taken for the stream to reach a steady state. This can be an important quantity when considering whether streaming may be established when a beam is being moved slowly. In water, Starritt *et al* (1989) reported shortest streaming rise-times of the order of few hundred milliseconds for unscanned pulsed beams, and of a few seconds for scanned (imaging) beams. Working at higher frequencies (10 and 20 MHz), with narrower beams and at acoustic powers as low as 3.5 mW, Hartley (1997) recorded rise times in water of 200 ms, and in blood of 80 ms. These observations confirm that acoustic streaming can reasonably be expected to be established within the period of typical scanning dwell times.

In practice the flow may not exist as a simple streamline when formed in a large volume. Evidence from visualisation of streams (Starritt *et al* 1991) and laser anemometry (Mitome *et al* 1996), shows that a nominally steady stream may alter in position and magnitude, probably caused by thermal convection in the fluid volume. This can be induced, for example, by the transducer self-heating. Such instabilities are particularly noticeable when slow streaming in large beams is being generated, for example in physiotherapy beams. Conversely there is some evidence from laser anemometry studies that the stream at the focus of a pulsed beam has a disturbed axial component surrounded by streamline flow. Full theoretical descriptions of such flows are not yet possible.

Equation (3.11) gives an expression for the streaming velocity in a plane propagating wave. Some authors (Wu and Du 1993, Kamakura *et al* 1995, Nowicki *et al* 1997) have developed expressions appropriate for streaming in weakly focused beams, which adequately predict experimental measurements of streaming in water and other fluids. Since the radiation pressure gradient

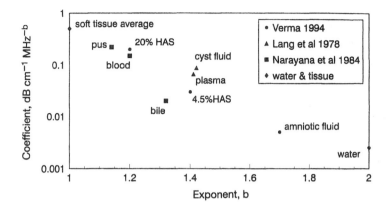

Figure 3.4. *Attenuation coefficient, α, for several biological liquids, expressed as α = af^b in dB cm^{-1} MHz^{-b}. The graph demonstrates the decrease of exponent from 2 (for water) to 1 (for soft tissue) as the attenuation at 1 MHz increases.*

depends on intensity, the streaming velocity tends to be a maximum at about the focus. The importance of nonlinear propagation effects becomes more relevant in focused beams (see Chapter 2). Under conditions of extreme nonlinearity, such as those found in diagnostic pulses propagating in water, the excess attenuation from shock propagation adds to that from the fluid alone. In such beams the streaming velocity in water is considerably greater than predicted using the simple formula (Starritt *et al* 1989). The degree of enhancement from this cause is less marked in other fluids, however, at least under the conditions appropriate in medical applications. This would seem to depend on the balance of loss mechanisms between that from the fundamental, and the excess loss from absorption of the harmonics generated by nonlinear propagation effects (see Chapter 2). Attenuation in biological fluids varies both in absolute value and in frequency dependence (see figure 3.4), and the balance between nonlinear propagation and thermo-viscous loss will be complex. In pulsed fields at 3.5 MHz (Zauhar *et al* 1998) and at 10 and 20 MHz (Hartley 1997) streaming velocities were observed to be lower in blood than in water for a given source power. Hartley also observed that the streaming velocity in blood increased with frequency, suggesting that nonlinear effects were not dominant under the conditions used in his experiment.

The studies referenced above have demonstrated that pulsed diagnostic beams with pulse lengths less than 1 μs can cause streaming velocities which can reach 10 cm s^{-1} in water. Furthermore, streaming can be generated in diagnostic imaging fields under some circumstances. It must be inferred from these observations that the radiation pressure gradient associated with the passage through a fluid of a single diagnostic pulse can be sufficient to

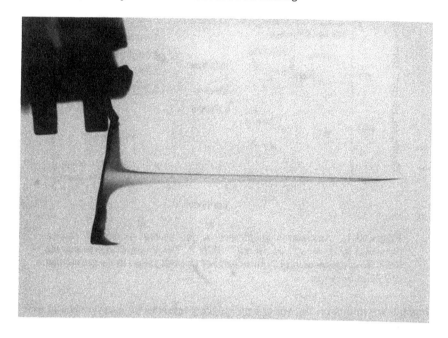

Figure 3.5. *Acoustic streaming demonstrated using thymol blue for a 3 MHz pulsed beam. (From Starritt et al (1991), with permission.)*

cause fluid displacement, since unless one pulse generates movement, there can be no integrated effect to form a stream.

3.5.1. Methods of measuring acoustic streaming

Fluid flow may be visualised and measured by a remarkably wide range of methods, many of which have been applied to the study of acoustic streaming. A useful review of methods for evaluating flow has been given by Merzkirch (1987). If it is required only to visualise the stream, a number of marker methods have been used. These include the sheet illumination of the stream with laser light, in combination with seeding the fluid with fine particles (Mitome *et al* 1993, Hertz 1993) (see figure 3.2). Starritt *et al* (1989) used a simple dye method, colouring the water with potassium permanganate. A useful visualisation method uses the pH sensitive dye thymol blue which is orange-yellow at low pH and becomes blue above $pH = 9.6$. Local pH is altered by electrolysis adjacent to a platinum electrode across which the stream passes (Starritt *et al* 1991). An example of a stream formed in water from a medical ultrasonic transducer is shown in figure 3.5. Recirculation mixing re-equilibrates the pH, so preventing the accumulation of the blue colour, allowing continual studies of streaming to be carried out over an extended period.

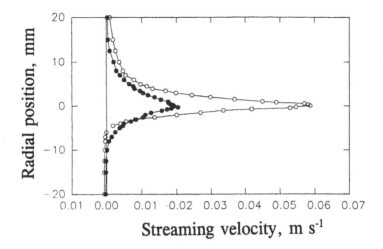

Figure 3.6. *Acoustic stream profiles measured using a laser velocimeter at the focus of a 5 MHz weakly focused beam. Total acoustic power was 150 mW in both cases. • continuous wave, low amplitude beam; ○ strongly shocked pulsed beam, 1.4 μs pulse length, 10.1 kHz prf, 1.20 W cm^{-2} time-average intensity. The beam width at the focus was 2.5 mm.*

Marker and dye methods are difficult to quantify however, requiring sequential image capture and off-line analysis. Several measurement methods have been explored to quantify streaming velocity. Hot-film anemometry was successfully used by Starritt *et al* (1989) for studies in water, giving spatial resolution of the order of 1 mm. However the method requires careful calibration, since the response to velocity is nonlinear, and it cannot be used reliably in any fluid which may contaminate the surface of the sensor; this includes most fluids of biological interest. Laser anemometry is a very powerful technique for flow measurement in transparent liquids, and has been used to explore acoustic streaming in water and some transparent biological solutions (Duck *et al* 1993, Mitome *et al* 1993, Hertz 1993). The method can achieve spatial resolution better than 1 mm and speed resolution of the order of 1 mm s^{-1}. An example of two stream profiles at the focus of a 5 MHz beam is shown in figure 3.6. Enhancement of the stream velocity from the use of high amplitude, nonlinear pulses is demonstrated. Nevertheless the equipment is fairly expensive, and again the observations are limited to transparent liquids, so excluding studies in blood.

Ultrasound Doppler methods present an obvious and attractive alternative approach. These have been implemented either by using the same beam to cause the stream as to measure it (Hartley 1997, Nowicki *et al* 1997) or by using a separate continuous-wave Doppler probe (Zauhar *et al* 1998). These approaches have advantages of cheapness and availability, but suffer from poor spatial resolution and limited sensitivity to low streaming velocities.

At present, though, all reports of streaming measurement in blood have used ultrasonic methods. The final method which should be easily available within a medical environment is the use of magnetic resonance. Such methods remain to be explored, but offer the ability to evaluate flow in three dimensions, with very high sensitivity, within opaque substances, and with adequate spatial resolution.

3.6.   OBSERVATIONS *IN VIVO* OF RADIATION PRESSURE EFFECTS

3.6.1.   Streaming

Streaming *in vivo* seems to be a relatively well noted incidental observation in clinical diagnostic practice, but one which is poorly documented (Duck 1998). Streaming has been noted in cyst fluid in breast, ovary and testicle, within ventricular haemorrhage in an infant, and within liquefied vitreous humour. Streaming has been suggested as a diagnostic aid to distinguish fluid from solid lesions (Nightingale *et al* 1995). Streaming itself *in vivo* would not appear to be hazardous. Nevertheless it demonstrates the presence of radiation pressure gradients which deserve study as a mechanism for biological effect within cellular structures.

3.6.2.   Observed biological effects apparently related to radiation pressure

Literature reports which relate a bio-effect to radiation pressure, rather than to a thermal or cavitation cause, fall broadly into two categories. The first category includes those which suggest a biological outcome directly from compression or tension. Lizzi *et al* (1981) reported transient blanching of the choroid of a rabbit eye prior to the onset of thermal damage, in work at intensities intended to cause retinal lesions. The blanching was noticed more readily at higher pulse amplitudes, and is therefore consistent with the view that radiation stress caused compression of the blood vessels. More recently, in studies directed at the effects of ultrasound exposure on cardiac function, Daleki *et al* (1997) reported that direct exposure of a frog heart to a long ultrasound pulse caused a reduction in central blood pressure. This could be mimicked by exposing an absorber on the outer myocardial surface to the same beam. It was concluded that radiation force exerted by either ultrasound exposure or external application gave similar outcomes.

The second class of reports relate to neuro-sensory reactions to ultrasound. Daleki *et al* (1995) noted that the pressure sensors in the skin can detect pulsed ultrasound either directly, or via an acoustic absorber. The equivalence of sensation between these two methods suggests a radiation force effect. At sufficiently high pulse intensities, it is possible to stimulate other sensory receptors, which include hot and cold skin receptors and pain

Table 3.3. *Summary of pulsing regimes required to generate acoustic bio-responses apparently caused by radiation pressure.*

Effect	Pulse length	Number of pulses
Choroid blanching (Lizzi)	100 μs	1
Tactile sensation (Daleki)	1 ms	repetitive
Cardiac response in frog (Daleki)	5 ms	1
Fluid movement (Starritt)	0.5 μs	1

receptors both at the skin and at depth (Gavrilov 1984). However in this series of studies, carried out over a range of frequencies, the threshold for response was most strongly correlated with particle displacement, rather than intensity or radiation pressure. Several authors have explored the apparent direct stimulation of the auditory nerve by ultrasound. Exposure of the auditory nerve can apparently result in the sensation of hearing the pulse repetition frequency, although direct bone conduction cannot be excluded. Magee and Davies (1993) stimulated the auditory nerve using a pulsed Doppler beam via the foramen magnum, apparently avoiding this criticism.

Table 3.3 summarises the pulsing regimes used to generate some of these responses.

3.7. DISCUSSION

The complete evaluation of the interaction of the radiation stress field with tissue is challenging. The stress field varies both spatially and temporally, and it may be that spatial and temporal gradients in the stress field may have greater importance in affecting tissues than time averaged or spatially averaged values. The variation of tissue attenuation can also give rise to a local radiation pressure gradient causing local strain (see Chapter 4). Radiation stress is experienced only during the time a pulse is passing through tissue and there is no stress between pulses. A proper evaluation of the radiation stress field within tissues caused by diagnostic pulses, and its outcome, is a substantial challenge.

The magnitude of the stress in modes such as pulse echo and pulsed Doppler depends on the pulse average intensity rather than the time average intensity. Traditionally, bio-effects which show a dependence on time average intensity are interpreted as being thermal in origin and those which depend on pulse average intensity or pulse amplitude are explained in terms of cavitation. Once we start to consider radiation stress as a possible mechanism it becomes more difficult to separate the dependence on exposure factors. Although radiation stress effects are experienced only during a pulse, and are therefore dependent on pulse amplitude or pulse average intensity, some of the outcomes, like streaming and some of the sensory effects, rely on an integration of the stresses. In this respect they resemble thermal effects. However the time scales can be very different. Acoustic streaming for example is established in time scales of less than a second, whereas thermal effects tends to develop slowly, generally requiring several minutes to reach equilibrium.

Table 3.3 shows the minimum pulse lengths and number of pulses required to produce the radiation stress effects described in the papers reviewed above. It shows that some effects can be produced and observed following a single pulse of ultrasound, while others require the stress to be repeated over a number of pulses. Choroid blanching occurred with a single pulse about 100 μs in duration and the cardiac response in frogs was also seen with a single long pulse, 5 ms in duration. In order to sense ultrasound on the skin a repetitive stress is required and similarly, while local fluid movement must be induced by a single pulse, there needs to be a repeated effect before it manifests itself as bulk streaming.

It is increasingly recognised that radiation force effects provide a possible explanation for ultrasound bio-effects which appear to be non-thermal and non-cavitational in nature. In adult tissue the radiation forces are highly unlikely to be significant compared with the tensile strength of the tissue. However embryonic tissue does not have the structural strength which it develops in later fetal and adult life. It is not known what is the stress threshold for permanent cell displacement in such weakly bound tissues. Until these matters are more fully explored, it is appropriate to continue to exercise caution in the use of highest exposures during ultrasound examinations during the earlier stages of pregnancy.

REFERENCES

Beissner K 1985 On the time-average acoustic pressure *Acustica* **57** 1–4
—— 1986 Two concepts of acoustic radiation pressure *J. Acoust. Soc. Am.* **79** 1610–2
Beyer R T 1950 Radiation pressure in a sound wave *Am. J. Phys.* **18** 25–9
—— 1978 Radiation pressure—the history of a mislabeled tensor *J. Acoust. Soc. Am.* **63** 1025–30
—— 1997 *Nonlinear Acoustics* (Woodbury, NY: Acoustical Society of America) (reprint of the 1974 book by US Department of the Navy)
Borgnis F E 1953 Acoustic radiation pressure of plane compressional waves *Rev. Mod. Phys.* **25** 653–64
Brillouin L 1964 *Tensors in Mechanics and Elasticity* (Tr R O Brennan) (New York: Academic)
Daleki D, Child S Z, Raeman C H and Carstensen E L 1995 Tactile perception of ultrasound *J. Acoust. Am.* **97** 3165–70
Daleki D, Raeman C R, Child S Z and Carstensen E L 1997 Effects of pulsed ultrasound on the frog heart III: the radiation force effect *Ultrasound. Med. Biol.* **23** 275–85
Duck F A 1998 Acoustic streaming and radiation pressure in diagnostic applications: what are the implications? *Safety of Diagnostic Ultrasound, Progress in Obstetrics & Gynaecology Series* eds S B Barnett and G Kossoff (New York: Parthenon) pp 87–98

Duck F A, MacGregor S A and Greenwell D 1993 Measurement of streaming velocities in medical ultrasonic beams using laser anemometry *Advances in Nonlinear Acoustics* ed H Hobaek (Singapore: World Scientific) pp 607–12

Dyer H J and Nyborg W L 1960 Ultrasonically induced movements in cells and cell models *IRE Trans. Med. Electron.* **ME-7** 163–5

Eckart C 1948 Vortices and streams caused by sound waves *Phys. Rev.* **73** 68–76

Gavrilov L R 1984 Use of focused ultrasound for stimulation of nerve structures *Ultrasonics* **22** 132–8

Hartley C J 1997 Characteristics of acoustic streaming created and measured by pulsed Doppler ultrasound *IEEE Trans. Ultrasonics Ferroelectr. Freq. Control* **44** 1278–85

Herrey E M J 1955 Experimental studies on acoustic radiation pressure *J. Acoust. Soc. Am.* **27** 891–6

Hertz T G 1993 Applications of acoustic streaming *PhD dissertation* (Lund University, Sweden)

Hueter T F and Bolt R H 1955 *Sonics* (New York: Wiley)

Kamakura T, Matsuda K, Kumamoto Y and Breazeale M A 1995 Acoustic streaming induced in focused Gaussian beams *J. Acoust. Soc. Am.* **97** 2740–6

Lang J, Zana R, Gairard B, Dale G and Gros Ch M 1978 Ultrasonic absorption in the human breast liquids *Ultrasound Med. Biol.* **4** 125–30

Lee C P and Wang T G 1993 Acoustic radiation pressure *J. Acoust. Soc. Am.* **94** 1099–109

Lizzi F L, Coleman D J, Driller J, Franzen L A and Leopold M 1981 Effects of pulsed ultrasound on ocular tissue *Ultrasound Med. Biol.* **7** 245–52

Magee T R and Davies A H 1993 Auditory phenomena during transcranial Doppler insonation of the basilar artery *J. Ultrasound Med.* **12** 747–50

Markham J J 1952 Second-order acoustic fields: streaming with viscosity and relaxation *Phys. Rev.* **86** 497–502

Merzkirch W 1987 *Flow Visualisation* 2nd edition (London: Academic)

Mitome H, Ishikawa A, Takeda H and Kyoma K 1993 Effects of attenuation of ultrasound as a source of driving force of acoustic streaming *Advances in Nonlinear Acoustics* ed H Hobaek (Singapore: World Scientific) pp 589–94

Mitome H, Kozuka T and Tuziuti T 1996 Measurement of the establishment process of acoustic streaming using laser Doppler velocimetry *Ultrasonics* **34** 527–30

Narayana P A, Ophir J and Maklad N F 1984 The attenuation of ultrasound in biological fluids *J. Acoust. Soc. Am.* **76** 1–4

Nightingale K R *et al* 1995 A novel ultrasonic technique for differentiating cysts from solid lesions: preliminary results in the breast *Ultrasound Med. Biol.* **21** 745–51

Nowicki A, Secomski W and Wojcik J 1997 Acoustic streaming: comparison of low-amplitude linear model with streaming velocities measured by 32-MHz Doppler *Ultrasound Med. Biol.* **23** 783–91

Nyborg W L 1953 Acoustic streaming due to attenuated plane waves *J. Acoust. Soc. Am.* **25** 68–75

—— 1998 Acoustic streaming *Nonlinear Acoustics* eds M F Hamilton and D T Blackstock (New York: Academic) pp 207–31

Rayleigh, Lord 1902 On the pressure of vibrations *Phil. Mag.* **3** 338–46

Starritt H C, Duck F A and Humphrey V F 1989 An experimental investigation of streaming in pulsed diagnostic ultrasound fields *Ultrasound Med. Biol.* **15** 363–73

Starritt H C, Duck F A and Humphrey V F 1991 Forces acting in the direction of propagation in pulsed ultrasound fields *Phys. Med. Biol.* **36** 1465–74

Tjøtta S 1959 On some non-linear effects in sound fields with special emphasis on the generation of vorticity and the formation of streaming patterns *Arch. Math. Naturvidensk* **55** 1–68

Tjøtta S and Naze Tjøtta J 1993 Acoustic streaming in ultrasound beams *Advances in Nonlinear Acoustics* ed H Hobaek (Singapore: World Scientific) pp 601–6

Verma P K 1994 personal communication

Wu J and Du G 1993 Acoustic streaming generated by focused Gaussian and finite amplitude tonebursts *Ultrasound. Med. Biol.* **19** 167–76

Zauhar G, Starritt H C and Duck F A 1998 Studies of acoustic streaming in biological fluids with an ultrasound Doppler technique *Br. J. Radiol.* **71** 297–302

CHAPTER 4

ULTRASONIC PROPERTIES OF TISSUES

Jeffrey C Bamber

INTRODUCTION

The properties that in practice are used for providing information about the structure of biological tissues, and which contribute to the physically complicated process of formation of medical ultrasound images, are acoustic speed, impedance, absorption, scattering and attenuation. In addition, for acoustic pressure amplitudes used in many medical applications, the nonlinearity of sound wave propagation becomes important. Knowledge of these characteristics, and their variation with frequency, amplitude, temperature, age, pathology, etc, is important for a thorough understanding and optimum use of present and potential ultrasonic diagnostic techniques. They also play an important role in determining the nature and magnitude of the biological effects of ultrasound, which form the basis both for ultrasound therapeutic procedures and concern for the continued safe use of medical ultrasound. The natural variation of the acoustical characteristics of biological media is often very broad and this range is further broadened by difficulty in making accurate measurements. As a consequence it is also difficult to specify in detail the wave–medium interaction mechanisms that are responsible for the observed acoustical characteristics.

4.1. BASIC CONCEPTS

4.1.1. Attenuation, absorption, scattering and reflection

In biological tissues both scattering (which includes refraction and reflection) and absorption contribute to the attenuation of a plane sound wave

57

propagating in the positive z-direction:

$$u(z, t) = u_0 e^{-\alpha z} e^{i\omega(t-z/c)} \qquad (4.1)$$

where c is the speed of sound of angular frequency ω, t is time and the wave amplitude, u_0, is modified by a constant factor, α, per unit path length, z. Practical situations, however, rarely involve perfectly plane waves and there are almost always additional losses (or gains) of acoustic intensity due to the diffraction field of the sound source. Diffraction corrections, to compensate for diffraction losses or associated diffraction phase changes, are required for accurate measurement of attenuation, scattering or sound speed.

Scattering and absorption by biological media are described in terms of bulk coefficients, μ_s and μ_a, which are intensity cross-sections per unit volume [1]. If it were possible to identify individual scatterers or absorbers (which is almost never the case) then the bulk scattering coefficient would be (in the absence of multiple scattering) $\mu_s = \sum_i n_{si}\sigma_{si}$, given that the number density distribution of individual scatterer cross-sections is described by $n_{si}(\sigma_{si})$. A similar relationship would hold for the absorption coefficient and for a plane wave of power W, the total power increment either absorbed or scattered, $dW_s + dW_a$, in a path length dz, would be $\mu_s W\,dz + \mu_a W\,dz$. Integration of this relationship gives

$$W = W_0 e^{-(\mu_s+\mu_a)z}. \qquad (4.2)$$

The sum of the intensity scattering and absorption coefficients $(\mu_s + \mu_a)$ is called the intensity attenuation coefficient, μ. When expressed in terms of the ratio of two measured powers,

$$\mu = -\frac{1}{z} \log_e \frac{W}{W_0} \qquad (4.3)$$

μ typically takes units of cm^{-1}. The amplitude attenuation, absorption and scattering coefficients (α, α_a, and α_s, respectively) are expressed similarly as the ratio of two measured wave amplitudes and also have units of cm^{-1}, so that $\mu = 2\alpha$. If, however, the ratios of either power or amplitude are expressed in terms of decibels then μ and α become numerically equal, taking units of dB cm^{-1}. It is common to use the symbol α when expressing the attenuation coefficient in dB cm^{-1}, which is therefore equal to 4.343 times μ in cm^{-1} and 8.686 times α in cm^{-1}.

For individual scatterers the differential scattering cross-section, $\sigma_{ds}(\theta, \varphi)$, defines the ratio of the power scattered in a particular direction, $W_s(\theta, \varphi)$, to the incident intensity, I, per unit solid angle Ω:

$$\sigma_{ds}(\theta, \varphi) = \frac{d\sigma_s}{d\Omega}(\theta, \varphi) = \frac{d}{d\Omega} \frac{W_s(\theta, \varphi)}{I} \qquad (4.4)$$

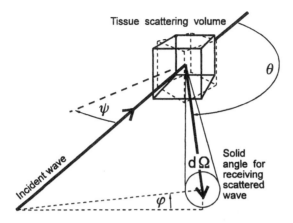

Figure 4.1. *Geometrical variables used when defining scattering coefficients (adapted from [91]).*

where figure 4.1 defines the geometrical variables. For a medium in which the scattering is incoherent (i.e. when there is no measurable spatial correlation in the scattering properties) the power scattered will be proportional to the number of scatterers, or the volume of the medium. For a medium of discrete scatterers a differential scattering coefficient might then be defined, similarly to the scattering coefficient, as $\mu_{ds}(\theta, \varphi) = \sum_i n_{si}\sigma_{dsi}(\theta, \varphi)$. In general, however, the scattering strength as a function of scattering angle (θ, φ) is described directly by the differential scattering coefficient $\mu_{ds}(\theta, \varphi) = (d\mu_s/d\Omega)(\theta, \varphi)$, defined as the power scattered per unit incident intensity, per unit volume, per unit solid angle. An important special case of this is the backscattering coefficient, $\mu_{bs} = (d\mu_s/d\Omega)(\pi, 0)$.

The scattering coefficient and the differential scattering coefficient are related by

$$\mu_s = \int_0^{4\pi} \mu_{ds}(\theta, \varphi)\, d\Omega \qquad \text{(note: } d\Omega = \sin\theta\, d\theta\, d\varphi\text{)}. \qquad (4.5)$$

For scattering with azimuthal symmetry this reduces to

$$\mu_s = 2\pi \int_0^{\pi} \mu_{ds}(\theta) \sin\theta\, d\theta. \qquad (4.6)$$

If the scattering is completely isotropic (spherical symmetry) this reduces further to

$$\mu_s = 4\pi \mu_{ds} = 4\pi \mu_{bs}. \qquad (4.7)$$

Unlike μ_s, the units for $\mu_{ds}(\theta, \varphi)$ and μ_{bs} may only be length^{-1} solid angle^{-1} (usually cm^{-1} sterad^{-1}); there is no decibel equivalent.

There are coherent and incoherent components to the scattering from tissues, the relative contributions depending on the randomness of the scattering medium and the ratio of either the wavelength or the size of the scattering volume to the typical spacing of scattering structures. Patterns of scattering observed as a function of measurement variables such as angle, orientation, frequency or spatial position, commonly display average large-scale trends on which are superimposed finer fluctuations due to interference effects. If the sound is scattered incoherently (i.e. from many randomly distributed scattering sources) then these finer fluctuations, although deterministically related to the scattering structure, represent multiplicative noise as far as measuring an average scattering coefficient is concerned. The probability of the scattered amplitude is then described by a Rayleigh distribution with a standard deviation 0.52 of the desired mean value and the typical scale of the fluctuations in the scattering coefficient being determined by the parameters of the experiment (e.g. beam width, pulse length) rather than by the tissue structure. Semi-periodic interference fluctuations due to coherent scattering may be observed if there is underlying regularity to the tissue structure. Alternatively, the properties of the fine-scale interference fluctuations in the scattering pattern may become more characteristic of the tissue structure if they arise from coherent scattering structure that has a spatial scale approaching, or greater than, the size of the scattering volume. Thus information about the scattering structure is potentially available from either component of the scattering pattern: the average trend due to incoherent scattering or the local fluctuations due to the coherent component, but not the local fluctuations due to the incoherent scattering.

Interference, or diffraction, patterns may be observed as fluctuations in the values of various 'instantaneous' differential scattering coefficients with changing scattering angle (θ, φ), wavelength, or tissue orientation angle, ψ. On the other hand, in order to study the average angular or orientation dependence of scattering, or indeed to use such data to estimate μ_s and its frequency dependence, it is necessary to compute the average of a number of statistically independent scattering patterns obtained by altering some suitable independent (or assumed to be independent) variable, such as the spatial position within the tissue or, for isotropic tissues, tissue orientation. This process is invariably a compromise and usually results in loss of resolution of some kind, whether this be spatial, spectral, angular or ability to resolve spatial anisotropy.

Reflection is essentially geometrical scattering from structures much larger than the wavelength. For plane interfaces Snell's law of refraction applies. The intensity reflection coefficient, R, for a plane wave incident at an angle θ on a plane interface between two fluids of characteristic acoustic impedance Z_1 and Z_2, (where $Z_1 = \rho_0 c$) and emerging at an angle of refraction ϕ, is the ratio of the reflected to the incident intensity and is equal to

$$R = \left(\frac{Z_2 \cos\theta - Z_1 \cos\varphi}{Z_2 \cos\theta + Z_1 \cos\varphi} \right)^2 \qquad (4.8)$$

but this expression will not adequately describe the situation at interfaces involving solids, where conversion between longitudinal and shear waves may take place.

4.1.2. Speed of sound

Although both group and phase speeds may be defined, and sound speed is wave-amplitude dependent, the field of medical ultrasonics has adopted the simplistic viewpoint of a single valued sound speed which can be measured for each medium, given by

$$c = \sqrt{1/\rho_0 \kappa} \qquad (4.9)$$

where ρ_0 is the mean density and κ is the adiabatic bulk compressibility.

4.1.3. Nonlinearity

The assumption that density and acoustic pressure are linearly related is an approximation resulting from the use of only the first term in a series expression between the two. Extension of the theory to the next order of small quantities by inclusion of the next (quadratic) term results in a second order, or nonlinear, relationship between density and pressure. The phase speed may then be written

$$c_p = c \left(1 + \frac{B}{2A} \frac{u}{c} \right)^{2A/B+1} \qquad (4.10)$$

where u is the particle velocity, A is $\rho_0 c^2$, and the ratio B/A is known as the nonlinearity parameter, which varies with the medium and with temperature [2–4]. The dependence of sound speed on particle velocity, and hence acoustic pressure, described by equation (4.10) produces a wave which changes shape, becoming progressively steeper during the fall from the high- to the low-pressure cycles, as it propagates. These and other matters associated with nonlinear propagation are covered more fully in Chapter 2.

4.1.4. Transducer diffraction field

The acoustical characteristics of biological media are defined in terms of plane wave propagation (see chapter 1) but diffraction from the source often contributes substantially to the measured values, which must therefore be corrected to yield (plane wave) values that are independent of the

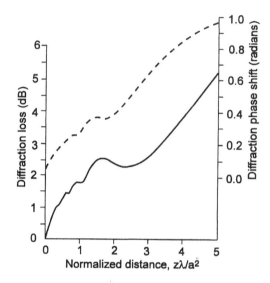

Figure 4.2. *Diffraction loss and phase shift from a circular piston source of radius a, seen by an identical receiver, as a function of normalised distance (units of λ/a^2) between the two (adapted from [85, 86]).*

measurement system. Diffraction corrections may be applied to any acoustic field parameter (intensity, pressure, etc) as a function of both space (three dimensions) and time (or frequency spectrum). For a single frequency component, the apparent magnitude and phase of the signal received in a loss-less medium by a plane circular transducer of identical size and shape to a transmitting transducer decay and fluctuate with transducer separation (figure 4.2). Diffraction loss (or 'gain') curves of this kind will vary with the size and shape of either of the transducers, and with the displacement phase/amplitude profiles of either the source or the receiver (i.e. their focusing/apodisation characteristics). Furthermore, modern medical ultrasound imaging systems use electronically focused transducer arrays in which the received aperture size, apodisation and focal length may all vary with distance.

4.1.5. *Pulse-echo imaging, speckle and echo texture*

A B-scan is formed from the magnitude of the backscattered signal ($|A|$) as a function of the spatial position of the scattering volume. It depicts the spatial variation in backscattering coefficient, smoothed by the shape of the resolution cell or point spread function (determined by the lateral extent of the beam and axial extent of the pulse envelope) and modified by a number of factors which include the source acoustic spectrum, the diffraction field of the source, the attenuation of sound in the tissue and the geometry of the receiver. However, superimposed on this average scattering magnitude are fine-scale fluctuations due to interference effects. Mathematically the B-scan is obtained from the magnitude of a coherent convolution between a spatially variant, object dependent radio frequency (RF) point spread function $h(x, y, z)$ and the tissue backscattering impulse response $T(x, y, z)$ [5, 6]:

$$|A|(x, y, z) = |h(x, y, z) * T(x, y, z)|. \qquad (4.11)$$

$T(x, y, z)$ is related, by whichever model for scattering is favoured, to the spatial rate of change in density and/or compressibility in the medium. Biological media contain structures over a very wide range of spatial frequencies and the echo magnitude for a single image point will result from the coherent sum of echoes due to a number of structures present within the resolution cell surrounding that point. Such interference effects strongly influence the fine detail, or texture, of B-scan images. For any uniform region of the medium containing a large number density (more than about 10 per resolution cell) of small ($\ll \lambda$), randomly distributed (incoherent) scattering sources, the image magnitude, although deterministic, varies apparently randomly with position. The result, known as a speckle pattern, is the multiplicative noise discussed in section 4.1.1 and has (in the absence of nonlinear signal processing) predictable statistical properties. Any departure from this situation, such as the introduction of an isolated coherent scatterer, the existence of distributed coherent scattering from a non-random (periodic) distribution of scatterers, a reduction of the number density of scatterers, or the existence of a transition from a region of one backscattering coefficient to another, will result in a change in these statistical properties. The probability density function, $p(|A|)$, of fully developed speckle is a Rayleigh distribution [7]:

$$p(|A|) = \frac{2|A|}{\langle A^2 \rangle} \exp \left\{ -\frac{A^2}{\langle A^2 \rangle} \right\} \qquad (4.12)$$

where $\langle A^2 \rangle$ is the mean-square scattering amplitude. The second order statistical properties of speckle are described by the fact that the autocorrelation function of speckle is determined only by the autocorrelation function of the interrogating point spread function, in turn determined by the transducer aperture, centre frequency, focal length and bandwidth. The average speckle cell sizes (or speckle correlation length), in the directions parallel to (S_{cz}) and perpendicular to (S_{cx}) the pulse propagation, are given for the focal distance, z_f, of an aperture of diameter D and a Gaussian pulse with half-power bandwidth, Δf, by

$$S_{cz} = 1.37/\Delta f \qquad S_{cx} = 0.87 \lambda z_f / D \qquad (4.13)$$

where f is in MHz and dimensions in millimetres. Estimation of scattering (and attenuation) coefficients from backscattered signals, which is a form of speckle reduction, requires averaging. For spatial averaging, equation (4.13) defines the distances between samples required for independent estimates. Due to the phase integrating nature of the receiver (see section 4.1.6), however, the speckle in the near field (Fresnel zone) and beyond the focal zone of a focused aperture possesses a much shorter correlation length that is difficult to predict analytically [8]. The phenomenon of speckle also occurs in other coherent systems such as imaging with laser light and radar. For the acoustic case, however, opportunities exist for interference effects to occur from sources other than the simple convolutional model defined by equation (4.11), e.g. multiple scattering, phase reversal at scattering boundaries and wave refraction or distortion due to sound speed inhomogeneity of the medium.

4.1.6. *Receiver phase sensitivity*

Unlike optical detectors, which are energy sensitive, acoustic receivers are phase sensitive. For scattered waves received under near-field conditions (e.g. from outside the focal plane), or if the wave propagates through an aberrating medium, phase variations across the receiver will cause destructive addition across the wave, resulting in further fine-scale variation (i.e. speckle) and underestimation of the signal magnitude. This may cause underestimation of scattering coefficients, overestimation of attenuation, loss of signal in imaging and additional variations in estimated sound speed.

4.2. MEASUREMENT METHODS

The reader is referred to previous reviews (e.g. [1, 4, 9, 91]) for more complete descriptions of methods of measuring ultrasonic propagation properties of tissues.

4.2.1. *Measurement of the absorption coefficient*

A thermocouple probe may be used to measure that portion of the ultrasonic energy absorbed locally and transformed into thermal energy [4, 9–11]. The method may be applied *in situ*, *in vivo* and in small local structures such as the mouse spinal cord [12, 13] and is specific to a very local region of tissue. A thermocouple junction of diameter small relative to the ultrasonic wavelength is implanted in the sample under study and exposed to short bursts of sound. A rise in temperature during the first 0.1 s, due to viscous motion absorption by the wires, is to be ignored, but at frequencies below about 300 kHz this tends to obscure accurate observation of the subsequent, relatively linear, rate of temperature rise which is due to absorption in the sample. Calculation of μ_a is from the density, specific heat and acoustic intensity using an iterative technique in which an initial value, calculated using an approximate measured incident intensity, I, is used as a value for μ to estimate the true value of I knowing the path length in the sample. Successive better estimates of μ_a and I are then obtained. It is difficult to produce broad plane waves and small thermocouples for high frequency work and convergence of μ_a does not occur if μ is too large. Measurements up to 7 MHz have been made in tissues, with a total uncertainty of the order of 10–15% [14]. An alternative approach employs the rate of cooling of the tissue around the thermocouple following a short (less than 0.1 s) sound pulse [15].

4.2.2. *Measurement of the attenuation coefficient*

The attenuation coefficient, μ, may be measured using systems that are *narrow band* or *broad band* [1]. A further distinction is between *fixed path* and *variable path* instruments; the path length being that between the transmitter and receiver. *Variable path* systems, which measure the rate of change of received signal with the position of the receiver/reflector, provide absolute values of μ but diffraction corrections are required, which are proportionately worse at low frequencies (below 3 MHz the accuracy becomes worse than ±5%). Above 12 MHz millilitre sample volumes can be measured without appreciable diffraction corrections. Accuracies of ±0.5% are possible with good temperature control, homogeneity, calibration of electronic amplification, distance measurement and transducer/reflector alignment. Variable path methods have been used to study solutions and liquids (e.g. water [16]), or homogenised tissues, at frequencies up to about 200 MHz.

The *substitution* [17] or *insertion* [18] techniques are *fixed path*, making diffraction corrections less important and alignment tolerances less stringent, but the measurements are relative to the attenuation coefficient of a reference liquid (often water). In the *substitution method*, suitable for measurements down to 0.3 MHz (errors as low as ±2%) on liquids or ground tissues in suspension, the transmitting and receiving transducers are mechanically linked so that they move simultaneously: one through the reference liquid, the other through the test medium. The two media are separated by an acoustically transparent window and only the proportion of the acoustic path length occupied by the reference and test liquids is varied. The *insertion technique*, which has been applied to solid tissues at frequencies from below one megahertz to many hundreds of megahertz, involves using equation (4.3) to determine μ from the ratio of the powers in the signal received with and without the tissue between the transmitter and receiver. A reference medium separates the transducers from the sample, as shown in figure 2.3, reducing the error due to diffraction losses by reducing the relative change in overall path length when the tissue is inserted. Neglecting forward scattering

$$\mu = \mu_w + \frac{1}{2d_z}\{\ln W' - \ln W + 4\ln(1 - R)\} \tag{4.14}$$

where μ_w is the attenuation coefficient of the reference medium (e.g. water), d_z is the tissue thickness, W and W' are the powers received with and without the tissue present, and R is the reflection coefficient at the tissue/water interface. It is difficult to accurately cut and mount parallel-sided tissue slabs and measurement errors can be up to ±10%. These may be reduced by repeating measurements on multiple tissue slabs of different thickness or by using cutting and mounting procedures designed to reduce specimen thickness errors to ~±2% [19].

Ultrasonic interferometers and reverberation chambers need to be calibrated with a reference liquid but very small volumes (e.g. 10 ml) of liquids or suspensions may be studied, with accuracies of the order ±10% [20], over a wide frequency range (e.g. 0.2 to 10 MHz or higher).

Broad band techniques use spectral analysis of modulated acoustic signals to obtain μ as a function of frequency within the bandwidth of the system [1]. For example, using equation (4.3), $\mu(f)$ may be obtained from the logarithm of the power spectrum of the pulse transmitted through the specimen divided by the power spectrum of the reference pulse. When combined with the fixed path insertion technique, these methods are popular for studying solid tissues but the experimental difficulties are easily underestimated and understanding of the errors is incomplete. Diffraction corrections, while small, vary with frequency and may be important when measuring media with very low attenuation. Spectral processing necessitates the use of phase sensitive receivers which suffer from phase cancellation artefacts and the use of short pulses with relatively high peak acoustic pressures may result in additional losses due to nonlinear propagation.

In vivo measurement presents more problems. Complete through-propagation is possible for only a few organs (e.g. the female breast) but if regions of tissue sufficiently large and homogeneous can be located, estimates of μ may be obtained from the average rate of decrease of the signal backscattered by the tissue with depth of propagation [1]. If commercial ultrasound imaging instrumentation is employed corrections must be applied for nonlinear signal processing [21]. These are *variable path* methods and therefore require (frequency dependent) diffraction corrections, but the diffraction loss as a function of distance (see section 4.1.4) is unknown since frequency dependent attenuation and the inhomogeneous nature of the propagation medium will alter the diffraction field, and the diffraction loss depends on the nature of the receiving transducer which in this case is determined by the scattering structure. The use of different models for the diffraction loss produces different results for the attenuation estimate. Fixed path methods therefore offer greater advantages for *in vivo* work and may be implemented by a method analogous to the substitution technique: the pulse-echo transducer is coupled to the tissue by a water bath and is moved towards or away from the tissue so as to ensure that each estimate of the backscattered signal is obtained at the same position in the diffraction field [22].

References [1] and [91] provide discussions of errors in attenuation measurement. Errors present even for homogeneous media include those associated with: the forward scattered part of μ_s, which causes underestimation of μ to a degree dependent on measurement geometry (which could be 20% for soft tissues or higher for strongly scattering tissues such as lung or bone), the tissue path length, the frequency f, linearity of voltage signals corresponding to W and W', linearity of transduction or sound propagation, reflection at specimen boundaries, signal loss in the reference medium, and differences in sound speed between the tissue and coupling medium (diffraction losses and refraction or phase cancellation due to misalignment or non-parallelism of the specimen).

The acoustical inhomogeneity of tissues gives rise to additional errors due to refraction, diffractive scattering and phase cancellation. Phase cancellation losses arise because the power-dependent signal is obtained from the voltage signal squared, proportional to the integral of acoustic pressure across the receiver, whereas its true value is the integral of the squared pressure distribution. If the phase of the wave received is not constant over the receiver then W will be underestimated. Distorted wavefronts may be caused by variations in tissue sound speed or specimen thickness. The error may be eliminated by using power sensitive receivers such as the radiation force balance [23] or acoustoelectric receivers [24], and reduced by the use of thin, evenly cut specimens, a coupling medium whose sound speed is matched to that of the specimen, a narrow sound beam in the vicinity of the specimen, and a receiver of size and/or distance from the specimen adjusted according to the spatial coherence of the distorted wave. A focused transmitter and receiver, with the specimen at the joint focal plane, helps to reduce phase cancellation without increasing refraction losses.

The conditions under which tissues are measured may substantially influence their observed acoustical characteristics. Reviews of this topic may be found elsewhere [1, 91] but the important factors may include the ambient temperature, time since death, temperature of storage, medium of storage, chemical preservation (fixation), freezing, medium of measurement (e.g. pH, concentration), unwanted gas inclusion or production, and inherent biological variation within or between specimens. Factors which may influence the relationship between the acoustical characteristics *in vitro* and *in vivo* are: blood flow and other tissue movement, blood pressure/volume, difference between blood and the medium of measurement, surrounding tissues and organs (e.g. tension supplied by connection), temperature, tissue orientation (for anisotropic tissues) and any factors specific to the death of particular cells and tissues.

Performance of an attenuation measurement system may be assessed by measurement of a standard material which has known, or widely agreed upon, characteristics. For absolute measurements in liquids the data for water [16] are often employed, while for relative measurements the sound speed and attenuation coefficient of castor oil [1, 25, 91] are suitably similar to those of many soft tissues in the low MHz range of frequencies.

4.2.3. *Measurement of sound speed*

Prior reviews provide summaries of the methods of measuring sound speed [2, 4, 9, 27, 91]. Most attenuation measurement methods are also capable of providing simultaneous data on sound speed. They have similar limitations with regard to their applicability to tissues versus liquids, and cover similar frequency ranges to those described in section 4.2.1. The most common absolute measurement methods are variants of a general transit-time, or time-of-flight (TOF), measurement, initially developed as a variable path technique for liquids [28]. Sound speed is obtained either from the variation in TOF with path length or from the TOF for a known fixed path length. There have been many approaches to determining the TOF; in the 'sing-around' system, separate transmitting and receiving transducers are used, the received pulse is shaped to be fed back as a trigger for the transmitting transducer and the resulting pulse repetition period provides a self-adjusting measurement of TOF. Accuracies of $\sim \pm 0.1\%$ have been claimed using such methods [29], or even better accuracies using continuous wave ultrasonic interferometers [30].

For solid tissues relative methods are more suitable, the reference medium usually being water or saline. A common approach is another variation of the pulse TOF method, using the insertion geometry described by figure 4.3, either with a single transducer plus reflector or with separate transmitting and receiving transducers. The average (group) speed of sound in the tissue, c_t, may then be calculated from

$$\frac{1}{c_t} = \frac{1}{c_w} - \frac{\Delta t}{2d_z} \tag{4.15}$$

where c_w is the speed of sound in the reference medium at known temperature and Δt is the measured time shift in the position of the received sound pulse with and without the tissue specimen in place. If long pulses are employed velocity dispersion can be observed. It is also possible to measure phase speed and dispersion in short pulse TOF systems by subtracting the phase spectrum of the received sound pulse (obtained by Fourier transformation) from that of the pulse which travelled through reference

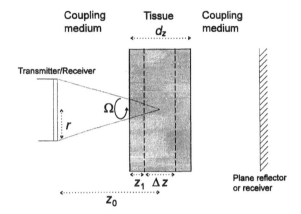

Figure 4.3. *Typical experimental arrangement for attenuation and sound speed measurement by the insertion technique, and backscattering measurement by the substitution method. Here a single transducer acts as both transmitter and receiver but a separate receiver may also be used, in place of the reflector (adapted from [1]).*

medium only [31]. The substitution method [17] described in section 4.2.1 can provide phase speed by noting the variation in the difference in phase between the received signal and that of the source as relative path length is varied. Similar approaches to phase comparison, although with fixed relative path lengths, are employed when making relative measurements of sound speed using acoustic microscopes.

For *in vivo* measurement, whenever conventional transmission measurement is impossible, methods that use signals backscattered by the tissue include (i) the use of misregistration between multiple images of the same tissue structure to estimate the beam refraction that occurs when sound enters the body from a known coupling medium [32]; (ii) statistical estimation of TOF over a path including scattering from a region of tissue defined by overlapping beams from transmitting and receiving transducers [33, 35]; and (iii) finding the sound speed which, when used to reconstruct a phased array image, produces the sharpest image of a specified structure [34]. Most of these methods can provide accuracies and precisions of ±0.5–1% under laboratory conditions but *in vivo* tissue inhomogeneity is a problem.

Errors are reviewed elsewhere [27, 91] but note that in the relative measurement of figure 4.3 a path length error of ±10% translates to a sound speed error of less than ±1%, if the reference medium is water and its temperature is known. Also, pulse time-of-arrival (TOA) can be difficult to define; it is common to employ the time when the received signal is first detectable above noise or the occurrence of one of the zero crossings, but these measures will be in error if the pulse changes shape after propagation through the tissue. Account must be taken of attenuation-related pulse amplitude changes and pulse stretching due to frequency dependent attenuation [27]. TOA of the centre of the pulse, e.g. estimated using cross-correlation methods, is thought to be better although earliest TOA is advantageous if there is multipath propagation. Phase cancellation may seriously affect TOF measurements and provides further justification for correlation techniques. Diffraction phase corrections, made in variable path and interferometric systems at low frequencies, may or may not be negligible for the insertion technique. The effects of measurement conditions, where they are known, follow a related pattern to their effects on attenuation [27, 91].

Water [30], saline [36] and water–ethanol mixtures [37] may be used for checking the accuracy and precision of a sound speed measurement [27, 91].

4.2.4. *Measurement of scattering*

A complete specification of ultrasonic scattering by tissue, reviewed elsewhere [38–40, 91], requires measurement of many different scattering coefficients, some of which were defined in section 4.1.1, and measurement of the variation of any scattering coefficient with one of the experimental variables (frequency, scattering angle, orientation, spatial position, etc) involves averaging over one or more of the other variables.

Both direct and indirect estimates of $\mu_s(f)$ are possible. Indirect measurements have resulted from:

• the difference between the attenuation coefficient μ and the absorption coefficient μ_a [15, 41], although both measurements have rarely been made on the same specimens and the result is likely to be inaccurate;

• subtracting from a measurement of μ for an intact tissue, the value measured after grinding the tissue to remove the scatterers [42] (but it is difficult to do this without altering μ_a [27, 91]);

• making use of changes in μ_{bs} and μ, measured simultaneously, when the tissue is modified in some way [91]. If the modification does not alter the angular scattering distribution μ_{bs} will be proportional to μ_s. This is particularly useful if it can be assumed that μ_a is constant, in which case the ratio of μ_s/μ may be obtained from the ratio of the fractional change in μ to that in μ_{bs}.

Direct measurement of $\mu_s(f)$ would involve integrating over all angles the measured average angular dependence of $\mu_{ds}(\theta, \varphi)$. In practice measurements are made over a limited range of angles and equations (4.6) or (4.7) used even though the level of isotropy of scattering is in general unknown [43–45]. The measurement of scattering coefficients requires correction for attenuation and for the scattering volume, which may vary with scattering angle. A more established literature is associated with μ_{bs} but, although it is simpler to measure, there are still a variety of approaches.

For the arrangement in figure 4.3, the power backscattered by the tissue, W_{bs}, may be derived in a similar manner to the 'weather radar equation' [46] and divided by the power, W_{ref}, received from a plane interface of reflection coefficient, R_{ref}, positioned perpendicular to the sound beam at distance z_0 [91]. This substitution technique yields

$$\mu_{bs} = \frac{W_{bs}}{W_{ref}} \frac{2\mu R_{ref}}{\Omega (1 - R)^2 \, e^{2\mu_w(z_1 + \Delta z/2)} \, e^{-2\mu z_1} (1 - e^{-2\mu \Delta z})} \tag{4.16}$$

where Ω is the solid angle subtended at the receiver by the centre of the scattering volume at distance z_0, μ_w is the intensity attenuation coefficient of the coupling medium (water), μ is the intensity attenuation coefficient of the tissue, z_1 is the distance from the tissue surface to the beginning of the scattering volume, Δz is the length of the scattering volume (defined by convolving the acoustic pulse with the signal gate) and the incident beam is assumed to have the highly simplified structure of a constant cross-section (along the length of the scattering volume) of uniform magnitude and phase within specified boundaries, outside which the incident intensity is zero. This technique is designed to correct for diffraction losses, the transmitted acoustic power and the cross-sectional area of the scattering volume; many versions have been derived since the first measurement of μ_{bs} for blood [47], the least restrictive appearing to be that of reference [38].

Many of the errors in scattering measurements are associated with the normalisation in the substitution method, which is an area of discrepancy between investigators [44, 48, 49, 91]. Equation (4.16) describes a simple approach and its deficiencies include the fact that the incident beam directivity is not a rectangular function. The effect of this depends on the nature of the reflector, in particular the degree of coherence in the scattering, but for a plane interface at the focal plane of a concave transducer the error is thought to be of the order of 1 dB [91]. In principle, however, complete specification of the transmitted field can be used to normalise for reliable estimates of μ_{bs} at almost any position in the field of either focused or unfocused sources [38].

Other potential uncertainties are associated with factors analogous to those in attenuation measurement, e.g. linearity of received signal amplification, uncertainty in attenuating tissue path length, beam refraction by the specimen and phase cancellation. In addition μ_{bs} estimates are affected by uncertainty in the value of the area of the receiving transducer aperture, and the values of c or μ for the tissue specimen. Phase cancellation is reduced by ensuring that observations are made under far-field conditions but even energy sensitive detection would require averaging of diffractive fluctuations, with a sufficient number of uncorrelated instantaneous estimates. Errors for tissues are not known but may be as high as $\pm 70\%$ [91] and scattering measurements may be even more severely influenced by the condition of the tissue than attenuation [1, 91].

Tests of scattering measurement systems are less straightforward than for attenuation, due to the lack of availability of an ideal standard material. As a plane reflector a water/CCl$_4$ interface may be used, for which equation (4.8) shows R to be −48 dB at 20°C (although strongly temperature dependent), and random distributions of microscopic glass spheres in gelatine are popular incoherent scattering models for which μ_{ds} can be computed as a function of both frequency and angle.

4.2.5. Measurement of nonlinearity

There are two main methods for determining B/A. The *finite amplitude method* [50–53] is based on the measurement of the amplitude of the second harmonic generated during propagation of a sinusoidal wave. Two transducers are used, as in variable path attenuation measurement. The second harmonic pressure $p_2(z)$, as a function of propagation distance z, is

$$p_2(z) = \frac{\pi f(2 + B/A)}{2\rho_0 c^3} p_1^2(0)z \exp\left[-\left(\alpha_{a1} + \frac{\alpha_{a2}}{2}\right)z\right] \mathrm{DIFF}(z) \qquad (4.17)$$

where $p_1(0)$ is the acoustic pressure output of the transmitting transducer at the fundamental frequency, α_{a1} and α_{a2} are the absorption coefficients of the measured medium at the fundamental and second harmonic frequencies, and $\mathrm{DIFF}(z)$ is the diffraction correction. By analogy with the insertion method of attenuation measurement, normalisation for $p_1(0)$ and $\mathrm{DIFF}(z)$ is achieved by comparison of $p_2(z)$ determined for the sample with that of a reference medium (10% NaCl solution) of known B/A and an acoustic impedance similar to that of the sample. The *thermodynamic (or 'pressure jump') method* requires measurement of the rate of change in sound speed in the sample with ambient pressure, p, when the pressure is decreased rapidly (at about 5–10 MPa s^{-1}) to approximate an adiabatic depressurisation [53, 54]

$$B/A = 2\rho_0 c \left(\frac{\partial c_p}{\partial p}\right). \qquad (4.18)$$

The finite amplitude and thermodynamic methods have sources of error in common with those for attenuation and sound speed estimation, respectively. The total systematic error is $\sim\pm 8\%$ for the finite amplitude method and $\sim\pm 5\%$ for the thermodynamic method (but worse if samples are inhomogeneous) [55]. For liquids, relative thermodynamic measurements of B/A can be achieved with a precision better than $\pm 0.3\%$ [56]. Temperature appears to be important but there appears to be no significant difference between *in vivo* and *in vitro* values [57].

4.3. ULTRASONIC PROPERTIES OF TISSUES

This section provides a summary of material in previous reviews [1, 4, 27, 58–63, 91].

4.3.1. Absorption and attenuation

4.3.1.1. Molecular species. As illustrated in figure 4.4, μ_a depends strongly on the structural complexity of the medium. Although body tissues contain over 60% water the contribution of the absorption by water to μ in tissues is negligible at low megahertz frequencies but, because of the f^2 dependence, this contribution may prove to be more substantial above 100 MHz. Small molecules, such as amino acids in aqueous solution, barely modify $\mu_a(f)$ of water but when assembled into whole biopolymers, such as proteins, they contribute significantly to μ_a. For biopolymers, μ_a may increase with molecular weight, but not always and the dependence is not well understood [64]. Absorption is often characteristic of interaction between macromolecules, having a nonlinear concentration dependence and being increased by chemical fixatives that promote cross-linkages.

The frequency dependence of ultrasonic absorption for biopolymers (typically between f^1 and $f^{1.4}$) suggests a distribution of time constants for relaxation processes, but the specific relaxation mechanisms and their number are unknown. Relaxation is the process by which the acoustic wave may repeatedly perturb some physical or chemical equilibrium in the medium. The wave energy is cyclically redistributed between translational, molecular vibrational and structural, or lattice vibrational and structural states, but with a time lag for return to the original energy distribution, so that wave energy degrades to heat. The frequency dependence of a single relaxation process has a general form described by

$$\mu_R = \frac{2A_R f^2}{1 + (f/f_R)^2} \tag{4.19}$$

where A_R is a constant (the relaxation amplitude) and f_R is the relaxation frequency corresponding to the time constant for the process. At frequencies well below f_R, $\mu_R \propto f^2$, which describes the behaviour of water at medical frequencies due to the 'classical' processes of shear viscosity and thermal conductivity. Absorption in solutions of biopolymers has been modelled by summing multiple relaxation processes and the classical absorption due to shear viscosity, η, (neglecting thermal losses):

$$\mu = 2f^2 \left(\frac{16\pi^2 \eta}{3\rho_0 c^3} + \sum_i \frac{A_{Ri}}{1 + (f/f_{Ri})^2} \right). \tag{4.20}$$

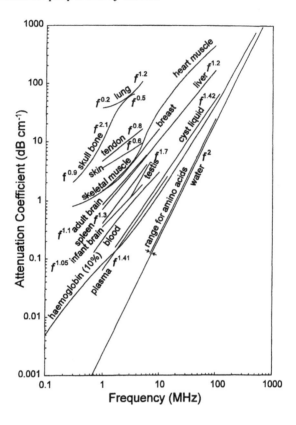

Figure 4.4. *General trends for the variation of the attenuation coefficient with frequency and tissue (after [1], which also contains the references to the original sources of data).*

It is thought that at physiological pH the dominant relaxation mechanisms arise from perturbing the hydration layers of macromolecules, i.e. structural relaxation between the number of water molecules at any instant in a free or bound state.

4.3.1.2. Tissues. Despite the difficulty, from published data, of distinguishing the biological variation of μ of many tissues from systematic effects, the following observations may be made.

For most tissues, the contribution of scattering to attenuation is unknown. Scattering has not been observed at medical frequencies from some uniform media such as amniotic fluid, aqueous humour, vitreous humour, and cyst liquids (although not all). The very high value for μ in inflated lung (see figure 4.4) is predominantly due to scattering by gas bubbles and some models attribute the frequency dependence and high value in bone, which may be between 2 and 20 times that for soft tissue, to scattering

Table 4.1. *Selection of published estimates for the contribution of scattering to attenuation.*

Method	Species	Direct or indirect	μ_s/μ (%)	Freq. (MHz)	Reference
(i) Liver					
Angular distribution of μ_{ds}	Human	Direct	19.00	4–7	[44]
Angular distribution of μ_{ds}	Calf	Direct	2.00	3–7	[43]
Measurement of μ_a and μ	Beef	Indirect	18.00	5.60	[15]
Changes in μ_{bs} and μ	Human	Indirect	9–16	4.00	[1]
Change in μ on destruction of tissue structure (known overestimation, see text)	Beef	Indirect	30.00	1–10	[42]
(ii) Blood					
From μ_{bs} assuming isotropic scattering	Human	Indirect	0.1–0.6	5–15	[90, 44]
Angular distribution of μ_{ds}	Human	Direct	0.1–0.5	4–7	[44]

from anisotropic grain-like structures. Bone is discussed in greater detail in Chapter 16 of this volume.

An intact cellular structure appears to be responsible for about 19% of the relatively low attenuation coefficient of blood and high level structure accounts for about 30% of attenuation in liver [42]. However in neither case are these necessarily the contributions of scattering to attenuation, since high level structure may contribute to absorption through mechanisms such as viscous relative motion, where damping of the sound wave may occur due to relative motion between a structure and the surrounding viscous medium, when the density of the structure is different from that of the medium. There is a cyclic transfer of momentum between the structural inhomogeneities and the embedding medium, which takes place at a finite rate resulting in a frequency dependence of absorption not unlike that for relaxation phenomena but where the equivalent of the relaxation frequency is itself a function of frequency. The frequency dependent function is specific to the shape of the inhomogeneity and has been evaluated only for simple shapes such as spheres and cylinders. A frequency dependence for this source of attenuation in tissue may be obtained by summing a number of processes involving a distribution of sizes and shapes of inhomogeneity. It is thought that viscous motional losses could account for 10% of attenuation in blood and that only 1% at maximum can be accounted for by longitudinal wave scattering [65]. In solid soft tissue the evidence for the contribution of scattering to attenuation is inconsistent. In normal liver, the tissue most often studied, viscous relative motion losses have been calculated to be as high as 60% of μ. Table 4.1 summarises published results from both direct and indirect estimates of μ_s/μ.

The frequency dependence of μ, as shown in figure 4.4, is often described over local frequency ranges by the model $\mu = bf^m$. Most soft tissues and body liquids exhibit a value of m a little greater than unity and this has been observed, for some tissues, to extend to $f > 100$ MHz. At still higher frequencies one might expect the f^2 dependence of μ_a by water to begin to dominate. The wide variation in published results for any particular soft tissue does not in general permit an adequate evaluation of competing hypotheses, in terms of scattering or absorption models, for the observed frequency dependence of attenuation. A possible exception is muscle, for which the data have been said to suggest the existence of a strong relaxation region at $\sim$40 kHz, in addition to mechanisms at other frequencies.

In soft tissues μ is a complex function of temperature, T. For $T = 6$–40°C, and $f > 2$ MHz, $d\mu/dT$ is negative and decreases both with T and f. As f decreases one finds a frequency at which $d\mu/dT \sim 0$, below which $d\mu/dT$ may be positive. Above 40°C, $d\mu/dT$ becomes positive and irreversible, even at high frequencies, consistent with heat denaturation of macromolecules. This behaviour is not understood, being partially consistent with either molecular relaxation or viscous relative motion.

The structural protein collagen has a particularly high value of elastic modulus and is thought to play a key role in determining attenuation and scattering characteristics of many tissues. The data in figure 4.4 exhibit the trend that increasing μ correlates with increasing collagen content. For parenchymal tissues, which in their non-diseased state contain relatively little collagen, μ depends mainly on the overall protein content, i.e. is inversely related to the water content. Various diseases and wounds may cause a local increase in collagen content which, in the absence of an increase in fluid content, invariably results in an increase in attenuation and scattering. For some tissues and/or disease conditions, fat content is also important. Sound is scattered strongly at fat/non-fat interfaces due to a low sound speed in fat. If the interfaces are present in sufficient number, such as in normal human breast or fatty diseased liver, then they may contribute to raising μ and μ_s in these tissues. Other diseases, some types of cancer for example, raise the water content and consequently lower μ and μ_s although the situation is complicated since tissue changes usually involve the simultaneous alteration of many such properties.

4.3.2. Sound speed

Figure 4.5 provides examples of the ranges of c that have been measured in various mammalian tissues, with some non-biological media for comparison. Excluding lung and bone (left of the figure) the total range in tissues is $\sim \pm 10\%$ of the mean, and spatial variations are generally neglected in the design of most medical imaging systems. The large differences in values for c and ρ_0 on either side of boundaries between a soft tissue and gas

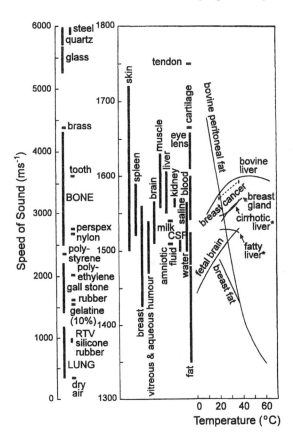

Figure 4.5. *Ranges of values for speed of ultrasound in tissues for temperatures 20–37°C (left and middle) and examples of the temperature dependence of sound speed (far right). Sources of data as reviewed in [27], except where indicated by *, in which case they are from [54] (after [91]).*

or bone give rise to very high levels of scattering which, combined with the associated high μ, make it difficult (often impossible) to obtain useful ultrasonic images through bone, lung or other gas.

For soft tissues in the middle section of figure 4.5, variation in c is determined largely at the molecular level, rather than by the presence of intact cells or higher level structure. Frequency dispersion (studied up to 100 MHz) is very small, the largest reported variation being greater than 0.5% increase in c per decade, which is close to that predicted from both relaxation theory and general relationships between attenuation and dispersion [66]. The temperature dependence, however, is strong enough to have accounted for some of the variability of early results *in vitro*. For fat dc/dT is strongly negative and for non-fatty tissues it is positive, which gives rise to unpredictable temperature dependencies in complex tissues such as breast, a minimum in sound speed in adult brain at 15°C [66], a slight negative temperature dependence in fatty liver [54] and a strong temperature dependence of scattering from fat/non-fat interfaces. For fatty tissue c has also been observed to drop abruptly with temperature in association with a solid/liquid phase transition, although it is not clear that melting caused the rapid change in c [92]. Heat denaturation or coagulation of tissues produces a small but measurable decrease in sound speed [67]. The shapes of the curves of c versus T for non-fatty tissues, with a maximum in the region of 40–70°C are similar to those for water and aqueous salt solutions, whereas fat behaves more like most other liquids. The low sound speed for fat at body temperature makes it an acoustically important tissue component, accounting for some of the observed variation in c between tissues. Increasing sound speed also correlates with increasing protein content, particularly the structural protein collagen present in skin, cartilage and tendon, and with decreasing water content. For tissues such as liver, which contain only moderate amounts of collagen, normal and pathological variations in sound speed are largely explained by simultaneous variation of both the water and fat content. *In vivo* measurement of c in an organ such as the liver may enable unique identification of increased fat content as the cause of increased attenuation or scattering which might otherwise be due to fibrosis. High water content is believed to explain the results for fetal brain (figure 4.5, right).

The high sound speed and attenuation in skull bone is accompanied by a relatively large dispersion, believed to be associated with scattering. Bone may also propagate transverse waves and be highly anisotropic, dispersion of longitudinal sound speed varying ~1–12%, over 1–3 MHz, depending on the bone type and direction of propagation. Transverse wave speed also varies with direction but falls in the range 1800–2200 ms^{-1}.

Muscle is also anisotropic but there are differing observations for whether c is highest along or across the fibres [27], and different authors have observed sound speed to increase and decrease with muscle contraction (although an increase has only been observed *in vivo* where factors such as blood content vary during contraction).

4.3.3. Scattering

The scattering properties of tissues have been reviewed by various authors [5, 68, 91]. Most published results are for μ_{bs} (figure 4.6) and the data exhibit a wide variation for some tissues, making it difficult to evaluate models for their scattering structure.

The 'simplest' of the scattering tissues is blood, for which good agreement has been obtained between experimental results and theoretical predictions based on modelling blood as a collection of small (radius, r, about 3 μm) fluid (aqueous haemoglobin solution) spheres of the same volume as red blood cells. Experiments with red cell 'ghosts' have demonstrated that

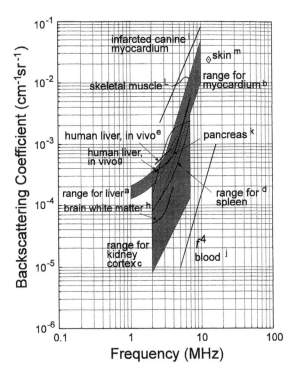

Figure 4.6. *Examples of trends for variation of ultrasonic backscattering coefficient with frequency and biological medium. Key: a, b, c, d, i, j, k, l, as reviewed in [68] but excluding calf liver; e [48]; g [87]; h [73]; i [88]; m [88] (adapted from [91]).*

there is negligible contribution to scattering from the cell membrane. When $\lambda > \sim 20r$, Rayleigh scattering theory applies and [70]

$$\sigma_{ds}(\theta) = \frac{V_e^2 \pi^2}{\lambda^4} \left[\frac{\kappa_e - \kappa_f}{\kappa_f} + \frac{3(\rho_e - \rho_f)}{2\rho_e + \rho_f} \cos \theta \right]^2 \qquad (4.21)$$

where V_e is the volume of a red cell, κ_e, ρ_e and κ_f, ρ_f are the compressibility and mass density of red cells and the surrounding fluid respectively. Thus the scattered power will vary as f^4 (observed in flowing blood or red cells suspended in saline) and the square of the red cell volume (observed by utilising the species dependence of red cell size). The predicted angular scattering distribution (predominantly backward, from the sum of the monopole compressibility and dipole density terms) agrees well with measured distributions although a model that includes the effect of fluid viscosity on the dipole term, provides a better fit [71]. There is a linear relationship between μ_{bs} and red cell concentration only for haematocrit <8%. Dilute blood is exceptional in that σ_s and σ_{ds} for individual scatterers can be calculated [68]. For normal whole blood, red cells do not behave as either independent or small scatterers:

- σ_{bs} per cell decreases until it peaks at haematocrit $= 13\%$;
- red cells aggregate into larger units that scatter more strongly, according the V^2 dependence. Aggregation is reduced by shearing forces so that σ_{bs} is inversely related to flow rate; hence venous blood is more echogenic than arterial blood;
- μ_{bs} correlates with both erythrocyte sedimentation rate and the concentration of plasma proteins associated with clotting;
- for whole blood the frequency dependence is reduced from f^4 at low flow rates.

Turbulent flow also increases the scattering from blood and blood clots are more echogenic than unclotted blood. Blood structure is spatially correlated and may thus, at any instant, act like solid tissues to produce partially coherent scattering, to a degree dependent on the haematocrit [70].

Identification and modelling of the scattering structure of tissues more complex than blood has proven difficult. A popular model for the scattering structure of tissues is the inhomogeneous continuum model in which mass density and compressibility are assumed to fluctuate continuously by small amounts about their mean values, ρ_0 and κ_0, and in a manner described statistically by spatial autocorrelation functions (ACF) in both quantities. Both Gaussian and exponential ACFs have been employed. A further simplification, but one that is disputed, is that fluctuations in density are negligible by comparison with those in compressibility. An alternative assumption, that ρ and κ fluctuations are inversely related, leads (for a Gaussian ACF in both ρ and κ) to [44]

$$\mu_{ds}(\theta) = \frac{\pi^8 \bar{a}^3}{\lambda^4} \left[\frac{\sqrt{\langle |\kappa - \kappa_0|^2 \rangle}}{\kappa_0} - \frac{\sqrt{\langle |\rho - \rho_0|^2 \rangle}}{\rho_0} \cos \theta \right]^2$$
$$\times \exp[-2\pi^2 \bar{a}^2 (1 - \cos \theta)] \qquad (4.22)$$

where $\bar{a}$ is the correlation distance of the ACF, describing the scale of a single characteristic scattering structure (inversely related to mean spatial frequency). The frequency dependence of backscattering from myocardium (figure 4.6) for example appears to be quite well described by a fit somewhere in the range $f^{2.7}$–$f^{3.3}$, consistent with a relatively simple characteristic structure close to the cellular level. Liver has a complex structure: μ_{bs} increases with f from $\sim f^0$ at 1 MHz to about f^3 at 7 MHz, suggesting that a simple model with one correlation length (which predicts a decreasing frequency dependence) is inadequate. Composite models with two or three correlation lengths give a better fit [72] and would also seem to be appropriate for other tissues such as brain [73] and kidney [74]. It

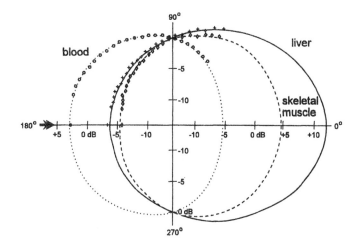

Figure 4.7. *Polar plots of the angular dependence of the differential scattering coefficient at 6 MHz for blood, liver and skeletal muscle [44].*

therefore seems that in such tissues (figure 4.6) most of the scattering at low frequencies (1–2 MHz) arises from relatively large structures, such as small vessels and liver lobules (∼1 mm). As f increases the slope of the frequency dependence increases and eventually approaches that for blood, indicating that at higher frequencies the scattering from the cellular structure (∼20 μm) dominates. It may be anticipated that at higher frequencies the frequency dependence of scattering should begin to decrease again, unless scattering from subcellular structures begins to dominate the total scattered power.

The average angular scattering pattern of tissues [43, 44] has also been used to estimate the characteristic size of the scattering structures and the results are consistent with the above observation that the size of the dominant scattering structure decreases with increasing frequency. Figure 4.7 shows angular scattering data [44] from which it was estimated, by fitting equation (4.22), that at 6 MHz 'effective scattering structures' exist within blood, skeletal muscle (perpendicular to the fibres) and liver with correlation lengths of 8 ± 1 μm, 74 ± 5 μm and 61 ± 8 μm, and with density fluctuations of 50%, 28% and 15% of the compressibility fluctuations, respectively.

Methods of measuring the various kinds of μ_{ds} have given rise to approaches to parametric imaging that permit a model-based analysis of tissue structure; by fitting, to the experimental data, the variation of μ_{ds} with a given measurement variable predicted from a given model of the tissue structure, descriptions of the tissue may be provided in terms of parameters of the model. Within this approach, effective 'scatterer size' is available from the average (incoherent) frequency or angular dependence of μ_{ds} [75, 76, 93]. Effective 'scatterer spacings' have been estimated from

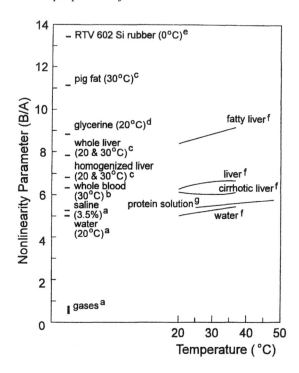

Figure 4.8. *Examples of the nonlinearity parameter B/A, at atmospheric pressure. Key: a [3]; b [50]; c [55]; d [89]; e [85]; f [54]; g [53] (after [91]).*

interference (diffraction) patterns associated with the coherent component of scattering, as a function of frequency [77], tissue orientation [73, 78] and spatial position [79, 80]. For liver, observed fluctuations are consistent with scatterer spacings ∼1 mm. 'Structural anisotropy' is available from the average (incoherent) orientation (ψ) dependence of μ_{ds}; the outer region of the kidney, the cortex, has a radial pattern of collecting systems and exhibits strong anisotropy of μ_{bs}, as does scattering from muscle [81]. Such scattering has been modelled using a two-dimensional anisotropic Gaussian ACF and has assisted in the identification of two dominant scattering structures in the renal cortex: blood vessels/tubules and glomeruli [75].

From the above discussion it may be noted that differences in the scattering coefficients for different tissues (figure 4.6) may arise from a combination of at least five tissue properties: the compressibility and density variations that cause the scattering; the characteristic size of the scattering structures; their characteristic spacing (number density); their degree of alignment; and their preferential orientation if strongly aligned. Tissue echogenicity is often related to connective tissue content since the compressibility and density of structural proteins such as collagen and elastin are very different from those of other tissue components [82]. Strong scattering is therefore observed from skin, which is rich in relatively large (50–100 μm) strongly aligned bundles of collagen fibres, and collagen-rich pathology such as infarcted myocardium. At the other extreme, blood is an isotropic cellular scatterer containing no structural protein. Because of the low sound speed in fat at body temperature strong scattering occurs in tissues such as breast, which contains large fat/non-fat interfaces, and backscattering from diffusely diseased liver is highly correlated with percentage fat content. Increased water content, diluting the number density of scatterers, is associated with reduced backscatter in liver and non-fibrous liver tumours [78]. Lack of alignment and the small size of collagen fibres appears to account for reduced backscattering, relative to dermis, in some highly collagenous skin tumours [83]. The temperature dependence of scattering coefficients is unknown but heat coagulation of liver slightly decreases the backscattering coefficient [67].

4.3.4. Nonlinearity

Examples of reported values for B/A are provided in figure 4.8. For water from 0 to 40°C, solutions of macromolecules and most tissues, B/A increases slightly with increasing sound speed and temperature. B/A for fat, however, is greater than for aqueous media and other tissues (cf. sound speed, where the reverse is true) and, as temperature increases, B/A and c vary in an inverse manner for fatty liver [54]. This behaviour is similar to that seen in general for pure non-biological media where B/A is inversely correlated to sound speed [3, 84]. For solutions of biological compounds B/A depends on the concentration and on the type of solute but is independent of molecular weight. Solute–solvent interactions are therefore regarded as the most likely source of nonlinearity in such media, rather than inter- or intramolecular interactions, with the solute hydration shell structure playing an important role [56]. B/A has been shown to result from a linear combination of contributions from various chemical components, and simultaneous measurement of c and B/A enables chemical composition to be predicted in terms of the volume fractions of water, fat and residual components such as proteins and carbohydrates [54]. Values for whole and homogenised liver suggest that B/A is influenced by the presence of large scale structure in tissue [53]; after subtracting the contribution due to water the remaining contributions to B/A appear to be 26% from cell–cell adhesive force, 20% from hepatocyte cellular structure and 15% from secondary/tertiary protein structure.

REFERENCES

[1] Bamber J C 1986 Attenuation and absorption *Physical Principles of Medical Ultrasonics* ed C R Hill (Chichester: Ellis Horwood) pp 118–99

[2]   Beyer R T and Letcher S V 1969 *Physical Ultrasonics* (New York: Academic Press)

[3]   Beyer R T 1974 *Nonlinear Acoustics* Naval Ship Systems Command, US Department of the Navy (republished by the Acoustical Society of America, 1997)

[4]   Dunn F, Edmonds P D and Fry W J 1969 Absorption and dispersion of ultrasound in biological media *Biological Engineering* ed H P Schwan (New York: McGraw-Hill) pp 205–332

[5]   Dickinson R J 1986 Reflection and scattering *Physical Principles of Medical Ultrasonics* ed C R Hill (Chichester: Ellis Horwood) pp 225–77

[6]   Jensen J A 1991 A model for the propagation and scattering of ultrasound in tissue *J. Acoust. Soc. Am.* **89** 182–90

[7]   Wagner R F, Smith S W, Sandrik J M and Lopez H 1983 Statistics of speckle in ultrasound B-scans *IEEE Trans. Sonics Ultrasonics* **30** 156–63

[8]   Flax S W, Glover G H and Pelc N J 1981 Textural variations in B-mode ultrasonography: a stochastic model *Ultrasonic Imaging* **3** 235–57

[9]   Goss S A *et al* 1979 Elements of tissue characterization, part II. Ultrasonic propagation parameter measurements *Ultrasonic Tissue Characterization II* ed M Linzer, NBS Spec. Publ. 525 (Washington, DC: US Govt Printing Office) pp 43–51

[10]   Fry W J and Fry R B 1954 Determination of absolute sound levels and acoustic absorption coefficients by thermocouple probes–theory and experiment *J. Acoust. Soc. Am.* **26** 294–317

[11]   Fry W J and Dunn F 1962 Ultrasound: analysis and experimental methods in biological research *Physical Techniques in Biological Research* vol 4, ed W L Nastuk (New York: Academic Press) pp 261–394

[12]   Dunn F 1962 Temperature and amplitude dependence of acoustic absorption in tissue *J. Acoust. Soc. Am.* **34** 1545–7

[13]   Frizzel L A, Carstensen E L and Davis D 1979 Ultrasonic absorption in liver tissue *J. Acoust. Soc. Am.* **65** 1309–12

[14]   Goss S A, Cobb J W and Frizzell L A 1977 Effect of beam width and thermocouple size on the measurement of ultrasonic absorption using the thermocouple technique *1977 Ultrasonics Symp. Proc., IEEE Cat. No. 77CH 1264-1SU* (New York: IEEE) pp 206–11

[15]   Parker K J 1983 Ultrasonic attenuation and absorption in liver tissue *Ultrasound Med. Biol.* **9** 363–9

[16]   Pinkerton J M M 1947 A pulse method for the measurement of ultrasonic absorption in liquids: results for water *Nature* **160** 128–9

[17]   Schwan H P and Carstensen E L 1952 Ultrasonics aids diathermy experiments *Electronics* July, p 216

[18]   Esche R 1952 Untersuchungen zur Ultraschallabsorption in tierischen Geweben und Kunststoffen *Akustische Biehefte* **1** 71–4

[19]   Bamber J C and Bush N L 1991 Quantitative imaging of acoustical and histological properties of excised tissues *Acoustical Imaging* vol 18, ed H Lee and G Wade (New York: Plenum) pp 17–25

[20]   Goss S A and Dunn F 1980 Ultrasonic propagation properties of collagen *Phys. Med. Biol.* **25** 827–37

[21]   Parker K J and Waag R C 1983 Measurement of ultrasonic attenuation within regions selected from B-scan images *IEEE Trans. Biomed. Eng.* **BME-30** 431–7

[22]   Ophir J and Mehta D 1988 Elimination of diffraction error in acoustic attenuation estimation via axial beam translation *Ultrasonic Imaging* **10** 129–52

[23] Marcus P W and Carstensen E L 1975 Problems with absorption measurements of inhomogeneous solids *J. Acoust. Soc. Am.* **58** 1334–5

[24] Busse L J and Miller J G 1981 Detection of spatially nonuniform ultrasonic radiation with phase sensitive (piezoelectric) and phase insensitive (acoustoelectric) receivers *J. Acoust. Soc. Am.* **70** 1377–86

[25] Dunn F and Breyer J 1962 Generation and detection of ultra-high frequency sound in liquids *J. Acoust. Soc. Am.* **34** (6) 775–8

[26] Dunn F and Fry W J 1961 Ultrasonic absorption and reflection by lung tissue *Phys. Med. Biol.* **5** 401–10

[27] Bamber J C 1986 Speed of sound *Physical Principles of Medical Ultrasonics* ed C R Hill (Chichester: Ellis Horwood) pp 200–24

[28] Pelham J R and Galt J K 1946 Ultrasonic propagation in liquids: I. Application of pulse technique to velocity and absorption measurements at 15 megacycles *J. Acoust. Soc. Am.* **14** 608–14

[29] Van Venrooij G E P M 1971 Measurement of sound velocity in human tissue *Ultrasonics* **9** 240–2

[30] Del Grosso V A and Mader C W 1972 Speed of sound in pure water *J. Acoust. Soc. Am.* **52** 1442–6

[31] Verhoef W A, Cloostermans M J T M and Thijssen J M 1985 Diffraction and dispersion effects on the estimation of ultrasound attenuation and velocity in biological tissues *IEEE Trans. Biomed. Eng.* **BME-32** 521–9

[32] Robinson D E, Chen F and Wilson L S 1982 Measurement of velocity of propagation from ultrasonic pulse-echo data *Ultrasound Med. Biol.* **8** 413–20

[33] Haumschild D J and Greenleaf J F 1983 A crossed beam method for ultrasonic speed measurement in tissue *Ultrasonic Imaging* **12** 168

[34] Hayashi N *et al* 1985 In vivo measurement of sound speed in normal and abnormal livers using a high resolution ultrasonic scanner *Proc. 4th Meeting World Federation for Ultrasound in Medicine and Biology* ed R W Gill and M J Dadd (Sydney: Pergamon Press) p 520

[35] Ophir J 1986 A beam tracking method for estimation of ultrasound propagation speed in biological tissues *IEEE Trans. Ultrasonics Ferroelectr. Freq. Contr.* **33** 359–67

[36] Kinsler L E *et al* 1982 *Fundamentals of Acoustics* 3rd edn (New York: Wiley) p 397

[37] Giacomini A 1947 Ultrasonic velocity in ethanol–water mixtures *J. Acoust. Soc. Am.* **19** 701–2

[38] Madsen E L 1993 Method of determination of acoustic backscatter and attenuation coefficients independent of depth and instrumentation *Ultrasonic Scattering in Biological Tissues* ed K K Shung and G A Thieme (Boca Raton, FL: CRC Press) pp 205–50

[39] Reid J M 1993 Standard substitution methods for measuring ultrasonic scattering in tissues *Ultrasonic Scattering in Biological Tissues* ed K K Shung and G A Thieme (Boca Raton: CRC Press) pp 171–204

[40] Waag R C and Astheimer P 1993 Measurement system effects in ultrasonic scattering experiments *Ultrasonic Scattering in Biological Tissues* ed K K Shung and G A Thieme (Boca Raton, FL: CRC Press) pp 251–90

[41] Pohlhammer J and O'Brien W D Jr 1981 Dependence of the ultrasonic scatter coefficient on collagen concentration in mammalian tissue *J. Acoust. Soc. Am.* **69** 283–5

[42] Pauly H and Schwan H P 1971 Mechanism of absorption of ultrasound in liver tissue *J. Acoust. Soc. Am.* **50** 692–9

[43] Campbell J A and Waag R C 1984 Measurements of calf liver ultrasonic differential and total scattering cross sections *J. Acoust. Soc. Am.* **75** 603–11

[44] Nassiri D K and Hill C R 1986 The differential and total bulk acoustic scattering cross sections of some human and animal tissues *J. Acoust. Soc. Am.* **79** 2034–47

[45] Burke T M, Madsen E L and Zagzebski J A 1987 A preliminary study on the angular distribution of scattered ultrasound from bovine liver and myocardium *Ultrasonic Imaging* **9** 132–45

[46] Battan L J 1973 *Radar Observation of the Atmosphere* (Chicago: University of Chicago Press)

[47] Sigelmann R A and Reid J M 1973 Analysis and measurement of ultrasound backscattering from an ensemble of scatterers excited by sinewave bursts *J. Acoust. Soc. Am.* **53** 1351–5

[48] Zagzebski J A *et al* 1993 Quantitative backscatter imaging *Ultrasonic Scattering in Biological Tissues* ed K K Shung and G A Thieme (Boca Raton, FL: CRC Press) pp 451–86

[49] Ueda M and Ozawa Y 1985 Spectral analysis of echoes for backscattering coefficient measurement *J. Acoust. Soc. Am.* **77** 38–47

[50] Law W K *et al* 1981 Ultrasonic determination of the nonlinearity parameter B/A for biological media *J. Acoust. Soc. Am.* **39** 1210–2

[51] Cobb W N 1983 Finite amplitude method for the determination of the acoustic nonlinearity parameter B/A *J. Acoust. Soc. Am.* **73** 1525–31

[52] Cain C A *et al* 1986 On ultrasonic methods for measurement of the nonlinearity parameter B/A in fluid-like media *J. Acoust. Soc. Am.* **80** 685–8

[53] Zhang J and Dunn F 1987 In vivo B/A determination in a mammalian organ *J. Acoust. Soc. Am.* **81** 1635–7

[54] Sehgal C M *et al* 1986 Measurement and use of acoustic nonlinearity and sound speed to estimate composition of excised livers *Ultrasound Med. Biol.* **12** 865–74

[55] Law W K *et al* 1985 Determination of the nonlinearity parameter B/A of biological media *Ultrasound Med. Biol.* **11** 307–18

[56] Sarvazyan A P *et al* 1990 Acoustic nonlinearity parameter B/A of aqueous solutions of some amino acids and proteins *J. Acoust. Soc. Am.* **88** 1555–61

[57] Zhang J *et al* 1991 Influences of structural factors of biological media on the acoustic nonlinearity parameter B/A *J. Acoust. Soc. Am.* **89** 80–91

[58] Carstensen E L 1979 Absorption of sound in tissue *Ultrasonic Tissue Characterization II* (NBS Spec. Publ. 525) ed M Linzer (Washington, DC: US Govt Printing Office) pp 29–40

[59] Chivers R C and Parry R J 1978 Ultrasonic velocity and attenuation in human tissue *J. Acoust. Soc. Am.* **63** 940–53

[60] Dunn F and O'Brien W D 1978 Absorption and dispersion *Ultrasound: its Application in Medicine and Biology* ed F J Fry (Amsterdam: Elsevier) p 393

[61] Goss S A, Johnston R L and Dunn F 1978 Comprehensive compilation of empirical ultrasonic properties of mammalian tissues *J. Acoust. Soc. Am.* **64** 423–57

[62] Goss S A, Johnston R L and Dunn F 1980 Compilation of empirical ultrasonic properties of mammalian tissues II *J. Acoust. Soc. Am.* **68** 93–108

[63] Wells P N T 1975 Absorption and dispersion of ultrasound in biological tissue *Ultrasound Med. Biol.* **1** 369–76

[64] Kremkau F W and Cowgill R W 1984 Biomolecular absorption of ultrasound. I. Molecular weight *J. Acoust. Soc. Am.* **76** 1330–5

[65]   O'Donnell M and Miller J G 1979 Mechanisms of ultrasonic attenuation in soft tissue *Ultrasonic tissue characterization II* (NBS Spec. Publ. 525) ed M Linzer (Washington, DC: US Govt Printing Office) pp 37–40

[66]   Kremkau F W *et al* 1981 Ultrasonic attenuation and propagation speed in normal human brain *J. Acoust. Soc. Am.* **70** 29–38

[67]   Bush N L, Rivens I, ter Haar G R and Bamber J C 1993 Acoustic properties of lesions generated with an ultrasound therapy system *Ultrasound Med. Biol.* **19** 789–801

[68]   Shung K K 1993 *In vitro* experimental results on ultrasonic scattering in biological tissues *Ultrasonic Scattering in Biological Tissues* ed K K Shung and G A Thieme (Boca Raton, FL: CRC Press) pp 291–312

[69]   Yuan Y W and Shung K K 1986 The effect of focusing on ultrasonic backscattering measurements *Ultrasonic Imaging* **8** 212–130

[70]   Mo L Y L and Cobbold R S C 1993 Theoretical models of ultrasonic scattering in blood *Ultrasonic Scattering in Biological Tissues* ed K K Shung and G A Thieme (Boca Raton, FL: CRC Press) pp 125–70

[71]   Shung K K and Reid J M 1977 The acoustical properties of deoxygenated sickle cell blood and hemoglobin S solution *Ann. Biomed. Eng.* **5** 150–8

[72]   Bamber J C 1979 Theoretical modelling of the acoustic scattering structure of human liver *Acoust. Lett.* **3** 114–19

[73]   Nicholas D 1982 Evaluation of backscattering coefficients for excised human tissues; results, interpretation and associated measurements *J. Ultrasound Med. Biol.* **8** 17–28

[74]   Insana M F *et al* 1990 Describing small-scale structure in random media using pulse-echo ultrasound *J. Acoust. Soc. Am.* **87** 179–92

[75]   Insana M F and Brown D G 1993 Acoustic scattering theory applied to soft biological tissues *Ultrasonic Scattering in Biological Tissues* ed K K Shung and G A Thieme (Boca Raton, FL: CRC Press) pp 75–124

[76]   Lizzi F and Feleppa E J 1993 *In vivo* ophthalmological tissue characterization by scattering *Ultrasonic Scattering in Biological Tissues* ed K K Shung and G A Thieme (Boca Raton, FL: CRC Press) pp 393–408

[77]   Waag R C *et al* 1976 Tissue macrostructure from ultrasound scattering *Proc. Conf. on Computerised Tomography in Radiology* (St. Louis: American College of Radiology) pp 175–86

[78]   Bamber J C *et al* 1981 Acoustic properties of normal and cancerous human liver *Ultrasound Med. Biol.* **7** 121–44

[79]   Fellingham L L and Sommer F G 1984 Ultrasonic characterization of tissue structure in the in vivo human liver and spleen *IEEE Trans. Sonics Ultrasonics* **31** 418–28

[80]   Garra B S 1993 In vivo liver and splenic tissue characterization by scattering *Ultrasonic Scattering in Biological Tissues* ed K K Shung and G A Thieme (Boca Raton, FL: CRC Press) pp 347–91

[81]   Mottley J G and Miller J G 1988 Anisotropy of the ultrasonic backscatter of myocardial tissue. I. Theory and measurements in vitro *J. Acoust. Soc. Am.* **83** 755–61

[82]   Fields S and Dunn F 1973 Correlation of echographic visualizability of tissue with biological composition and physical state *J. Acoust. Soc. Am.* **54** 809–12

[83]   Bamber J C *et al* 1992 Correlation between histology and high resolution echographic images of small skin tumours *Acoustical Imaging* vol 19, ed H Ermert and H P Harjes (New York: Plenum) pp 369–74

[84]   Madigosky W M *et al* 1981 Sound velocities and B/A in fluorocarbon fluids and in several low density solids *J. Acoust. Soc. Am.* **69** 1639–43

[85] Seki H, Granato A and Truell R 1956 Diffraction effects in the ultrasonic field of a piston source and their importance in the accurate measurement of attenuation *J. Acoust. Soc. Am.* **28** 230–8

[86] Papadakis E P 1966 Ultrasonic diffraction loss and phase change in anisotropic materials *J. Acoust. Soc. Am.* **40** 863–76

[87] O'Donnell M and Reilly H F 1985 Clinical evaluation of the B'-scan *IEEE Trans. Sonics Ultrasonics* **32** 450–7

[88] Foster F S *et al* 1984 The ultrasound macroscope: initial studies of breast tissue *Ultrasonic Imaging* **6** 243–61

[89] Bjørnø L 1975 Non-linear ultrasound—a review *Ultrasonics Int. Conf. Proc.* (Guildford: IPC Science and Technology Press) pp 110–5

[90] Shung K K *et al* 1977 Angular dependence of scattering of ultrasound from blood *IEEE Trans. Biomed. Eng.* **24** 325

[91] Bamber J C 1997 Acoustical characteristics of biological media *Encyclopedia of Acoustics* ed M J Crocker (New York: Wiley) pp 1703–26

[92] Johnson S A *et al* 1977 Non-intrusive measurement of microwave and ultrasound induced hyperthermia by acoustic temperature tomography *Ultrasonics Symp. Proc.* IEEE Cat. No. 77 Ch1264-1SU (New York: IEEE) pp 977–82

[93] Mercer J 1997 Ultrasound scatterer size imaging of skin tumours—potential and limitations *PhD Thesis*, University of London

PART 2

TECHNOLOGY AND MEASUREMENT IN DIAGNOSTIC IMAGING

CHAPTER 5

TRANSDUCER ARRAYS FOR MEDICAL ULTRASOUND IMAGING

Thomas L Szabo

INTRODUCTION

The transducer is the indispensible part of a diagnostic ultrasound imaging system. Phased and linear arrays focus and electronically scan elastic waves into the body, and gather the echoes from tissues for generating real-time images of the body. Over the last 30 years, transducers have continued to evolve: providing images of better clarity detail, and fidelity; going where no array has gone before (new applications and orifices), and producing new information about organ function and tissue viability.

The role of the transducer in the imaging process is examined here in a comprehensive way. Beginning with individual piezoelectric elements, in the first half, we shall cover array structures, design challenges, transducer modelling and materials, and achievable performance. In the second half, the beam-forming functions of an array in conjunction with an imaging system are examined. Array configurations, beam steering and focusing, apodisation, achievable resolution and non-ideal array performance are also included. References are provided to more detailed works for those wishing to pursue any topic in more depth.

5.1. PIEZOELECTRIC TRANSDUCER ELEMENTS

5.1.1. A basic transducer model

Ultrasound transducers for medical purposes (Hunt *et al* 1993) employ piezoelectric elements that convert electrical signals to mechanical forces and vice versa. The basic characteristics of a piezoelectric transducer can be described by a simple model. The transducer is essentially a capacitor that converts electrical signals into acoustic waves. This ability can be expressed as a piezoelectric stress equation (Auld 1990),

$$S = c^D T - hD \qquad (5.1)$$

where S is stress; T is strain; D is dielectric displacement; c^D is elastic stiffness constant obtained under conditions of constant D; and h is a piezoelectric constant. The piezoelectric element can be thought of as a singing capacitor of area A and thickness d, $C_0 = \varepsilon^S A/d$, as illustrated in figure 5.1, where C_0 is the clamped capacitance and ε^S is the clamped (zero strain) dielectric constant. When a voltage, V, is applied across the two electroded faces of the capacitor, oppositely signed stresses appear on these sides through the piezoelectric effect. These stresses can be represented as impulsive forces on each electroded face:

$$S(t) = (hC_0 V/2A)[-\delta(t - d/2) + \delta(t + d/2)] \qquad (5.2)$$

under a clamped condition of zero strain caused by the loading of each face by media. In this example, the load impedance Z_A is equal to the transducer acoustic impedance. Through a Fourier transform of this stress relation, the stress frequency response is

$$S(\omega) = -\mathrm{i}(hC_0 V/2A) \sin(\pi(2n + 1)\omega/2\omega_0) \qquad (5.3)$$

where the fundamental resonant frequency is $f_0 = v/(2d)$; v is the acoustic velocity between the electrodes, $v = \sqrt{c^D/\rho}$; the angular frequency is $\omega_0 = 2\pi f_0$ and $n = 0, 1, 2, \dots$. Both equations (5.2) and (5.3) are demonstrated by figure 5.1. Note that only odd harmonic frequencies are generated.

The electrical input impedance, $Z_T(\omega)$ of a piezoelectric transducer can be represented by a capacitive reactance in series with a radiation resistance, R_A, and a radiation reactance, X_A

$$Z_T(\omega) = R_A(\omega) + \mathrm{i}[X_A(\omega) - 1/\omega C_0]. \qquad (5.4)$$

To first order, we can determine the radiation resistance from the frequency response of the stress found earlier. The real electrical power, W_E, into the transducer for an applied voltage V and current I is

$$W_E = II^* R_A/2 = |I|^2 R_A/2 \qquad (5.5)$$

where the current is $I = \mathrm{i}\omega Q = \mathrm{i}\omega C_0 V$. The total acoustic power into the medium on both sides of the ceramic delivered into the surrounding medium of specific impedance $Z_A = \rho v A$ is

$$W_A = SS^*/2Z_A + SS^*/2Z_A = |T(\omega)|^2/Z_A = |hC_0 V \sin(\pi\omega/2\omega_0)|^2/Z_A. \qquad (5.6)$$

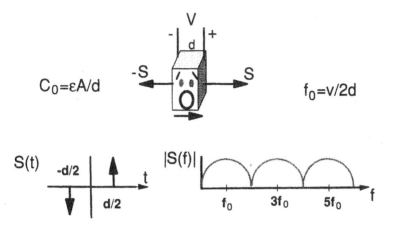

Figure 5.1. *Simplified model of piezoelectric element of thickness d and area A with voltage induced stresses at electroded sides and corresponding stress frequency response.*

By equating the powers above from equations (5.5) and (5.6), R_A can be found as

$$R_A(\omega) = R_{A0} \operatorname{sinc}^2(\omega/2\omega_0) \tag{5.7}$$

in which $\operatorname{sinc}(x) = \sin(\pi x)/(\pi x)$. Since $h = k_T\sqrt{c^D/\varepsilon^S}$ (k_T is a coupling coefficient), R_{A0} can be shown to be

$$R_{A0} = \frac{2k_T^2}{\pi\omega_0 C_0}. \tag{5.8}$$

The radiation reactance can be determined by causality as the Hilbert transform of R_A (Nalamwar and Epstein 1972)

$$X_A(\omega) = R_{A0}\frac{[\sin(\pi\omega/\omega_0) - \pi\omega/\omega_0]}{2(\pi\omega/2\omega_0)^2}. \tag{5.9}$$

In figure 5.2, the transducer impedance is dissected into its parts, R_A, X_A, X_C and X_A+X_C. Let $X_C = 1/(\omega_0 C_0)$ and $X = X_A+X_C$. At the centre frequency of 5 MHz the radiation resistance is R_{A0} and $X_A = 0$, and $X = -1/(\omega_0 C_0)$.

5.1.2. Transducer elements as acoustic resonators

The three-dimensional nature of the piezoelectric element results in the inevitable coupling of the main vibrational mode along the poling axis to those in the other two perpendicular directions. As the piezoelectric material is squeezed between the electrodes, it bulges out at the sides; therefore, the vertical extensional mode is coupled to lateral dilational modes.

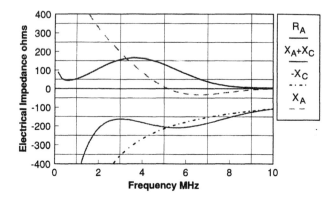

Figure 5.2. *Transducer electrical impedance for a 5 MHz reso-nant frequency PZT-5H beam mode transducer with an impedance of 30×10^6 kg m^{-2} s^1 for backing and water loading.*

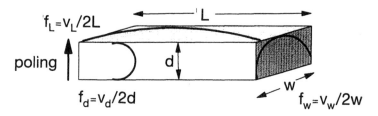

Figure 5.3. *Diagram showing a piezoelectric element as a multimode acoustic resonator with three fundamental resonant frequencies which are intercoupled.*

The piezoelectric element can be considered to be a three-dimensional acoustic resonator with each direction capable of supporting fundamental and harmonic resonant frequencies as illustrated in figure 5.3. Furthermore, these modes are coupled more tightly together as the three dimensions become closer in size. Fortunately, for most array elements, the elevation dimension, L, is much greater than the others, so that its resonant frequency is very low and far below the imaging frequency band. The properties of a typical array element resonator depend mainly on the width to thickness ratio, w/d (Sato *et al* 1979, Kino and DeSilets 1980).

Two extreme geometries are given by figure 5.4. The thickness mode in which $w \gg d$ and $L \gg d$, applicable to large mechanical transducers (with a large area normal to the poling axis), and the beam mode for plank-shaped elements where L is the greatest dimension, $w \gg d$ and $L \gg d$. Both the electromechanical coupling constant and the acoustic velocity are slowly varying functions of the ratio w/d. Cases of interest for these extreme geometries are listed in table 5.1 for a widely used piezoelectric ceramic, PZT-5H. In practice, the w/d ratios are small so that coupling constants and

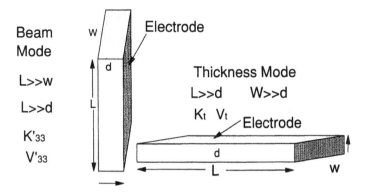

Figure 5.4. *Piezoelectric elements of two extreme geometries: the beam and thickness modes.*

Table 5.1. *Constants for Vernitron PZT-5H for thickness and beam modes.*

Mode	Electro-mechanical coupling coefficient k	Wave speed v (m s^{-1} × 10³)	Acoustic impedance Z (kg m^{-2} s^{-1} × 10⁶)	Relative dielectric constant $\varepsilon^S/\varepsilon_0$
Thickness	$k_T = 0.5$	$v_T = 4.56$	$Z = 34.7$	$\varepsilon_T/\varepsilon_0 = 1470$
Beam	$k'_{33} = 0.693$	$v'_{33} = 3.97$	$Z'_{33} = 29.7$	$\varepsilon_{33}/\varepsilon_0 = 1508$

acoustic velocities are closer to the beam values; therefore, small elements benefit from a higher coupling constant compared to large thickness mode transducers (Souquet *et al* 1979, Fukumoto *et al* 1981). The actual values for the piezoelectric ceramic are determined either by experimental methods (Szabo 1982) or by the more accurate models described later.

5.1.3. Transducer array structures

Because a piezoelectric element is bidirectional, efforts have been made to optimise the conversion efficiency and bandwidth in the forward direction towards the tissue. On the rear face, an absorbing backing material is used to reflect energy into the forward direction and to broaden bandwidth. The disparity between the acoustic crystal impedance $Z_c = Z'_{33} = 30 \times 10^6$ kg m^{-2} s^{-1} from table 5.1 and normalised tissue or water impedance ($Z_w = 1.5 \times 10^6$ kg m^{-2} s^{-1}) can be reduced by quarter-wave (length) matching layers. Each matching layer of characteristic impedance, Z_1, loaded by an impedance, Z_2, transforms it into an input impedance, $Z_{in} = Z_1^2/Z_2$, at its resonant frequency. For example, a single matching layer impedance would be chosen to have a value $Z_{ml} = \sqrt{Z_c Z_w} = 6.4 \times 10^6$ kg m^{-2} s^{-1},

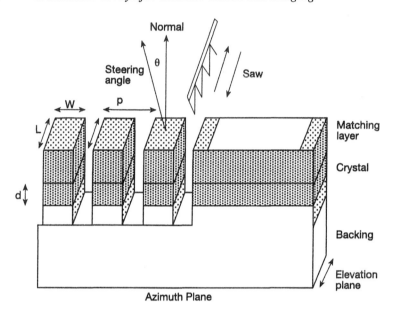

Figure 5.5. *The geometry of a multilayer structure being diced into array elements by a saw.*

where Z_c is normalised crystal impedance and Z_w is normalised water impedance.

A typical array structure begins as the sandwich of layers shown in figure 5.5. During construction, the sandwich is diced into individual elements and final layers and a mechanical lens (not shown) are added as toppings. Here acoustic port 1 of the model corresponds to the top direction in this illustration. In order to better describe the performance of this structure, more comprehensive transducer models are required.

5.1.4. Transducer models

By convention, a one-dimensional transducer equivalent circuit model for a piezoelectric element is viewed as a three-port device like that of figure 5.6. This representation provides electrical analogues of voltages and currents for acoustic force and particle velocity. As in the simpler model described earlier, the piezoelectric element has an electrical port where voltage is applied and two acoustic ports on either electroded face. By reciprocity, this three-port model is equally valid for either transmit or receive operation. In order to represent one element of the diced transducer array structure, an acoustic port is loaded by transmission line sections representing each matching or bonding layer and the electrical port is connected to electrical matching networks and terminations as demonstrated in figure 5.7. Input variables for the overall transducer model typically

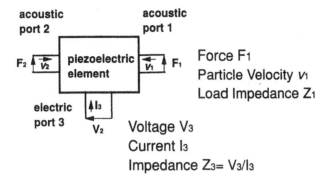

Figure 5.6. *One-dimensional equivalent circuit model of a piezoelectric element as a three-port device.*

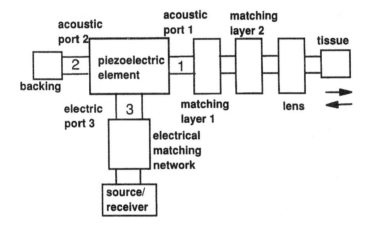

Figure 5.7. *One-way one-dimensional equivalent circuit model of whole transducer including backing, electrical matching and matching layers and lens.*

consist of the following: for the piezoelectric element, the area A, thickness d, effective velocity v_e, coupling constant k_e and impedance Z_c, clamped dielectric constant ε^S, acoustic loss, and dielectric loss tangent; and for each layer, the acoustic characteristic impedance, velocity, layer thickness and loss. Typical output parameters of interest are the overall electrical input impedance, frequency response or insertion loss, an impulse response, and the pulse/spectrum resulting from different excitation functions which are obtained by a Fourier approach using fast Fourier transforms (Mequio *et al* 1983).

What are the specific equivalent circuit models that lie within the box of figure 5.6? Many models have been proposed, but the two used most frequently are the Mason (Berlincourt *et al* 1964, Mason 1948) and KLM

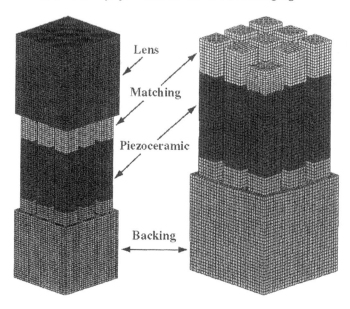

Figure 5.8.   *Two finite element simulations of 2D arrays; the one on the left is capped by a lens and the one on the right is without a lens. Simulations generated by PZFLEX program, courtesy of Weidlinger Associates.*

(Leedom *et al* 1971) thickness mode models. While the structures of these models differ, they were derived from the same elastic wave equations and therefore give exactly the same numerical results. The KLM model uses the centre of the crystal as the point of acoustic excitation which some find analytically more appealing, even though physics shows that stresses are created at the crystal boundaries. Both one-dimensional models work reasonably well for first order design purposes, with modifications made for array elements where effective parameters appropriate for the element geometry are substituted for thickness mode values. A third model has the capability to link the transducer to nonlinear and active electronic drivers and receivers. This model (Hutchens and Morris 1984) uses an approximate Mason model written as an input file to the SPICE program, software written for electrical circuit and electronic simulation and design (Vladimerescu 1984).

Finite element modelling (FEM) (Wojcik *et al* 1993, 1996) is being used to simulate transducer array performance more realistically, including complex modal interactions. This type of modelling is useful either for previewing array behaviour prior to construction or for analysing spurious modes. For two-dimensional arrays, spaced on half-wavelength centres, the w/L ratios approach unity, so that acoustic resonant modes are coupled strongly. A simulation of a 2D array with and without a lens is illustrated in figure 5.8. Realisation of more accurate modelling for higher array

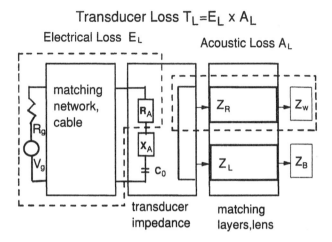

Figure 5.9. *Diagram showing transducer loss as the product of electrical and acoustic loss factors.*

performance depends on the development and full characterisation of new and existing materials. In addition, simplified design methods accounting for complex array modal interactions are needed.

5.1.5. Transducer design

The starting point of design is usually from a one-dimensional model with effective parameters appropriate for the geometry selected. The design process is made more challenging by fixed constraints such as the size of the element, the required bandwidth, an impulse response of certain minimum length, the need for a lens with losses, and electrical requirements such as a cable and electronic interfaces. The overall combination of requirements for one-way transducer loss is represented schematically in figure 5.9 for the transmit case. The challenge of design is how most efficiently to couple energy from the source V_g to the tissue represented by load Z_w while simultaneously meeting all given constraints. The problem can be viewed in the following two parts: maximising the electrical loss factor E_L which is the ratio of power delivered to R_A to that available from the source; and maximising the acoustic loss factor A_L, the power transfer from R_A to Z_w.

From the KLM model, the radiation resistance at resonance can be expressed in a simple form which is a generalisation of equation (5.8) (DeSilets *et al* 1978),

$$R_A(\omega_0) = \frac{4k_T^2}{\pi \omega_0 C_0} \left[\frac{Z_c}{Z_L + Z_R} \right] \qquad (5.10)$$

where Z_c is the acoustic crystal impedance; Z_L, the impedance at acoustic port 2 and Z_R, the impedance load seen at acoustic port 1. This relation is

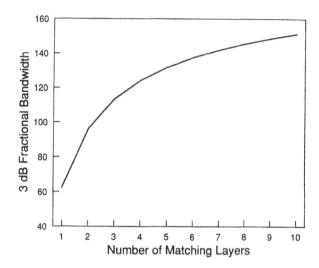

Figure 5.10. *One-way −3 dB bandwidths generated by maximally flat criteria and an impedance mismatch ratio of* $30/1.5 = 20$.

useful in explaining relationships among essential parameters. Usually the right port is closely matched to Z_c through quarter-wavelength matching layers. If both acoustic port load impedances are made equal to Z_c, broad bandwidth is obtained at the expense of half the power lost in the backing at port 2. At the other extreme, air-backing puts 100% of the acoustic power in port 1, but with narrower bandwidth. As the size of the element decreases, C_0 decreases, and R_A increases, making electrical matching to the element more difficult. Acoustic loss at resonance acts like a power splitter,

$$A_L(\omega_0) = \frac{Z_R}{Z_L + Z_R}. \tag{5.11}$$

This relation can be used to estimate the impact of quarter-wave matching and backing impedance combinations. For the basic model when $Z_L = Z_R = Z_c$, $A_L = 1/2$.

For multiple matching layers, several schemes have been proposed for selecting layer impedances (DeSilets *et al* 1978, Matthaei *et al* 1980), including binomial (Pace 1979), maximally flat, Tchebycheff, and others (Souquet *et al* 1979). The maximum fractional bandwidth for the maximally flat approach (Matthaei *et al* 1980) is shown in figure 5.10 as a function of the number of matching layers; in practice, somewhat less is achieved, depending on design constraints. Once matching layer impedances are selected, finding the appropriate low loss materials of the desired impedance is difficult; special composite mixtures are often required.

Most arrays are made with ferroelectric ceramics such as PZT (lead zirconate titanate). Other piezoelectric ceramics such as modified lead

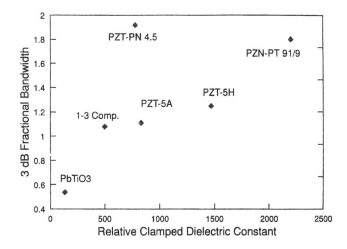

Figure 5.11. *−3 dB bandwidth and relative dielectric constants for piezoelectric materials with the upper right corner being best for arrays.*

titanate (PbTiO$_3$) (Takeuchi *et al* 1982) have been formulated to have lower coupling to lateral modes. Composite materials have been developed to combine the high piezoelectric coupling of ceramics with materials of lower acoustic impedance (Smith *et al* 1984). One of the most successful of these combinations is the 1–3 composite (Gururaja *et al* 1985), in which rods of PZT, embedded in an epoxy matrix, provide the high coupling constant of the rod mode (close to that of the beam mode) with lower overall acoustic impedance. A recent research area is the development of higher coupling single crystals such as PZT-PN (Lopath *et al* 1997, Saitoh *et al* 1997).

Piezoelectric materials can be compared by their electrical Q (Q_e) and relative dielectric constants. Large dielectric constants increase clamped capacitance and improve the electrical impedance matching of small elements. For an inductively tuned transducer with acoustic loading equal to Z_c on both sides and $R_{A0} = R_g$,

$$\Delta f_3/f_c = 1/Q_e = \omega_0(R_{A0} + R_g)C_0 = 8k_T^2/\pi. \tag{5.12}$$

Fractional bandwidth from this equation and ε are plotted in figure 5.11 for several materials of interest (Fukumoto *et al* 1981, Lopath *et al* 1997, Saitoh *et al* 1997). For best array performance, materials with the highest combination of fractional bandwidth and relative dielectric constant, in the upper right corner of figure 5.11, are preferred.

Imaging arrays are more than a collection of individual elements which happen to be neighbours. Because of array construction, elements are mechanically connected and interactions occur (Larson 1981). Cross-coupling can affect element directivity and contribute unwanted spurious

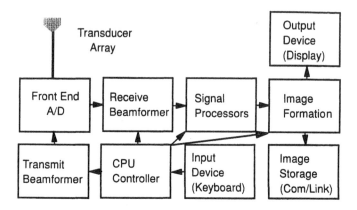

Figure 5.12.   *Block diagram for a diagnostic ultrasound imaging system.*

signals that end up in the image as clutter. Constructive and destructive interference of spurious modes can alter the main pulse width and tail, narrow the corresponding spectrum and add unwanted spectral peaks and valleys. The second phase of transducer design is the minimisation of unwanted effects by accounting for the three-dimensional nature of the array elements and their construction as well as safety considerations (Ziskin and Szabo 1993, We *et al* 1995). Because of multiple constraints, non-resonant design methods are employed (Szabo 1984).

The basic types of array are phased, linear and curved linear, with many variations for different clinical applications. One example is the transoesophageal array (Bom *et al* 1993) which is small enough to put down a throat with remote rotational and flexural movements. Some arrays are shaped to fit other orifices such as the transrectal and endovaginal arrays, while others are for surgical environments and for intravascular imaging.

5.2. IMAGING

Modern diagnostic ultrasound imaging systems employ digital architectures to achieve more flexibility and control over beam-forming and signal and image processing, calculations, and system input/output requirements. A block diagram of a typical ultrasound imaging system with electronic focusing is given in figure 5.12. One or more computer controllers orchestrate the internal processing, beam-forming and image manipulation in a highly interactive manner responsive to the system user. The formation of an image begins with the application of excitation signals to the array from the transmit beamformer in the front end. After a beam is launched, pulse echo signals return from tissue and enter the array elements after which they are beamformed and signal processed according to line position and

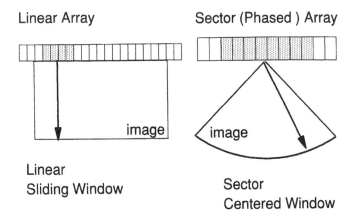

Figure 5.13. *Two basic imaging formats: linear and sector.*

Figure 5.14. *Illustration of single transmit focus and the use of zones for multiple focus splicing.*

depth. The beam position is then sequenced through a pattern required by the selected image format. These lines are converted into grey-scale or colour line data which are processed and interpolated to form an image frame for display. More details on how imaging systems work can be found in the special issues of *Hewlett Packard Journal* (1983) listed in the References.

The two major image formats are linear and sector, as shown in figure 5.13. For linear arrays an electronically switched sliding configuration of active elements (an aperture or window) forms each acoustic image line. In a sector format, all, or a subset of, elements centred at the middle of the array create a sequence of image lines at different scan angles.

5.3. BEAM-FORMING

Electronic focusing and steering occur in the azimuth or imaging plane. Transmit and receive functions are illustrated diagramatically in figure 5.14. A transmit beam is launched along a selected direction with a single focal depth as shown by the narrowing of the −6 dB beam contour (size

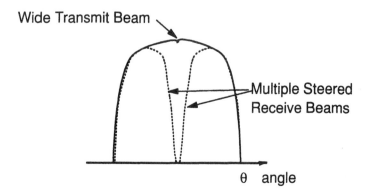

Figure 5.15.

Figure 5.15. *Parallel processing consists of a wide transmit beam with two receive beams steered at two different angles within the transmit beam to enhance frame rate.*

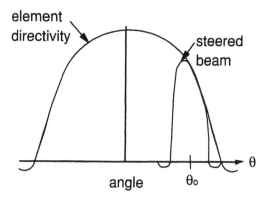

Figure 5.16. *Wide angle directivity of a single element which affects the shape of a steered beam at a wide steering angle.*

exaggerated from illustration) in the focal region. Additional transmit foci can be bought at the expense of slower frame rate by dividing the total scan depth into a number of zones, transmitting in each one, and by splicing the focal regions together. On reception of the pulse echo signals from each transmit direction, dynamic receive focusing is applied at each depth.

Electronic transmit beam-forming is achieved by delaying transmit excitation waveforms in a cylindrical or spherical profile so that the resulting pressure waveforms from each element will add constructively at the focal point (Macovski 1983). A linear delay shift provides steering capability. On reception, returning echoes undergo similar focusing and steering delays to focus the beam at each time (depth) along an acoustic line. By widening the transmit beam and by offsetting the receive line position from the

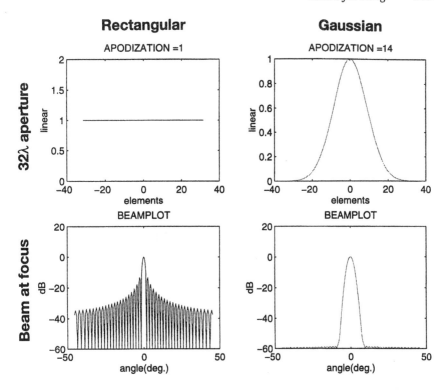

Figure 5.17. *On the left is an unapodised flat function with its corresponding beamshape at a focus. On the right is shown a truncated Gaussian apodisation which has a beamplot with a wider main beam and lower sidelobes for a 32-wavelength, 64-element-long array.*

transmit direction, multiple receive beams can be generated simultaneously, as depicted in figure 5.15. This process, called 'parallel processing', increases the image frame rate.

Each element has its own directivity (Macovski 1983) which has a broad angular extent compared to a focused beam. Because the elements in an array are nearly identical, the effect of element directivity is to 'roll off' or taper the amplitude of the focused beam at wide steering angles as illustrated in figure 5.16.

Apodisation is a means of altering the surface displacement of a transducer through amplitude weighting or shading ('t Hoen 1982). Apodisation functions are tapered gradually towards the ends of the aperture as exemplified by a truncated Gaussian function. This is compared to an unapodised flat amplitude function in figure 5.17. At the focal length, the beamshape is the Fourier transform of the transducer aperture apodisation function as illustrated by the beams of figure 5.17 for a single frequency excitation. This figure demonstrates that apodisation (Kino 1987) involves

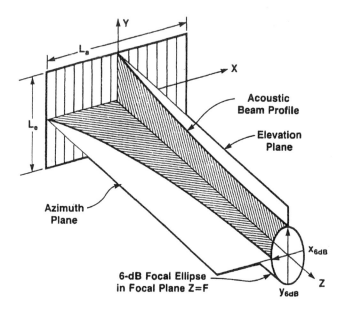

Figure 5.18. *Focused beam tapering to a focal plane ellipse with azimuth and elevation planes depicted.*

a trade-off between main lobe detail resolution (−6 dB beamwidth, for example) and sidelobe suppression or contrast resolution (−40 dB beamwidth).

The imaging or azimuth plane is not consistently thin but has a non-uniform slice thickness in the elevation direction, as illustrated in figure 5.18. Focusing in the elevation plane is accomplished most often with a single fixed focus mechanical lens. Round trip measurements of beams are made by reflections from a small point target; one way beams are usually measured with hydrophones and, recently, by Schlieren imaging (Zanelli and Kadri 1994).

Electronic focusing can also be used for the elevation plane. In figure 5.19, an ordinary 1D array is shown at the top of the figure. For example, this array might have 64 elements spaced half a wavelength apart. At the bottom right of the figure is a 2D array (Smith *et al* 1995) with the same spacing, but with 64 rows and 64 columns of elements so the number of elements and system channel interconnects are $64 \times 64 = 3600$.

If the complete focusing and steering flexibility of a 2D array can be sacrificed somewhat, a 1.5D array can provide intermediate performance, hence its name, '1.5D'. The 1.5D array in the figure is undersampled in the elevation direction (typically at several wavelength intervals). For an example, let there be six samples in elevation and 64 in azimuth, resulting in $6 \times 64 = 384$ elements. This undersampling gives rise to multiple main

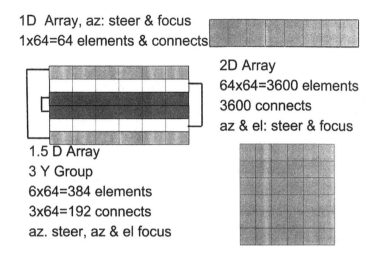

1D Array, az: steer & focus
1x64=64 elements & connects

2D Array
64x64=3600 elements
3600 connects
az & el: steer & focus

1.5 D Array
3 Y Group
6x64=384 elements
3x64=192 connects
az. steer, az & el focus

Figure 5.19. *Major configurations of arrays: 1D, 1.5D and 2D.*

Spatial Resolution at Focus

- Axial 1.8λ
- Azimuth $.9\lambda \ F_{AZ}/L_{AZ}$
- Elevation $.9\lambda \ F_{EL}/L_{EL}$

-6 dB Ellipsoid

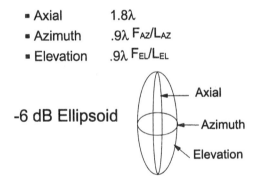

Figure 5.20. *Resolution cell in the form of a −6 dB ellipsoid centred on a coincident azimuth and elevation focus.*

beams called grating lobes whose effect is minimised by not steering the beam. No elevation steering means focusing symmetry which allows for hardwiring pairs of elements receiving the same focusing delays. In this example three groups of elements can be organized into three 'Y Groups' that reduce by half the number of connections to 192.

Three-dimensional ultrasound images are generated usually from a sequence of imaging planes that are mechanically scanned in certain patterns such as rotation, translation or angular scanning or fanning. The image planes are assembled through interpolation to form a 3D image matrix. Once the

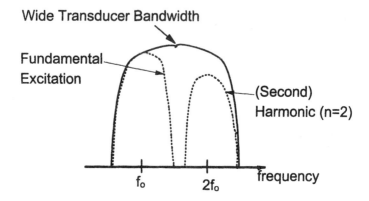

Figure 5.21.   *Harmonic imaging is made possible by a very wide bandwidth transducer which allows fundamental excitation and second harmonic reception.*

matrix is computed, arbitrary slice planes, or surface-rendered images or images with transparent surfaces can be created. Through synchronisation methods during acquisition, moving 3D images can be generated, frame by frame (Masotti and Pini 1993).

At each point in imaging space, the spatial impulse response or point spread function has a measurable resolution cell size. For example, a -6 dB beam contour ellipsoid can be used as a measure of resolution as illustrated by figure 5.20 for a coincident elevation and azimuth focal depth. In general, away from the focus, the time resolution is the smallest and the elevation resolution is the largest. For 3D imaging, consistently fine resolution is desired. 1.5D and 2D arrays are ways of providing elevation plane resolution which is improved over fixed mechanical lens focusing methods.

5.4.   OTHER IMAGING MODES

Until now, the discussion has centred on 2D (B-mode) imaging. Other modes include colour flow imaging, Doppler (pulsed, CW or power), and acoustic quantification; but space does not allow further discussion. (For Doppler see Chapter 6.)

A relatively new modality is harmonic imaging in which advantage is taken of the fact that water and the body tissue are nonlinear elastic media even under modest pressure amplitudes (Dunn *et al* 1981). Therefore, as focused beams propagate, finite amplitude distortion generates higher harmonic spectral components. Transducers of very wide bandwidths can capture not only the fundamental excitation frequency band but the second harmonic spectrum as well, as depicted in figure 5.21. Note that the second harmonic is purely a nonlinear propagation effect. A transducer, excited at its centre frequency under linear conditions, may be expected only to generate odd harmonics. Images created from the second harmonic are providing new clinical information, especially for assessing heart function with contrast agents (Goldberg 1993) and for providing improved views of patients otherwise difficult to image.

5.5. CONCLUSION

Transducer array design is complicated because of the increasing number of constraints and requirements as well as mechanical interactions within array structures. The growing use of finite element modelling in design is revealing that most materials are not adequately characterised for accurate simulations. New materials including composites can extend transducer performance. Despite these challenges, arrays are continuing to improve with trends towards wider bandwidths and higher centre frequencies. The development of 1.5D and 2D arrays provides new challenges. In addition, new clinical applications for transducers continue to emerge.

Another trend is the combining of imaging with quantitative measurement capabilities. New developments are in the areas of contrast and harmonic imaging which require very wide transducer bandwidths. A further area of potential improvement is imaging system methods for minimising or correcting for unwanted elastic imaging body effects.

REFERENCES

Auld B A 1990 *Acoustic Fields and Waves in Solids* vol 1, 2nd edn (Malabar, FL: Krieger) ch 8

Berlincourt D A, Curran D R and Jaffe H 1964 Piezoelectric and piezomagnetic materials and their function in transducers *Physical Acoustics* vol 1, part A, ed W P Mason (New York: Academic Press) pp 170–270

Bom N, Brommersma P D and Lance'e C T 1993 Probes as used in cardiology with emphasis on transesophageal and intravascular applications *Advances in Ultrasound Techniques and Instrumentation* ed P N T Wells (New York: Churchill Livingstone) pp 19–33

DeSilets C S, Fraser J D and Kino G S 1978 The design of efficient broadband piezoelectric transducers *IEEE Trans. Sonics Ultrasonics* SU-25 115–25

Dunn F, Law W K and Frizzell L A 1981 Nonlinear ultrasonic wave propagation in biological materials *1981 IEEE Ultrasonics Symp. Proc.* pp 527–32

Fukumoto A, Kawabuchi M and Sato J 1981 Design of ultrasound transducers using new piezoelectric ceramic materials *Ultrasound Med. Biol.* 7 275–84

Goldberg B B 1993 Ultrasound contrast agents *Advances in Ultrasound Techniques and Instrumentation* ed P N T Wells (New York: Churchill Livingstone) pp 35–45

Gururaja T R, Schulze W Aa, Cross L E, Newnham R E, Auld B A and Wang Y 1985 Piezoelectric composite materials for ultrasonic transducer applications, part 1:

Resonant modes of vibration of PZT rod-polymer composites *IEEE Trans. Sonics Ultrasonics* **SU-32** 481

Hewlett Packard Journal 1983 **34** special issues October 1983 and December 1983

Hunt J W, Arditi A and Foster F S 1983 Ultrasound transducers for pulse-echo medical imaging *IEEE Trans. Biomed. Eng.* **BME-30** 453–81

Hutchens C G and Morris S A 1984 A three port model for thickness mode transducers using SPICE II *1984 IEEE Ultrasonics Symp. Proc.* pp 897–902

Kino G S 1987 *Acoustic Waves: Devices, Imaging, and Analog Signal Processing* (Englewood Cliffs, NJ: Prentice-Hall)

Kino G S and DeSilets C S 1980 Design of slotted transducer arrays with matched backings *Ultrasonic Imaging* **1** 189

Larson J D 1981 Non-ideal radiators in phased array transducers *1981 IEEE Ultrasonics Symp. Proc.* pp 673–83

Leedom D, Krimholtz R and Matthaei G 1971 Equivalent circuits for transducers having arbitrary even-or-odd symmetry *IEEE Trans. Sonics Ultrasonics* **SU-18** 128–41

Lopath P D, Park S-E, Shung K K and Shrout T R 1997 Pb(Zn$_{1/3}$Nb$_{2/3}$)O$_3$/PbTiO$_3$ single crystal piezoelectrics for ultrasonic transducers *Proc. SPIE* **3037** 170–4

Macovski A 1983 *Medical Imaging Systems* (Englewood Cliffs, NJ: Prentice-Hall)

Mason W P 1948 *Electromechanical Transducers and Wave Filters* 2nd edn (New York: Wiley) pp 209–10, 399–404

Masotti L and Pini R 1993 Three dimensional imaging *Advances in Ultrasound Techniques and Instrumentation* ed P N T Wells (New York: Churchill Livingstone) pp 69–77

Matthaei G L, Young L and Jones E M T 1980 *Microwave Filters, Impedance-matching Networks, and Coupling Structures* (Dedham, MA: Artech House) ch 6 pp 255–354

Mequio C, Coursant R H and Pesque P 1983 Simulation of the acousto-electric response of piezoelectric structures by means of the Fast Fourier-Transform algorithm *Acta Electron.* **25** 311–23

Nalamwar A L and Epstein M 1972 Immitance characterization of acoustic surface-wave transducers *Proc. IEEE* **60** 336–7

Pace N G 1979 Impulse response of water-loaded air-backed piezoelectric disks *IEEE Trans. Sonics Ultrasonics* **SU-26** 37–41

Saitoh S, Kobayashi T, Shimanuki S and Yamashita Y 1997 Single element ultrasonic probe using PZN-PT single crystal *Proc. SPIE* **3037** 22–9

Sato J, Kawabuchi M and Fukumoto A 1979 Dependence of the electromechanical coupling coefficient on the width-to-thickness ratio of plank-shaped piezoelectric transducers used for electronically scanned ultrasound diagnostic system *J. Acoust. Soc. Am.* **67** 1609–11

Smith S W, Davidsen R E, Emery C D, Goldberg R L and Light E D 1995 Update on 2-D array transducers for medical ultrasound *1995 IEEE Ultrasonics Symp. Proc.* pp 1273–8

Smith W A, Shaulov A A and Singer B M 1984 Properties of composite piezoelectric materials for ultrasonic transducers *1984 IEEE Ultrasonics Symp. Proc.* pp 539–44

Souquet J, Defranould P and Desbois J 1979 Design of low-loss wide-band ultrasonic transducers for noninvasive medical application *IEEE Trans. Sonics Ultrasonics* **SU-26** 75–81

Szabo T L 1982 Miniature phased-array transducer modelling and design *1982 IEEE Ultrasonics Symp. Proc.* pp 810–4

—— 1984 Principles of nonresonant design *1984 IEEE Ultrasonics Symp. Proc.* pp 804–8

't Hoen P J 1982 Aperture apodization to reduce off-axis intensity of the pulsed-mode directivity function of linear arrays *Ultrasonics* 231–6

Takeuchi H, Jyomura S, Yamamoto E and Ito Y 1982 Electromechanical properties of (Pb,Ln)(Ti,Mn)O$_3$ ceramics *J. Acoust. Soc. Am.* **72** 1114

Vladimerescu A 1984 *Spice Version 2F.1 User's Guide* (Berkeley, CA: Department of Electrical Engineering and Computer Science, University of California)

Wojcik G L, DeSilets C, Nikodym L, Vaughan D, Abboud N and Mould J Jr 1996 Computer modeling of diced matching layers *1996 IEEE Ultrasonics Symp. Proc.* pp 1503–8

Wojcik G L, Vaughan D K, Abboud N and Mould Jr J 1993 Electrochemical modeling using explicit time-domain finite elements *1993 Ultrasonics Symp. Proc.* pp 1107–12

Wu J, Cubberley F, Gormely G and Szabo T L 1995 Temperature rise generated by diagnostic ultrasound in a transcranial phantom *Ultrasound Med. Biol.* **21** 561–8

Zanelli C I and Kadri M M 1994 Measurements of acoustic pressure in the non-linear range in water using quantitative schlieren *1994 IEEE Ultrasonics Symp. Proc.* pp 1765–8

Ziskin M C and Szabo T L 1993 Impact of safety considerations on ultrasound equipment and design and use *Advances in Ultrasound Techniques and Instrumentation* ed P N T Wells (New York: Churchill Livingstone) pp 151–60

CHAPTER 6

CURRENT DOPPLER TECHNOLOGY AND TECHNIQUES

Peter N T Wells

SUMMARY

The Doppler effect is the change in the frequency of a wave perceived by an observer when received from a moving source. Scattering of ultrasound by flowing blood can be characterised as a Gaussian random process. The velocity of blood flow can be estimated from the Doppler shift frequency, which generally falls in the audible range in medical applications. Continuous wave ultrasound cannot provide information about the distance to the target; by pulsing the ultrasound, distance can be estimated from the echo delay time. Doppler signals can be processed to determine the direction of flow. The frequency spectrum can be analysed by numerous methods, the most useful generally being the fast Fourier transform. Combining pulsed Doppler and real-time two-dimensional scanning results in duplex scanning, which allows the anatomical location of the Doppler sample volume to be identified. In colour flow imaging, the whole of the two-dimensional scan is superimposed by a colour map derived from moving targets within the scan plane. Frequency domain and time domain processing are possible. The colour can be coded either according to the velocity of target motion, or according to the power of echoes from moving targets. The information, having been acquired in contiguous two-dimensional scans, can be displayed in three dimensions. Contrast agents, consisting of gas microbubbles, can be used to enhance the Doppler signal amplitude. Contrast agents also scatter ultrasound at the second harmonic of the irradiating frequency and this provides a technique for clutter rejection.

6.1. THE DOPPLER EFFECT

The paper that made Christian Doppler famous, 'On the coloured light of double stars and certain other stars of the heavens', was read by him before the Royal Bohemian Society of Sciences in Prague on 25 May 1842 and published in the following year (Doppler 1843). Eden (1992) has written a fascinating account of these events. Suffice it to say here that Doppler's paper begins by recapitulating the wave theory of light and introduces the idea that the observed frequency increases when moving towards the source and vice versa. Using a rather clumsy mathematical derivation, Doppler showed that

$$f_o = ((c - v_o)/(c - v_s))f_s \qquad (6.1)$$

where f_o is the apparent frequency perceived by the observer, f_s is the frequency of the source, c is the wave speed and v_o and v_s are the vector velocities of the observer and the source respectively. Doppler went on erroneously to conclude that blue stars owe their colour to the fact that they are approaching the observer while receding stars are red. This argument depends on the false assumption that stars only emit white light and not ultraviolet or infrared light. Only later did it become established that blue stars appear so because they are hotter than red ones.

6.2.   THE ORIGIN OF THE DOPPLER SIGNAL

The detectability of ultrasound backscattered by blood ultimately depends on the signal-to-noise ratio. The suspended ultrasonic scatterers that are overwhelmingly dominant are the erythrocytes (the red cells). An individual erythrocyte is a biconcave disk, about 8 μm in diameter and with a maximum thickness of 2 μm. The ratio of the total erythrocyte volume to the total blood volume is known as the haematocrit; typically, this is 0.45 in normal adults, being slightly higher in men than in women.

   At the typical ultrasonic frequency of 3 MHz, the wavelength in blood is about 0.5 mm. Thus, an individual erythrocyte is two orders of magnitude smaller than the wavelength; it can be considered to be a Rayleigh scatterer, for which the backscattered power increases as the fourth power of the frequency (see Chapter 4). Brody and Meindl (1974) treated blood as a suspension of point scatterers in modelling the scattering process. This neglects the fact that the erythrocytes are quite closely packed, so they do not behave as uncorrelated scatterers but actually are strongly interacting. Angelsen (1980) avoided this problem, in his model, by treating blood as a continuous medium characterised by fluctuations in mass density and compressibility. The situation is further complicated by the tendency of erythrocytes to aggregate, forming rouleaux that can survive even under normal flow conditions (Machi *et al* 1983). Yuan and Shung (1989) concluded that scattering by porcine blood decreases with increasing shear rate. Although bovine blood does not exhibit this shear rate dependence, using it as a model for human blood allowed Mo and Cobbold (1986) to conclude that ultrasound backscattered by flowing blood can be characterised as a Gaussian random process.

Ultrasonic blood flow detection is only possible because ultrasonic scattering by blood is a stochastic process. As flowing blood interacts with a beam of ultrasound, it behaves like a random array of discrete targets, albeit limited in extent both temporally and spatially. There are limitations, however, in the assumption that the backscattered power is proportional to the number of red cells interacting with the ultrasound (Shung *et al* 1992).

The choice of the ultrasonic frequency that gives maximum signal-to-noise ratio for echoes detected from blood is determined by the compromise between the backscattering efficiency of blood (which increases with frequency) and the useful depth of penetration (which decreases with frequency). In practice, a frequency of around 3.5 MHz is generally used for a penetration of about 150 mm, but it can be increased to, say, 10 MHz when the penetration is reduced to about 20 mm. With pulsed Doppler systems, as described in section 6.3.2, it may be advantageous to use a somewhat lower ultrasonic frequency because this relieves the constraint imposed by the Nyquist criterion.

Happily, Doppler's equation is true for acoustic waves. The Doppler shift in the frequency of ultrasound backscattered by blood is a special case in which the source and the observer are stationary and the ultrasonic beam is folded back by the moving scattering ensembles so that the transmitting and receiving beams are effectively coincident. In this way, motion of an ensemble results in change in the distance that the ultrasonic wave travels between the transmitting and the receiving transducers. Described in another way, a moving scattering ensemble of erythrocytes acts both as a receiver of Doppler-shifted ultrasound as the distance between the ensemble and the transmitting transducer changes, and as a transmitter of this already Doppler-shifted ultrasound which is Doppler-shifted again as the distance between the ensemble and the receiving transducer also simultaneously changes.

It is actually the difference f_D between the frequencies of the observed and the transmitted ultrasound, known as the Doppler shift frequency, which is important. In general, from equation (6.1),

$$f_D = f_o - f_s = \left(\frac{c - v_o}{c - v_s} - 1 \right) f_s. \tag{6.2}$$

With ultrasound backscattered from blood, the path length changes at a rate equal to the velocity $2v_i$, where v_i is the vector velocity of the ensemble. Provided that $c \gg v_i$, equation (6.2) can be simplified to

$$f_D = 2f v_i / c \tag{6.3}$$

using the notation that motion towards the effectively coincident transmitting and receiving transducers is positive.

There is one further practical consideration. Often, there is an angle θ between the effective direction of the ultrasonic beam and the flow of blood. Moreover, the Doppler shift frequency is the measured quantity from which the absolute flow velocity v of the blood is to be calculated. Then, equation (6.3) becomes

$$f_D = 2fv(\cos\theta)/c. \tag{6.4}$$

It is now appropriate to calculate the Doppler shift frequency for some typical physiological situation. Consider, for example, the Doppler shift frequency corresponding to blood flow during systole in the ascending aorta. The peak blood flow velocity is around 150 cm s^{-1}. With an ultrasonic frequency of 3 MHz and coincident flow and beam directions ($\theta = 0°$), $f_D = 6000$ Hz. Thus, although it is no more than a lucky chance, the Doppler shift frequencies in this and virtually all other practical physiological situations happen to fall in the audible range.

6.3.   THE NARROW FREQUENCY BAND TECHNIQUE

6.3.1.   *The continuous wave Doppler technique*

A block diagram of a continuous wave Doppler system is shown in figure 6.1. Uninterrupted ultrasonic waves of constant frequency and amplitude are emitted into the patient by the transmitting transducer. The echoes detected by the receiving transducer have the same frequency as that of the transmitter if they are reflected by stationary structures. If they originate from flowing blood, however, they are shifted in frequency by the Doppler effect. Thus, the output signals from the radio-frequency (RF) amplifier have a strong component at the transmitted frequency. In addition, in the example shown in figure 6.1, where the ultrasonic beam intercepts an artery in which blood flows in the forward direction in systole and in the reverse direction during diastole, there are components in the upper and lower sidebands corresponding respectively to approaching and receding flows. These sidebands can be separated by phase quadrature detection (Nippa *et al* 1975), the circuit for which is identified in the block diagram in figure 6.1.

The operator can listen to these separated signals on headphones, as shown in figure 6.1, with the forward flow signals fed to one ear and the reverse flow signals to the other. Although this works well, often it is desirable to have a permanent recording of the signals for detailed examination and analysis. A convenient way in which this can be achieved involves the use of a heterodyne processor and frequency spectrum analyser, also included in the block diagram in figure 6.1. The forward (U) and reverse (L) flow signals are separately multiplied by a signal at a pilot frequency, chosen at least just

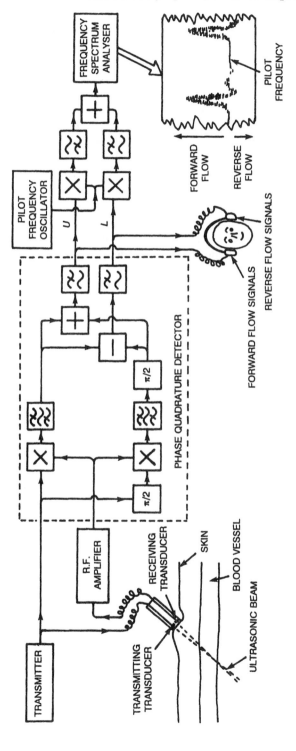

Figure 6.1. *A block diagram of a continuous wave Doppler system with directional phase quadrature detection providing audible forward (U) and reverse (L) flow signals, and a heterodyne processor and frequency spectrum analyser for visual display. From Wells (1994).*

to exceed the maximum reverse flow frequency. The output from the forward flow multiplier consists of a signal at the pilot frequency with the forward flow signals arranged symmetrically as upper and lower sidebands to the pilot frequency. A high-pass filter is arranged to eliminate the lower sideband. Similarly, the reverse flow signals are obtained as a lower-frequency signal sideband to the pilot frequency. Adding these two filtered signals results in a spectrum of frequencies in which the pilot frequency corresponds to zero flow velocity, frequencies above the pilot frequency are from forward flow and frequencies below are from reverse flow. Figure 6.1 shows how this spectrum can be displayed after frequency spectrum analysis, a process described in section 6.4.

Phase quadrature detection is very widely used to separate forward and reverse flow signals in ultrasonic Doppler instruments. For the circuit designer, success lies principally in designing the second of the $\pi/2$ phase-shift networks (see figure 6.1) so that its accuracy is maintained over the entire range of Doppler shift frequencies.

6.3.2. *The pulsed Doppler technique*

Pulsed Doppler systems use pulse-echo range gating to select Doppler signals from moving targets according to their distances from the ultrasonic probe (Wells 1969). Figure 6.2 shows the essential components of the circuit. The master oscillator runs at the frequency of the transmitted ultrasound. Clock pulses are obtained by dividing down from the master oscillator frequency and these clock pulses trigger the transmit sample-length monostable at the pulse repetition frequency of the system. This monostable generates a pulse that opens the transmit pulse gate for the length of time that ultrasound needs to be emitted by the transmitting transducer to produce the desired sample length (about 0.67 μs mm^{-1}). Echo signals from the receiving transducer are subjected to RF amplification before being fed to the phase quadrature detector and heterodyne processor, which function as explained in section 6.3.1. (Note that the pilot frequency can be chosen to maximise the frequency spectrum that can be accommodated within the Nyquist limit, which is explained later in this section.) Thus, following the transmission of each ultrasonic pulse, the output from the heterodyne processor consists of a wavetrain in which later time corresponds to increasing depth. The sample-depth monostable, which is also triggered by the clock pulse, introduces a time delay chosen by the operator to correspond to the echo delay time (about 1.33 μs mm^{-1}) from the beginning of the sample volume from which it is desired to collect the Doppler signals. The receive sample-length monostable then opens the receive pulse gate for a period equal to the transmitted pulse length, so that the output from the receive pulse gate is the Doppler signal sample. The amplitude of this signal sample is held in the sample-and-hold

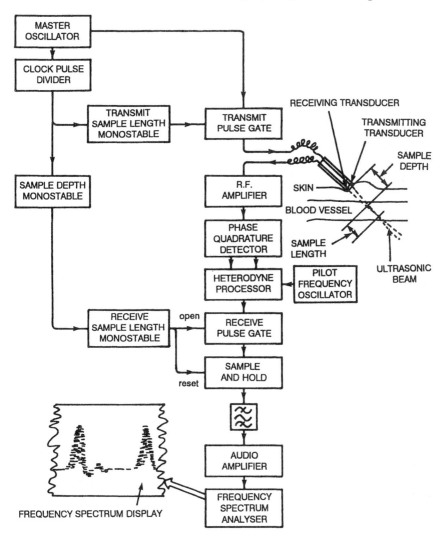

Figure 6.2. *A block diagram of a pulsed Doppler system providing depth and length adjustments for the sample volume and phase quadrature directional detection. From Wells (1994).*

circuit until it is updated by the sample derived from the next transmitted ultrasonic pulse.

All that is then necessary is for the output from the sample-and-hold circuit to be filtered (to eliminate both low-frequency high-amplitude signals from slowly moving strongly reflecting structures such as the walls of blood vessels, and high-frequency signals associated with the sampling transients) and amplified before frequency spectrum analysis.

According to the Nyquist criterion, there is no aliasing in the Doppler shift signals at the output from the sample-and-hold circuit provided that the pulse repetition frequency is more than twice the highest Doppler shift frequency.

6.4. FREQUENCY SPECTRUM ANALYSIS

A continuous frequency spectrum is produced by both continuous wave and, within the constraints imposed by sampling and the Nyquist criterion, pulsed Doppler methods. There are numerous techniques by which frequency spectrum analysis may be performed. Although nowadays often dealt with dismissively, the zero-crossing counter (Lunt 1975) is in fact quite a good estimator of the root-mean-square frequency. Other analogue methods for frequency spectrum analysis, such as bandpass filter banks (Light 1970) and time compression systems (Coghlan *et al* 1974) have largely fallen into disuse for Doppler applications.

Currently, Doppler signals of reasonable duration that do not slew too rapidly (examples of satisfactory Doppler blood flow signals are those obtained with stationary continuous wave or pulsed beams) are usually analysed by digital calculation of the fast Fourier transform (FFT). An excellent review of the method is presented in a tutorial paper by Challis and Kitney (1991). Assuming that the 'classical' Fourier transform can provide adequate resolution, it is generally the best option. (In the temporal domain, resolution in this context relates to the minimum duration of the signal segment that allows its frequency to be estimated with the required accuracy.) When this is not the case, one of the techniques of 'modern' spectral analysis should be used. Although these can be either parametric or non-parametric, it is the parametric approach using, for example, the autoregressive algorithm that has turned out to be the most promising with ultrasonic Doppler signals (D'Alessio *et al* 1984, Schlindwein and Evans 1990).

6.5. DUPLEX SCANNING

A duplex scanning system is one that enables two-dimensional ultrasonic pulse-echo imaging to guide the placement of an ultrasonic Doppler beam and thus to allow the anatomical location of the origin of the Doppler signals to be identified. Since its inception by Barber *et al* (1974), the common usage of the term 'duplex' has changed; what most people now consider to be a duplex scanner is one that has real-time imaging capability with either the imaging transducer or a separate transducer used to collect CW or pulsed Doppler signals, either simultaneously with imaging or sequentially.

Duplex scanners can be based on either mechanical or electronic array real-time imaging. The image frame rate is limited by the depth of the scanned field and the number of lines per frame. If the depth is, say 100 mm, the go-and-return pulse-echo delay time is about 130 μs per line. The corresponding maximum pulse repetition rate (equal to the number of image lines collected per second) is about 7700 Hz. This means that images made up of, say, 77 lines per frame can be collected at 100 frames per second, and so on.

For the collection of Doppler signals, the ultrasonic beam has effectively to be stationary and to dwell in the desired position sufficiently frequently and for a sufficiently long time during each acquisition period to allow adequate sampling of the moving targets. Inertia prevents the acceleration and deceleration of the moving parts of mechanical sector scanners that would be necessary for simultaneous collection of imaging and Doppler data, so the 'frozen' image frame approach is necessary unless a separate transducer is used for Doppler signal acquisition. Movement of the beam of an array scanner can be started and stopped instantaneously, so simultaneous operation is possible with this type of system.

6.6.   COLOUR FLOW IMAGING

6.6.1.   *Basic principles*

The basic principles of combining pulse-echo and Doppler two-dimensional images are illustrated in figure 6.3. Brandestini and Forster (1978) were the first to demonstrate the feasibility of this approach. They used a duplex scanner: two-dimensional grey-scale imaging was achieved in real time and the scan line of a 128-point pulsed Doppler system was swept relatively slowly through the imaged plane to 'paint' the flow signals into the vessel lumen using shades of red and blue. Thus, the colour flow imaging itself was not performed in real time; in practice, it was eventually possible to operate at four frames per second (Eyers *et al* 1981). Fundamentally, however, the multigated pulsed Doppler system used by Brandestini and Forster (1978) was limited in temporal resolution by the process of frequency estimation that it employed. Similar constraints applied to the instruments developed by Hoeks *et al* (1981) and Casty and Giddens (1984) for the study of flow velocity profiles.

It is the constraint on the maximum frame rate that is generally of most practical importance in colour flow imaging. With a simple scanning strategy in which the ultrasonic beam sweeps line by line in sequence through the scan plane, the beam has to dwell in each position for long enough to allow the lowest Doppler shift frequency that it is desired to detect to be determined. Simplistically, it is reasonable to assume that determining the

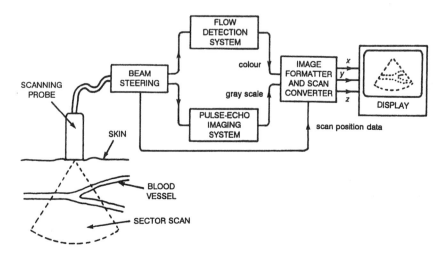

Figure 6.3. *A block diagram of a real-time two-dimensional colour flow imaging system. In this example, a phased or curved array transducer is used to produce a sector scan. Alternatively, a linear array transducer can be used to produce a scan in parallelogram format. The image formatter and scan converter accept the colour and grey-scale signals together with scan position data from the ultrasonic beam steering circuit. The scan converter may incorporate serial image memories to provide a ciné review facility. The operation of the entire system is synchronised by a clock pulse generator which, for simplicity, is not shown in this diagram. From Wells (1994).*

frequency of a signal requires a measurement time at least equal to the duration of one complete cycle of the signal. (During this time, of course, there are likely to be numerous samples at the pulse repetition frequency, so sampling quantisation is not a problem.) As an example, assume that it is desired to detect flow at velocities down to 10 cm s^{-1}, that the ultrasonic frequency is 3 MHz and $\theta = 0°$. This results in a Doppler shift frequency of 400 Hz; the corresponding period of one complete cycle is 2.5 ms. Thus, the ultrasonic beam has to dwell for 2.5 ms in each discrete position in the scan plane. Even if there are only 100 lines of colour information in the image, the time per frame is then 250 ms, corresponding to a frame rate of only 4 s^{-1}. This is hardly 'real time', and it is for this reason that more advanced frequency estimation and scanning strategies are used to increase the frame rate to produce images with clinically useful temporal resolutions.

The spatial resolution of colour flow imaging is controlled by the same beam and pulse properties as in grey-scale pulse-echo imaging. Generally, however, the beam is designed to be wider than for imaging (if only because a lower ultrasonic frequency is used) and, with Doppler detection at least, the pulse is made somewhat longer (to minimise spectral broadening).

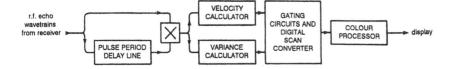

Figure 6.4. *A block diagram of an autocorrelation detector and associated gating, scan-converting and colour-processing circuits. The pulse period delay line inserts a time delay equal to the interval between consecutively transmitted ultrasonic pulses. The multiplier performs the process of auto-correlation. Notice the similarity between this frequency-domain (Doppler) processing approach and time-domain processing (see section 6.6.4) in which the multiplying autocorrelator is replaced in the circuit by a time-shifting cross-correlator. From Wells (1994).*

6.6.2. Autocorrelation detection

The basis of the autocorrelation detector (Kasai *et al* 1985) is that the echo wavetrain from stationary targets does not change with time, whereas sequential echo wavetrains from moving targets have corresponding changes in relative phase. The autocorrelation detector, illustrated in figure 6.4, produces an output signal that is derived from the relative phases of consecutive pairs of received echo wavetrains. Thus, the echo wavetrains themselves serve as references for phase comparison, rather than making phase comparison with a fixed-frequency reference signal. The RF echo wavetrains from the receiver are split into two circuit paths. One path leads directly to one input to a multiplier. The other path leads to the other input to the multiplier but through a delay line, which introduces a time shift exactly equal to the interval between the transmission of successive ultrasonic pulses. Hence, the output from the multiplier is the autocorrelation function of consecutive wavetrains; it has a constant amplitude unless pairs of consecutive wavetrains differ in phase. For this reason, the process is sometimes called 'pulse-pair-covariance' detection.

6.6.3. Other Doppler frequency estimators

Since the demonstration of the feasibility of real-time Doppler colour flow imaging with autocorrelation detection (see section 6.6.2), there has been increasing interest in other methods of frequency estimation in the search for enhanced performance.

The ideal estimator would be able accurately and consistently to determine the mean Doppler shift frequency, irrespective of its spectral distribution and the levels of the noise and stationary echo signals. All these factors, together with the duration of the time window, affect the bias and variance of the estimate. It is reasonable to expect that the variance increases with lower signal-to-noise ratios and shorter time windows. Brands and Hoeks (1992) have compared the performances of first- and second-order autoregressive estimators and a complex linear regression estimator in combination with a stationary-echo canceller to eliminate high amplitude signals from major tissue interfaces. A second-order autoregressive filter is able simultaneously to estimate the positions of two peaks in the spectrum, one of which can be assumed to correspond to the Doppler signals and the other to high-amplitude echoes from solid tissues. Both a first-order autoregressive filter and a complex linear estimator (which determines the slope of the regression line fitted through the unwrapped phase) can only estimate the position of a single peak in the spectrum. With realistically chosen conditions, the mean-frequency estimator based on the regression line through the unwrapped phase of the Doppler signal, together with an appropriate stationary-echo canceller, gives the best performance with short observation windows. For longer windows, however, the first-order autoregressive estimator is best. The performance of the second-order autoregressive filter is always poor in comparison with the other two.

6.6.4. *Time-domain processing*

Embree and O'Brien (1985) and Bonnefous and Pesque (1986) independently described how target-velocity information can be extracted from broad-band ultrasonic echoes by means of time-domain processing. The method depends on temporal tracking of the spatial position of individual coherent blood target ensembles. In principle, it can be applied either to the RF or to the video signals, although the two approaches require very different instrument designs.

6.6.4.1. *Cross-correlation detection.*   Blood flow velocity can be determined by a one-dimensional correlation between RF echo wavetrains collected from consecutively transmitted ultrasonic pulses. The time t required for an ultrasonic pulse to complete the round trip between a transducer and a scattering ensemble of blood cells situated a distance z from the transducer is given by

$$t = 2z/c. \tag{6.5}$$

If the pulse period (the reciprocal of the pulse repetition frequency) is equal to t_p and v_i is the vector velocity of the ensemble, the time shift τ of the echoes from consecutive pulses is given by

$$\tau = 2t_p v_i/c. \tag{6.6}$$

This time-domain formulation is clearly similar to the frequency-domain (Doppler) formulation given in equation (6.3). Moreover, the same cosine dependence between the angular direction of flow and that of the ultrasonic beam applies to both Doppler and time-domain processing. The cross-correlation can be achieved by shifting the relative time positions of the consecutive wavetrains and multiplying them together.

6.6.4.2. Speckle tracking. Perhaps the most obvious ultrasonic technique for studying structure motion is that which depends on tracking a strong quasispecular echo isolated from within sequential A scans by an appropriately positioned gate. This is the basis of M-mode echocardiography (Wells and Ross 1969). A natural extension of this approach is to track the speckle structure within a sequence of one-, two- or even three-dimensional images of blood flow. In this context, it is the demodulated (video) ultrasonic signal that gives rise to the image speckle, as distinct from the RF echoes from ensembles tracked by the cross-correlation technique discussed in section 6.6.4.1. The method was first demonstrated by Trahey *et al* (1987), using two sequential 10 MHz images of blood in a popliteal vein. If the sensitivity is sufficient to display the coherent speckle from the blood, the displacements of the individual blood ensembles can be tracked by a localised two-dimensional correlation search. Lines representing the velocity vectors can be superimposed on the image, so that velocity estimates are angle independent for in-plane flow.

6.6.5. Colour coding schemes

Besides possessing luminosity (brightness or shade), colours are regarded as possessing the attributes of hue and saturation. Hue is determined by the wavelength of the light. Saturation is the degree to which a colour departs from white and approaches a pure spectral colour. When a hue is desaturated with white light, it is classified as a tint. The wavelengths of the visible spectrum extend from about 740 nm (red) to 390 nm (violet). When suitably presented, the eye can discriminate between hues in the middle part of the spectrum with wavelength differences of 1.2 nm; about 130 steps of hue difference can be perceived across the visible spectrum. As a hue is diluted with white, about 20 tints can be identified.

Various schemes of colour coding are adopted in different colour flow imaging instruments and individual preferences vary. In general, however, increasing flow velocity towards the probe is colour coded red–orange–yellow–white; increasing reverse-flow velocity is displayed in dark blue–blue–light blue–white. These colour scales can be interpreted without the need to refer to a coding key.

In addition to colour coding according to velocity, information concerning flow disturbance, which is related to the variance of the estimate of velocity, can be simultaneously displayed by appropriate colour coding. This can be done, for example, by restricting the limits of the velocity map to the reds and blues and injecting an increasing amount of yellow as flow disturbance increases. The yellow converts the reds to oranges with forward flow and the blues to greens with reverse flow. Another use for green is to identify areas of equal velocity (or equal range of velocity) on the flow map. This is sometimes called 'green tagging'.

In another kind of colour-coded display, the red or blue luminosity (depending on personal preference) is used to indicate the power of the blood flow signal. By appropriately selecting the gain of the system, stationary structures are displayed with low luminosity, while blood flow appears proportionally brighter depending on the signal power. Although this kind of facility was available on some of the earliest commercial scanners (Jain *et al* 1991), it fell into disuse and has only recently been reintroduced (Rubin *et al* 1994). Modern colour flow-power images do not distinguish between forward and reverse flow; the colour luminosity is, presumably, related to the corresponding perfusion volume.

6.6.6. *Three-dimensional display*

Using the real-time imaging probe of a commercially available colour flow scanner, Pretorius *et al* (1992) described how they translated the scan plane at 5 mm intervals through the region of the carotid artery bifurcation and the jugular vein to collect a three-dimensional data set consisting of 33 images. The presence of colour was used to separate and thus to segment the grey-scale and blood flow data. By this means, rendered surface images of the blood vessels could be displayed, with the viewpoint under the control of the operator.

Similar techniques for image acquisition were used by Miyagi *et al* (1993) and Picot *et al* (1993). Like Pretorius *et al* (1992), Miyagi *et al* (1993) suppressed the grey-scale image to be left with a three-dimensional surface rendering of the blood-vessel image, which could be examined from any viewpoint selected by the operator. The three-dimensional image displayed by Picot *et al* (1993), however, retained the grey-scale information; the viewpoint could be changed and the image block could be cut in any desired plane.

6.7.   CONTRAST AGENTS AND SECOND HARMONIC IMAGING

The use of contrast agents to enhance the echogenicity of blood (Melany and Grant 1997) has obvious applications in improving the sensitivity of colour flow imaging. There is scope for the development of functional investigations based on the transit time of contrast agent boluses. Reception of the second harmonic signal backscattered from resonance bubble contrast agents (which have bubble diameters of the order of a few micrometres at low megahertz frequencies) provides significant clutter rejection and may also be a good approach to the measurement of perfusion (Schrope and Newhouse 1993). The use of contrast agents is discussed more fully in Chapter 12.

REFERENCES

Angelsen B A J 1980 A theoretical study of the scattering of ultrasound from blood *IEEE Trans. Biomed. Eng.* **27** 61–7

Barber F E *et al* 1974 Ultrasonic duplex echo-Doppler scanner *IEEE Trans. Biomed. Eng.* **21** 109–13

Bonnefous O and Pesque P 1986 Time domain formulation of pulse-Doppler ultrasound and blood velocity estimation by cross correlation *Ultrasonic Imaging* **8** 75–85

Brandestini M A and Forster F K 1978 Blood flow imaging using a discrete-time frequency meter *IEEE Ultrasonic Symp. Proc.* (New York: IEEE) pp 348–52

Brands P J and Hoeks A P G 1992 A comparison method for mean frequency estimators for Doppler ultrasound *Ultrasonic Imaging* **14** 367–86

Brody W R and Meindl J D 1974 Theoretical analysis of the cw Doppler ultrasonic flowmeter *IEEE Trans. Biomed. Eng.* **21** 183–92

Casty M and Giddens D P 1984 25+1 channel pulsed ultrasound Doppler velocity meter for quantitative flow measurements and turbulence analysis *Ultrasound Med. Biol.* **10** 161–72

Challis R E and Kitney R I 1991 Biomedical signal processing *Med. Biol. Eng. Comput.* **29** 1–17

Coghlan B A, Taylor M G and King D H 1974 On-line display of Doppler-shift spectra by a new time compression analyser *Cardiovascular Applications of Ultrasound* ed R S Reneman (Amsterdam: North-Holland) pp 55–65

D'Alessio T, Catanzariti A, di Guiliomaria C and Cavallaro A 1984 An algorithm to improve signal-to-noise discrimination in Doppler spectrum analysis *Ultrasonics* **10** 282–4

Doppler C 1843 Ueber das farbige Licht der Doppelsterne un einer anderer Gestirne des Himmels *Abhand. K.-bohm. Gessell.* **2** 465–82

Eden A 1992 *The Search for Christian Doppler* (Vienna: Springer)

Embree P M and O'Brien W D 1985 The accurate measurement of volume flow of blood by time domain correlation *IEEE Ultrason. Symp. Proc.* (New York: IEEE) pp 963–6

Eyers M K, Brandestine M, Phillips D J and Baker D W 1981 Color digital echo/Doppler image presentation *Ultrasound Med. Biol.* **7** 21–31

Hoeks A P G, Reneman R S and Peronneau P A 1981 A multi-gated pulsed Doppler system with serial data processing *IEEE Trans. Sonics Ultrasonics* **28** 242–7

Jain S P *et al* 1991 Influence of various instrument settings on flow information derived from the power mode *Ultrasound Med. Biol.* **17** 49–54

Kasai C, Namekawa K, Koyano A and Omoto R 1985 Real-time two-dimensional blood flow imaging using an autocorrelation technique *IEEE Trans. Sonics Ultrasonics* **32** 458–64

Light L H 1970 A recording spectrograph for analysing Doppler blood velocity signals in real time *J. Physiol.* **207** 42P–44P

Lunt M J 1975 Accuracy and limitations of the ultrasonic Doppler blood velocimeter and zero crossing detector *Ultrasound Med. Biol.* **2** 1–10

Machi J *et al* 1983 Relation of in vivo blood flow to blood echogenicity *J. Clin. Ultrasound* **11** 3–10

Melany M L and Grant E G 1997 Clinical experience with sonographic contrast agents *Semin. Ultrasound CT MRI* **18** 3–12

Miyagi Y, Masaoke H, Akamatsu N and Sekiba K 1993 Development of a three-dimensional color Doppler system *Med. Prog. Technol.* **18** 201–8

Mo L Y L and Cobbold R S C 1986 A stochastic model of the backscattered ultrasound from blood *IEEE Trans. Biomed. Eng.* **33** 20–7

Nippa J H *et al* 1975 Phase rotation for separating forward and reverse blood flow velocity signals *IEEE Trans. Sonics Ultrasonics* **22** 340–6

Picot P A *et al* 1993 Three-dimensional colour Doppler imaging *Ultrasound Med. Biol.* **19** 95–104

Pretorius D H, Nelson T R and Jaffe J S 1992 A 3-dimensional sonographic analysis based on color flow Doppler and gray scale image data *J. Ultrasound Med.* **11** 225–32

Rubin J M *et al* 1994 Power Doppler US: a potentially useful alternative to mean frequency-based color Doppler US *Radiology* **190** 853–6

Schlindwein F S and Evans D H 1990 Selection of the order of autoregressive models for spectral analysis of Doppler ultrasound signals *Ultrasound Med. Biol.* **16** 81–91

Schrope B A and Newhouse V L 1993 Second harmonic blood perfusion measurements *Ultrasound Med. Biol.* **19** 567–79

Shung K K, Cloutier G and Lim C C 1992 The effects of hematocrit, shear rate, and turbulence on ultrasonic Doppler spectrum from blood *IEEE Trans. Biomed. Eng.* **39** 462–9

Trahey G E, Allison J W and von Ramm O T 1987 Angle independent ultrasonic detection of blood flow *IEEE Trans. Biomed. Eng.* **34** 965–7

Wells P N T 1969 A range-gated ultrasonic Doppler system *Med. Biol. Eng.* **7** 641–52

Wells P N T 1994 Ultrasonic colour flow imaging *Phys. Med. Biol.* **39** 2113–45

Wells P N T and Ross F G M 1969 A time-to-voltage analogue converter for ultrasonic cardiology *Ultrasonics* **7** 171–6

Yuan Y-W and Shung K K 1989 Echoicity of whole blood *J. Ultrasound Med.* **8** 425–34

CHAPTER 7

THE PURPOSE AND TECHNIQUES OF ACOUSTIC OUTPUT MEASUREMENT

T A Whittingham

7.1. WHY MEASURE ACOUSTIC OUTPUTS?

Over the three decades or so in which diagnostic ultrasound has been in use, the magnitudes of temporal average intensity and other acoustic output quantities have increased considerably. The lack of any confirmed harmful effects from the earlier machines provides no guarantee that today's more powerful machines are free from potential hazard. Indeed, we shall see that from thermal considerations alone there are grounds for concern over the high outputs of which some modern machines are capable. It is important that clinicians and users have access to accurate data about the acoustic output of their machines if they are to make informed risk–benefit judgements.

7.2. ULTRASOUND DAMAGE MECHANISMS AND THEIR BIOLOGICAL SIGNIFICANCE

The two main damage mechanisms are generally considered to be heating and cavitation. There are other potential mechanisms involving mechanical forces, such as bulk acoustic streaming and standing wave radiation forces, but these are not generally considered to be as important and therefore will not be discussed further in this brief summary. Only the principal biological consequences of heating and cavitation are mentioned below; much more extensive reviews are available in the literature (Whittingham 1994, Barnett and Kossoff 1998). In addition, the mechanisms of heating and cavitation are described fully in Chapters 8 and 11 of this book, and streaming and radiation force mechanisms are discussed in Chapter 3.

7.2.1. Heating

Heating occurs due to absorption of ultrasound energy by tissue and to self-heating of the probe. The tolerance of tissues to temperature elevation depends on the type of tissue and the duration of the temperature rise. Tissues undergoing organogenesis are particularly at risk; this applies to the embryo in the first 8 weeks following conception and to the brain and spinal cord even after birth. The World Federation of Ultrasound in Medicine and Biology (Barnett 1998) has concluded that a temperature increase to the fetus in excess of 1.5°C may be hazardous if prolonged, while 4°C is hazardous if maintained for 5 minutes.

Ignoring for the moment the acoustic and thermal properties of the tissues, the greatest potential for heating exists at the point in the insonated field where the temporal average intensity I_{ta} is greatest. The value of I_{ta} at this point is known as the spatial peak time-averaged intensity, I_{spta}, and this quantity features in several regulations and standards. For convenience, I_{ta} is usually measured in water, but in some standards an attempt is made to allow for tissue attenuation. The 'derated' I_{ta} is the I_{spta} value, as measured in water, but reduced by a factor to simulate the effect of a medium with a stated attenuation coefficient, usually 0.3 dB cm^{-1} MHz^{-1}. Note that the derated I_{spta} is the largest value of derated I_{ta} in the ultrasound field, and this is not the same as the 'in water' I_{spta} with a derating factor applied.

The actual local rate of heat energy production (W cm^{-3}) is equal to $0.23\alpha I_{ta}$, where I_{ta} (W cm^{-2}) is the temporal average intensity in the tissue, and α (dB cm^{-1}) is the local ultrasonic absorption coefficient. Bone has a much higher absorption coefficient than any soft tissue, and is therefore an important potential target for heating effects. From the above it might be assumed that I_{spta} is a reliable predictor of worst case heating, under the assumption that the most strongly absorbing target could lie at the point of I_{spta}. Although this is sometimes true, it is not true in general.

The actual temperature rise will of course depend on the local specific heat capacity and the time of exposure, but it will also depend on the rate and distribution of heat production in neighbouring tissues, and the ease with which this heat can be transferred through the tissues. The latter can be described by a parameter known as the perfusion length (Swindell 1984), given by

$$L = \sqrt{\frac{k}{w s_b}} \qquad (7.1)$$

where k is the thermal conductivity of the tissue, w is the blood perfusion flow rate and s_b is the specific heat capacity of blood. The perfusion length is a few centimetres in relatively poorly perfused, or highly conducting tissues such as fat, bone or muscle, but only a few millimetres in well perfused or poorly conducting tissues such as heart, liver, kidney or brain. Where the perfusion length is smaller than any dimension of the insonated volume, heat transfer from surrounding tissue is small and the local I_{ta} is a reasonable indicator of heating potential. Where it is larger than any of the dimensions of the insonated volume, heat transfer within the insonated volume becomes important and the total temporal average acoustic power, W, transmitted by the probe may then be a better indicator of heating potential. A more complete discussion of the prediction of temperature rises in tissues perfused with vessels of more complex geometry may be found in Chapter 8.

In general then, neither the temporal average acoustic power nor the temporal average intensity can solely be used to compare the heating potential of different machines or control settings. A theoretically predicted temperature rise for a given acoustic field and clinical application is therefore a better guide in assessing thermal hazard than are basic acoustic field parameters such as W or I_{spta}. Calculations (Jago *et al* 1994) based on measured three-dimensional distributions of temporal average intensity from clinically used scanners and simple theoretical models of tissue layers, indicate that some diagnostic ultrasound machines are capable of generating worst case temperature rises in excess of 1.5°C in soft tissue, in all modes, including B-mode. In pulsed Doppler mode, temperature rises as high as 2.5°C were predicted for soft tissue and 6°C for bone. These temperature rises would be in addition to any pre-existing temperature elevation, for example due to fever.

A further cause of heating to consider is the probe itself, particularly in transvaginal applications. Examples of probe face temperatures of more than 10°C above the ambient air temperature were reported several years ago (Duck *et al* 1989). Manufacturers have addressed the problem, but this potentially important form of heating should always be checked.

7.2.2. Cavitation

This refers to the production and growth of bubbles by negative pressure excursions and to their subsequent behaviour, and is described in greater detail in Chapter 11. In the presence of an ultrasound wave, gas bubbles can form and grow in the half cycles for which the pressure is negative. Two forms of cavitation can be distinguished. In collapse (inertial) cavitation, pulses with very large variations in pressure can cause bubbles to form and collapse within one or two cycles, converting relatively large amounts of sound energy into heat and shock waves, and can create hazardous free radicals such as OH^-. Collapse cavitation can lead to tissue damage and cell destruction. As evidence of this, it may be noted that it is the mode of action of ultrasonic cleaning tanks. Stable cavitation is a less violent phenomenon requiring continuous waves, or very long pulses, of more moderate amplitude. Here, bubbles of a critical resonant size undergo large amplitude radial oscillations at the frequency of the applied ultrasound.

Smaller bubbles can grow until they reach the resonant size by a process known as rectified diffusion, whereby more gas is transferred into the bubble in negative pressure half cycles, when the bubble surface area is increased, than is transferred out again in the positive pressure half cycles, when the surface area is reduced. The resonant oscillations produce such vigorous swirling (microstreaming) of the liquid in the immediate vicinity of the bubble that free cells can be caught up in it and stressed. It has been postulated (Dyson 1991) that the acceleration of wound repair by physiotherapy ultrasound is due to an increase in the permeability of mast cell and T-cell walls brought about by such microstreaming.

Both forms of cavitation are more likely to happen if the peak negative pressure amplitude, p_r, is large and the frequency, f, is low. In fact, given a liquid environment with a plentiful supply of small 'nuclei' (tiny bubbles capable of growth into larger bubbles), Apfel and Holland (1991) have shown that the potential for the onset of collapse cavitation is proportional to a 'mechanical index' (MI) given by

$$\text{MI} = \frac{p_r}{\sqrt{f}}. \tag{7.2}$$

An account of the background and rationale for MI is given in Chapter 11.

Another quantity that is sometimes used as an indicator of the likelihood of cavitation is the spatial peak pulse-averaged intensity, I_{sppa}, which is the largest value of pulse average intensity (i.e. the intensity averaged over the duration of a pulse) to be found anywhere within the sound field. Note that although, at first sight, MI may appear to be a pressure parameter, and I_{sppa} a power or energy parameter, in fact MI^2 is proportional to the energy in the peak negative half cycle of the pulse and $\sqrt{I_{\text{sppa}}}$ is proportional to the rms pressure in the pulse.

Other factors which encourage cavitation include the availability of nuclei, agitation to encourage renewal of fresh nuclei and dissolved gas, a low viscosity liquid environment, a high pulse repetition frequency (prf), long pulse lengths and temperature elevation.

Damage from diagnostic levels of ultrasound has been demonstrated *in vivo* in animal tissues containing gas cavities, such as the lung and intestine (Child *et al* 1990, Dalecki *et al* 1995). There is no evidence of a hazard due to cavitation from diagnostic exposures *in vivo* in other tissues, but there is ample evidence of biological effects *in vitro* (see, for example, AIUM 1993 or Whittingham 1994). There are numerous liquid collections within and between tissues in which cavitation could occur provided suitable nuclei exist, including in or around critical targets such as the embryo or fetus, and the neonatal central nervous system. In view of the evidence of *in vivo* cavitation from physiotherapy and lithotripsy exposure, the demonstrated ability of diagnostic ultrasound to produce bubbles in gel (ter Haar *et al* 1989), and the difficulty of detecting isolated cavitational effects in tissue, there is a possibility, even a likelihood, that cavitational events do occur *in vivo*.

7.3. TRENDS IN ACOUSTIC OUTPUTS

Acoustic output measurements in the early years of diagnostic ultrasound (Hill 1969, Ziskin 1972, Carson *et al* 1978, Farmery and Whittingham 1978) reported values of intensities that were of the order of a few mW cm^{-2} or tens of mW cm^{-2} for B-mode, and up to a few hundred mW cm^{-2} for Doppler devices. Reported values of acoustic powers were typically 20 mW or less, although those for CW Doppler devices sometimes exceeded 100 mW. Reported values of acoustic outputs increased during the 1980s (Duck 1989). Peak negative pressures of up to 4 MPa were measured, and I_{spta} levels in Doppler mode were commonly in the range 1–2 W cm^{-2}. Acoustic powers reached 360 mW in B-mode and 500 mW in pulsed Doppler mode (Duck and Martin 1991, AIUM 1993).

A larger survey by Henderson *et al* (1994a) showed a continuing upward trend in reported acoustic output, particularly in worst case values of I_{spta} (i.e. the highest value of I_{spta} that could be obtained under any combination of scanner control settings). As shown in figure 7.1, reported median values of worst case I_{spta} increased by a factor of six for B-mode and two for colour Doppler (CD) mode, between surveys published in 1991 and 1994. In pulsed Doppler (PD) mode, the greatest reported I_{spta} value doubled to 9 W cm^{-2}. Some of this increase in reported output levels was probably due to improvements in measurement techniques and protocols for finding worst case outputs (Henderson *et al* 1993, 1994b), about which more will be said later. However, most of the increase is likely to be a real increase brought about by developments in scanner machine design.

For example, larger active apertures (more active elements) have been used in some scanner designs to form narrower beams, sometimes resulting in an increase in acoustic output power. Multiple transmissions down each scan line (each focused to a different depth zone) have reduced effective beam widths, and correspondingly higher line densities have also been introduced. These developments have improved lateral resolution, but they have also increased the number and frequency of pulses experienced by each field point. The provision of a B-mode 'write zoom' facility in more scanners has also been a factor. Unlike 'read zoom', which simply magnifies part of the stored image, write zoom involves the selected area being re-scanned at a higher frame rate and line density, again leading to an increase in the number and frequency of pulses in the scanned region. Another factor has been the use of higher ultrasonic frequencies for a given clinical application, again to improve spatial resolution, requiring increased transmission energies to compensate for the increased attenuation of the tissues. This has also

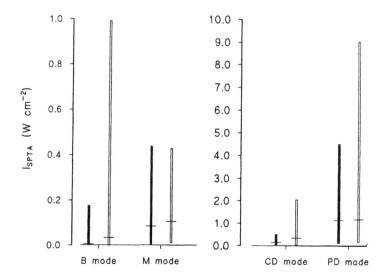

Figure 7.1. *Worst case I_{spta} levels reported by Duck and Martin (1991) (black) and Henderson et al (1994a) (open). Horizontal bars indicate median values.*

led to an increase in the pulse repetition frequency (prf), since the steeper rate of decrease of intensity with range at higher frequencies means less time is needed between transmissions to allow for unwanted weak returns from beyond the selected maximum depth. The significance of ultrasonic frequency and prf is indicated by the fact that the highest measured outputs for both I_{spta} and p_r in the 1994 survey were for 7.5 MHz transducers.

B-mode produces the lowest I_{spta} values because the beam is scanned across a region of tissue. The highest I_{spta} values are produced in pulsed Doppler mode, principally because the beam is held stationary rather than being scanned, and because very high prfs are employed in an effort to increase the Nyquist limit, i.e. the frequency at which aliasing of the Doppler-shifted frequency occurs. Colour Doppler modes tend to have I_{spta} values intermediate between those of pulsed Doppler and B-mode since, although the beam is scanned across the 'colour box', a sequence of several pulses is transmitted down each Doppler line at a high prf.

7.4.   REGULATIONS AND STANDARDS

In the UK, there are no mandatory limits on any acoustic output parameters from diagnostic ultrasound equipment. However, before an ultrasound scanner can be sold in the USA approval must be sought from the Food and Drug Administration (FDA), and its regulations (FDA 1997) impose some limits. The importance of the USA market means these limits generally

Table 7.1. *Limits set by FDA 510(k), Track 1. Derating involves reducing the in-water value by a factor which simulates the effect of 0.3 dB cm^{-1} MHz^{-1} attenuation in the medium.*

	Derated I_{spta} mW cm^{-2}	Derated I_{sppa} W cm^{-2}	MI = (derated p_{r})/$\sqrt{f}$ (p_{r} in MPa, f in MHz)
Peripheral vessel	720	190	1.9
Cardiac	430	190	1.9
Fetal and other	94	190	1.9
Ophthalmic	17	28	0.23

apply to machines sold in the UK and elsewhere. Some other countries, for example Japan, also have mandatory limits. There is also an International Electrotechnical Commission standard IEC 61157 (IEC 1992) with which manufacturers generally comply, which requires certain acoustic output parameters to be declared.

7.4.1. FDA 510(k) regulations

Manufacturers may choose to apply for FDA approval by either 'Track 1' or 'Track 3'. (There was an option called Track 2, but it is no longer available.)

FDA Track 1 requires that derated I_{spta} and either derated I_{sppa} or MI are below the limits shown in table 7.1. Note that these limits vary according to the application. The measurements of I_{sppa} and MI must be made at the position at which the 'derated pulse pressure squared integral (derated PPSI)' is a maximum. The PPSI is proportional to the pulse energy per unit area, and is found by integrating the value of p^2 over the pulse duration.

FDA Track 3 relaxes the limits for fetal, cardiac and all non-ophthalmic applications to those required by Track 1 for peripheral vascular applications, provided mechanical index (MI) and thermal index (TI) values are displayed. These indices are defined in the Output Display Standard of the American Institute of Ultrasound in Medicine/National Electrical Manufacturers Association (AIUM/NEMA 1998), and are discussed below. For ophthalmic applications the TI must be less than 1.0, the derated I_{spta} must be less than 50 mW cm^{-2} (compared to 17 mW cm^{-2} in Track 1) and the MI must be less than 0.23.

7.4.2. AIUM/NEMA Output Display Standard

This standard, usually called the 'ODS', requires that, if the machine is capable of producing a value for either MI or TI that is greater than 1.0, the index value must be displayed whenever it exceeds 0.4. It is not necessary to display the TI when in B-mode.

Thermal index (TI) is defined as the ratio of the total acoustic power to the acoustic power required to raise the tissue temperature by 1°C under defined assumptions. These assumptions include very simple tissue models and beam shapes. One of three different TIs is applicable, according to where the greatest heating is anticipated. TIS applies where it is likely to occur in soft tissue, TIB applies where bone near the focus of an unscanned beam is likely to be heated most, and TIC where it is likely to occur in bone close to the probe, e.g. as in transcranial scanning. Consideration of whether the operating mode is scanned or unscanned leads to four different formulae, shown below. In these formulae, W and W_3 are the in-water and derated (0.3 dB cm^{-1} MHz^{-1}) total acoustic powers respectively (mW); W_1 is the maximum in-water acoustic power transmitted by a 1 cm wide section of the probe face (mW); I_3 is the derated I_{spta} (mW cm^{-2}); f is the centre frequency (MHz); A is the radiating aperture (cm^2).

• Soft tissue only. (a) For scanned modes, or for unscanned modes that use an aperture of <1 cm^2: TIS $= W_1 f/210$. (b) For unscanned modes that use an aperture of >1 cm^2: TIS $=$ the maximum value of X as a function of range, where X is the lower of $W_3 . f/210$ and $I_3 . f/210$ at any range.

• Bone at focus. (a) For unscanned modes: TIB $=$ lower of $\sqrt{(W_3 I_3)}/50$ and $W_3/4.4$ measured where I_{spta} derated at 0.6 dB cm^{-1} MHz^{-1} is greatest. (b) For scanned modes: use TIS.

• Bone immediately next to probe: TIC $= W/40D$ where $D = \sqrt{(4A/\pi)}$.

Mechanical index (MI) is defined the spatial-peak value of the derated (0.3 dB cm^{-1} MHz^{-1}) peak negative pressure (MPa) divided by the square root of the pulse centre frequency (MHz$^{0.5}$). Note that, apart from in situations where bone is immediately next to the probe, the ODS specifies that TIS be used for scanned modes (e.g. B-mode or CD mode), irrespective of whether bone is present or not. Thus, for a colour Doppler scan of a second or third trimester fetus (in which bone will be present), TIS and not TIB is considered appropriate.

7.4.3. IEC 61157

IEC 61157 is an international standard that requires manufacturers to declare a number of acoustic output parameters if p_r, I_{spta}, or I_{ob} exceed certain threshold values. I_{ob} is the 'output beam intensity', defined as the total acoustic power divided by the 6 dB beam cross-sectional area at the transducer. The threshold for p_r (measured at the position of greatest PPSI) is 1 MPa; that for I_{spta} is 100 mW cm^{-2}; that for I_{ob} is 20 mW cm^{-2}. These thresholds are exceeded by virtually all commercial scanners. Parameters that must be declared include the three mentioned, plus total acoustic power, frequency, -6 dB beam width where the PPSI is greatest, and the modes and control settings that give maximum acoustic output values.

7.5. IS THERE A NEED FOR INDEPENDENT MEASUREMENTS?

It might be argued that the declaration of output parameters by the manufacturers, perhaps supported by the testing of representative machines at just one or two independent test houses, should be sufficient. Our experience is that manufacturers' reported data are not always reliable, and that considerable differences can sometimes exist between the outputs of different machines of the same make and model.

Finding worst case values of acoustic output parameters from modern complex scanners can be a difficult and elusive process. Measurement equipment is sometimes less than ideal, particularly in regard to hydrophone aperture size. In addition, record keeping lapses are always a possibility, resulting in software or hardware changes being made to individual machines without an accurate record being made in the scanner handbooks. It is perhaps not surprising, therefore, that there have been many instances when our own measurements have exceeded those declared by the manufacturer, often by a large factor. In an effort to highlight the problem, we published details of just three examples of I_{spta} measurements, chosen because they had been subject to close scrutiny by the manufacturers concerned (Jago *et al* 1995). Two of the three manufacturers accepted our higher figures (which were five times the declared values), while the third acknowledged that their outputs were probably higher than quoted in the equipment manual. One discrepancy was due to the manufacturer not taking sufficient account of multiple focal zones when using write zoom, as described above; the second was due to an obsolete high output software version mistakenly being supplied with the machine. These cases, and the scores of other unpublished examples we have on record, demonstrate that there is a need for independent acoustic output measurements to be made on every machine.

7.6. WHICH OUTPUT PARAMETERS SHOULD BE MEASURED?

If it is intended to check that equipment complies with a particular regulation or standard, it is clearly necessary to measure the worst case values of the specified parameters. In the case of the ODS, this involves measuring the power radiated by a 1 cm wide length of the probe, the centre frequency and derated values of p_{r}, I_{ta} and PPSI. Parameters specified in IEC 61157 include the 6 dB dimensions of the probe radiating area, the range and value of the greatest in-water PPSI, the 6 dB beam cross-section dimensions and the value of in-water p_{r} at that range, as well as W and in-water I_{spta}.

The difficulty of making these measurements at hospital sites for all probes and modes, under a wide range of control settings, means that we currently restrict on-site measurements to in-water values of p_{r}, I_{spta}, W, and the

6 dB beam cross-section dimensions at the range of I_{spta}. We also record the pulse waveforms, corresponding to worst case p_r and I_{spta}; this allows other pulse parameters such as centre frequency, pulse average intensity, pulse energy per unit area, p_c and pulse duration to be derived later. Although we only measure in-water values, these still provide some degree of a check on manufacturers' declared derated values, since derating the spatial peak in-water value should give a derated value which is no more than the declared derated value.

If a report is to be issued to the machine user, it is desirable to keep it brief and easy to understand. It is also desirable to give the recipient some criteria by which they can gauge the reported output values. Currently, our reports to users simply give worst case values of I_{spta} and p_r for all probes and all modes, and we have chosen to compare these with the limits recommended by the British Medical Ultrasound Society (BMUS 1988). Ultimately, it is likely that we shall calculate and report some form of thermal and mechanical indices as well. These may be those specified in the ODS or other future standards, or modified versions that are more amenable to a portable measurement system.

7.7. THE NEWCASTLE PORTABLE SYSTEM FOR ACOUSTIC OUTPUT MEASUREMENTS AT HOSPITAL SITES

Access to most scanning systems is limited by the need to cause minimal disruption to the normal clinical workload. This can be best achieved if measurements can be made at the site at which the scanner is installed. The measurement system must therefore be easily portable. It should be easy and quick to assemble at the measurement site as well as being light-weight and compact. The system described here is based on that of Martin (1986), and has been used successfully over a number of years to make many hundreds of measurements on almost the whole available range of diagnostic ultrasound scanners. A block diagram of the system is shown in figure 7.2. At the present time it is not capable of giving real-time read-outs of derated pressures and intensities, but work is in hand to make this possible by automatically monitoring the hydrophone depth and pulse centre frequency.

7.7.1. *The hydrophone and pre-amplifier*

Two types of hydrophone are candidates for pressure measurements on diagnostic ultrasound equipment: PVDF membrane hydrophones and needle hydrophones. The former consist of a thin film of polyvinylidene fluoride (PVDF) stretched across a 10 cm diameter supporting ring. Only a small (1 mm or 0.5 mm diameter) area in the centre is polarised, forming the

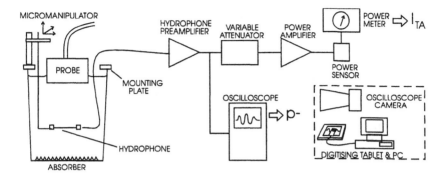

Figure 7.2. *Block diagram of the Newcastle portable system. An oscilloscope camera, digitising tablet and personal computer are also useful if an analogue oscilloscope is used.*

pressure sensing piezoelectric element. Printed circuit leads on the film connect electrodes on each side of this to the inner and outer conductors respectively of a coaxial lead, via terminals on the support ring. The support ring is wide enough to be well outside the beam in most applications. A 'bilaminar' version has the non-earthed electrode and lead protectively sandwiched between two PVDF films bonded together. Although having a greater thickness, and hence a lower resonance frequency and bandwidth and a higher reflection coefficient, bilaminar hydrophones are more suitable for use in non-deionised water and have better electrical shielding. Needle hydrophones consist of a small PZT or PVDF piezoelectric element mounted on the tip of a needle-like support. They are used with the needle pointing towards the source in order to minimise the effect of reflections from the support or, in the case of pulsed waves, to allow such reflections to be gated out. The Newcastle system uses a 50 μm thick bilaminar PVDF membrane hydrophone.

Nonlinear propagation effects can result in the generation of high frequency harmonics in the pulse waveform (Chapter 2). The international standard on measurement of ultrasonic fields IEC 61220 (IEC 1993) recommends that the sensitivity of the hydrophone and hydrophone amplifier should vary by less than ±6 dB over a frequency range extending to 3 octaves above the acoustic working frequency, or to 40 MHz, whichever is the smaller.

The resonance frequency for a 50 μm bilaminar hydrophone occurs at approximately 23 MHz. Consequently at frequencies below resonance the hydrophone sensitivity increases with frequency, by approximately 50% between 2 MHz and 20 MHz. It is possible to compensate for this by using a special hydrophone pre-amplifier with a complementary frequency response. The frequency response of the combination of a GEC Marconi Y-35-9724 hydrophone pre-amplifier with the 50 μm bilaminar hydrophone is flat to within 10% between 1 and 20 MHz and varies by less than a factor of two up to approximately 30 MHz. This combination, therefore, only meets the IEC bandwidth criterion for probes operating at frequencies below about 4 MHz. Hydrophones with higher resonance frequencies will be used when they become available. Cable resonances in the 70 cm cable used in the Newcastle system occur at approximately 65 MHz, and therefore do not cause a problem. Although the frequency response is reasonably flat, the calibration factor for the precise measured frequency is always used.

Ideally, the hydrophone element should be much smaller than the beam width. IEC 61102 (IEC 1991) and AIUM/NEMA UD2 (AIUM/NEMA 1992) standards specify that the effective diameter of the hydrophone element should be comparable to or smaller than $\lambda/2$. This corresponds to a diameter of <0.15 mm for a 5 MHz probe and <0.075 mm for a 10 MHz probe. Needle hydrophones down to 0.075 mm diameter are available, but they are difficult to mount in the small measurement tank of the portable system and have an uneven response at low frequencies. More relaxed criteria are specified for measurements where wavefronts can be considered to be reasonably planar. This applies to most practical situations. However, even these criteria still require hydrophone diameters below 0.5 mm. At the present time, membrane hydrophones with active element diameters of less than 0.5 mm are difficult to obtain. Algorithms are available to correct for spatial averaging effects when using hydrophone elements that are larger than ideal (IEC 1993, Smith 1989), although their accuracy is poor if the 6 dB beam width is less than the hydrophone diameter and they are not routinely applied.

It is important to reduce any error due to integrating noise associated with the hydrophone signal. Acoustic noise is reduced by lining the walls and base of the tank with acoustic absorber. Electrical noise and electrical interference from the probe are reduced by using a low noise pre-amplifier, connected to the bilaminar membrane hydrophone by coaxial cable.

7.7.2.   *Variable attenuator, power amplifier and power meter*

These components allow the temporal average intensity, I_{ta}, to be measured and displayed in real time as an analogue quantity, allowing rapid localisation of the point at which I_{ta} is a maximum and hence measurement of I_{spta}. A variable attenuator (Hatfield 2105) is used to keep the input voltage to the power amplifier (Mini circuits ZHL 32A) within the 600 mV peak to peak limit at which saturation occurs. The output from the power amplifier goes to a Marconi 6912 power sensor. This consists of a 50 Ω resistor, and a semiconductor thermocouple which produces a voltage proportional to the temporal average electrical power converted to heat in the resistor. This voltage is amplified and displayed by the meter (Marconi 6950). The sensor has a thermal time constant of approximately 30 ms. The meter time constant is variable up to 15 s and is set to a value greater than the scan repetition period, which can reach several hundred milliseconds at the lowest frame rates. The meter reading is therefore proportional to the temporal average of the square of the hydrophone output voltage, and hence to I_{ta}. The sensor and meter combination has a bandwidth from 30 kHz to 4.2 GHz and can measure electrical power levels from 1 μW to 100 mW. The electrical power of signals leaving the power amplifier is typically between 50 μW for a low output B-mode system and 60 mW for a high output pulsed Doppler system. Electrical noise at this stage is typically less than 1 μW.

7.7.3.   Oscilloscope

The oscilloscope displays the signal from the hydrophone amplifier so that p_r can be measured, and positions of peak pulse amplitude may be found. It also permits monitoring of any distortion of the ultrasound pulse. A delayed trigger facility is valuable for isolating individual pulses from B-mode and colour Doppler sequences. A Tektronix 2215 60 MHz analogue oscilloscope was used in the first system, but our second system uses a LeCroy 9350AM 500 MHz digital oscilloscope. The digital oscilloscope has a facility for calculating centre frequency and showing PPSI values in real time, as well as having much superior triggering facilities. It can also store digitised pulse waveforms for subsequent analysis if required.

7.7.4.   Oscilloscope camera, PC and digitisation tablet

If an analogue oscilloscope is used, an oscilloscope camera for recording pulse waveforms is useful. The pulse waveform may then be hand digitised on return to the laboratory, using a personal computer and a digitising tablet. This allows the centre frequency and other pulse parameters to be computed, and frequency dependent calibration factors to be applied automatically.

7.7.5.   The measurement tank

The portable measurement tank is a bucket, measuring 30 cm maximum internal diameter by 24 cm deep. It is larger than the dimensions of the largest ultrasound field to be measured and the bottom is covered with an acoustic absorber (carpet or astroturf) to reduce acoustic noise in the form of reflected ultrasound. An aluminium mounting plate around the top of the bucket provides a means of attaching the probe support and hydrophone positioning system. The main restriction on the size of the tank is the volume of distilled water needed to fill it. When transporting the equipment, 10–12 litres of water are carried in separate closed containers.

   Although the conductivity of tap water has little effect on bilaminar shielded hydrophones, distilled water is used in the measurement tank. One reason is that it allows the well documented acoustic impedance values, Z, for pure water to be used in the calculation of intensity (I) from pressure

(p) measurements, using $I = p^2/Z$. Another reason is that distilling also achieves some degree of degassing, helping to reduce field perturbations caused by bubbles adhering to the hydrophone or probe.

7.7.6.   The hydrophone positioning system

Movement of the hydrophone is achieved in the two horizontal directions perpendicular to the vertical beam axis by a micromanipulator with a reduction drive and a vernier millimetre scale. Movement in the vertical axial direction is controlled by a third micromanipulator, but without a reduction drive. The hydrophone is attached to the micromanipulator system via a clamp made in-house which supports it in a horizontal plane. The standard IEC 61220 states that the positioning of the hydrophone should be reproducible to ±0.1 mm.

7.7.7.   The probe mounting system

This has to accommodate and firmly support many different sizes and shapes of ultrasound probe. The use of retort stand clamps, attached to bars on the mounting plate, is a flexible and satisfactory method.

7.7.8.   Calibration and accuracy

The hydrophone is calibrated by the National Physical Laboratory (NPL), Teddington, every 12–18 months. The power meter is calibrated at similar intervals by the manufacturer. The amplifiers are supplied calibrated, and the gains are checked in-house annually. Accuracy depends on the characteristics of the ultrasonic field but overall the 95% confidence limit for p_r is approximately ±12%, while that for I_{spta} is approximately ±30%.

7.8.   THE NPL ULTRASOUND BEAM CALIBRATOR

The NPL Ultrasound Beam Calibrator is a comprehensive measurement system (Preston 1988) which has been developed at the NPL to enable the rapid determination of the acoustic output of most types of medical diagnostic ultrasonic equipment. It consists of a 21-element, 50 μm PVDF membrane hydrophone array mounted in a test-tank and connected to a fast data capture and presentation system. Elements are 0.4 mm in diameter and spaced 0.6 mm between centres (0.5 mm at 1.0 mm spacing in earlier versions). The pressure profile across the beam and the acoustic pressure waveform for any selected hydrophone element are presented in real time, together with the peak acoustic pressures (positive and negative), PPSI and pulse repetition rate. Other important acoustical parameters, including the acoustic working frequency and FDA derated values, are calculated. Measurement uncertainty depends on the particular application but it is typically ±11% for acoustic pressure parameters, and ±22% for intensity parameters (95% confidence level).

The system is powerful, but it is not as convenient as the Newcastle system for routine measurements at hospital sites. It is bulkier, and requires trigger signals that are often difficult to obtain from commercial scanners. The analogue method used in the Newcastle system is an easier way of measuring temporal average intensities, and it may be noted that the NPL proposes to use the analogue method in future versions of their Ultrasound Beam Calibrator.

7.9.   MEASUREMENT OF ACOUSTIC POWER

Unfortunately, at the present time there is no commercially available form of power measuring device that is both sufficiently sensitive and portable for measurements at hospital sites. However, the two designs shown in figures 7.3 and 7.4 have been found to be satisfactory, and suitable for in-house construction. They are both examples of radiation force balances, which use the relationship that, for an absorbing target,

$$\text{Force} = \frac{\text{Incident temporal average acoustic power}}{c} \tag{7.3}$$

where c is the speed of sound in the fluid around the target.

The balance (Perkins 1989) shown in figure 7.3 is a development of a design by Farmery and Whittingham (1978). It uses an air-filled, 45° conical reflector as a target, mounted within a cylindrical absorber to prevent reflections. The force on such a target is the same as that on an absorber perpendicular to the beam (see Chapter 3). Radiation force pushes the target back, generating a signal from an optical displacement detector. This signal is amplified and used to generate a current in an electromagnet coil, which in turn generates a counterbalancing force upon a permanent magnet, attached to the rear of the target. A meter reading of the coil current gives a direct measure of the magnetic restoring force, and hence of the radiation force. A modification of this design by Whittingham (unpublished) is the replacement of the conical reflecting target and surrounding absorbing cylinder by a flat absorbing target (figure 7.4). This is sufficiently wide to allow the direct measurement of power from linear arrays in scanning mode, not possible previously. The use of a flat target also allows the distance between the probe and the target to be reduced.

Both have sensitivities of around 1 mW, but at settings below about 100 mW full scale deflection, vibration transmitted from the surroundings can be a problem. Ten minutes or so should be allowed for any disturbance of

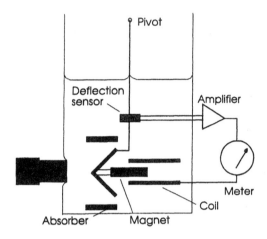

Figure 7.3.   *Principle of the Perkins balance.*

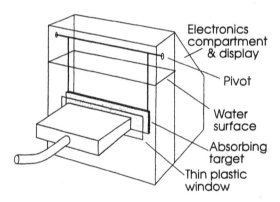

Figure 7.4.   *Modification (Whittingham) to accommodate wide transducers, e.g. linear arrays.*

the water, due to transportation for example, to settle down. If the ultrasonic field is strongly convergent (e.g. some strongly focused stationary beams), divergent (e.g. sector scanners in scanning mode), or obliquely incident on the target (e.g. angled Doppler or sector scan beams), then an appropriate 'cos θ' correction factor should be estimated and applied. With care, an accuracy of about 15% should be achievable. The balances may be calibrated either with a check source (a transducer and drive unit that delivers a beam of known acoustic power, available from the NPL) or by making provision for a detachable arm to which weights may be attached. The equation for radiation force given above means that, for calibration purposes, 1 W of acoustic power produces a force of 68 mg weight on an absorbing target.

Other methods of power measurement include calorimetry and integration of intensity across a plane perpendicular to the beam (planar integration). So far, calorimetric methods have lacked sufficient sensitivity. The planar integration method has excellent sensitivity but is really only suitable as a laboratory-based method.

7.10. FINDING WORST CASE VALUES

As there may be millions of different possible combinations of control settings on a modern scanning system, a systematic approach is clearly necessary. Unfortunately there are no foolproof rules, as controls and their effects vary between machines, often in unexpected ways. An important step is to spend some time identifying which controls affect the position of the spatial peak of the quantity, and which affect only its magnitude. This is much easier when the measurer has gained experience, but protocols have been published (Henderson *et al* 1993, 1994b) which can either be used directly, or form a useful starting point. Some guidance for finding worst case I_{spta} values, based on these protocols is given below. Finding worst case values of p_r and acoustic power is generally more straightforward.

In the following summarised protocols, small capitals denote a machine control.

7.10.1. Worst case I_{spta} of stationary beams, e.g. pulsed Doppler mode

The protocol may be summarised as:

(i) Set the OUTPUT POWER control to maximum and set the beam in the 'straight ahead' position, since this gives the greatest aperture, and hence is likely to give the greatest output. Start with the minimum possible settings of DEPTH OF FIELD and DOPPLER FOCUS depth.

(ii) Explore the field with the hydrophone to find the position of I_{spta}. Then try all combinations of GATE WIDTH, VELOCITY SCALE (prf) and DOPPLER FREQUENCY to find that which maximises the value of I_{spta} at this point. Make a note of this value.

(iii) Increase DOPPLER FOCUS depth until a change in the amplitude of the pulse can be seen on the oscilloscope. If the GATE is at the bottom of the image, increase the IMAGE DEPTH.

(iv) Move the hydrophone until the new position of I_{spta} is found and, leaving it there, again try all combinations of the controls mentioned in (ii) to maximise I_{ta} at this position. Make a note of the maximum value.

(v) Repeat steps (iii) and (iv) until all DOPPLER FOCUS depths settings have been selected. The largest I_{spta} value found is the worst case I_{spta}. Record this and the control settings that produced it.

7.10.2.   Worst case I_{spta} for scanned beam modes, e.g. B-mode

In contrast to stationary beams, the position of I_{spta} varies only slightly with FOCUS depth. This is due to the fact that the beams overlap in the scan plane so that, as far as temporal average intensity is concerned, there is a wide relatively uniform effective beam. Within this, the shape and focal position of any individual beam is relatively unimportant. In the elevation plane, of course, there is no beam overlap, so the range at which total power is concentrated most, and therefore where the I_{spta} is located, is close to that of the elevation plane focus. For sector probes and curvilinear arrays, the divergence of the effective beam means the I_{spta} is generally located between the elevation focus and the probe.

The protocol for B-mode is more complex and subject to more variations between machines than is that for pulsed Doppler. Consequently only a few general guidelines will be given here. Selecting the minimum WIDTH OF FIELD generally gives larger I_{spta} values, due to a higher frame rate and/or a higher line density. Generally, deep FOCUS settings give greater I_{spta} values, principally because of the larger apertures that are often employed. However, the effect of this can be offset by the fact that a deep FOCUS setting requires a large DEPTH OF FIELD, which means a reduced pulse repetition frequency. The effect of increasing the NUMBER OF FOCI can also be difficult to predict.

Start with the OUTPUT POWER control to maximum and select a single focus at the furthest range in the deepest DEPTH OF FIELD; progressively reduce the DEPTH OF FIELD (leave the FOCUS control alone, causing the focus to follow automatically at the furthest range in the image). Once the I_{spta} values start to fall, it is not necessary to continue investigating nearer FOCUS settings. If no maximum has been found before the smallest DEPTH OF FIELD has been selected, try all FOCUS settings within this field. Note the largest I_{spta} value obtained so far, advance the NUMBER OF FOCI by one and start again with the deepest DEPTH OF FIELD.

WRITE ZOOM may increase I_{spta} values considerably, the increase being greatest for the greater ZOOM BOX DEPTH settings, since greater ranges usually involve the greater apertures, and the minimum ZOOM BOX WIDTH setting, since this involves the greatest frame rate and/or line density. The effect of changes in ZOOM BOX HEIGHT that also change the number of foci within the box should be explored.

7.11.   CONCLUSIONS

It is possible for a medical physics department to measure the principal acoustic output characteristics of the clinical scanners in its area. There are several reasons for doing so. There has been a long term trend towards higher outputs, and this should be monitored. Some clinical machines are capable of harmful effects if not used prudently, with proper awareness of the modes and control settings that lead to high outputs. Reporting the results to clinical users allows them to become aware of how to reduce exposure, to make informed risk–benefit judgements based on their own machine, and to defend themselves in the event of any accusations relating to ultrasonic exposure. Manufacturers' efforts to moderate or reduce acoustic outputs will be given added impetus by increased customer awareness of, and interest in, actual values. Biomedical research will have the benefit of accurate knowledge about real clinical exposures.

REFERENCES

AIUM/NEMA 1992 *UD2: Acoustic Output Measurement Standard for Diagnostic Equipment* (Rockville, MD: American Institute of Ultrasound in Medicine)

AIUM/NEMA 1998 *Standard for Real-Time Display of Thermal and Mechanical Acoustic Output Indices on Diagnostic Ultrasound Equipment* 2nd edn (Rockville, MD: American Institute of Ultrasound in Medicine)

AIUM 1993 *Bioeffects and Safety of Ultrasound* (Rockville, MD: American Institute of Ultrasound in Medicine)

Apfel R E and Holland C K 1991 Gauging the likelihood of cavitation from short-pulse, low-duty cycle diagnostic ultrasound *Ultrasound Med. Biol.* **17** 179–85

Barnett S B (ed) 1998 WFUMB Symposium on Safety and Standardisation in Medical Ultrasound 1997 *Ultrasound Med. Biol.* in press

Barnett S B and Kossoff G (eds) 1998 *Safety of Diagnostic Ultrasound* (London: Parthenon)

BMUS 1988 *Br. Med. Ultrasound Soc. Bull.* no 50, August

Carson P L, Fischella P S and Oughton T V 1978 Ultrasonic powers and intensities produced by diagnostic ultrasound equipment *Ultrasound Med. Biol.* **3** 341

Child S Z *et al* 1990 Acoustic cavitation produced by microsecond pulses of ultrasound *Ultrasound Med. Biol.* **16** 817–25

Dalecki D, Raeman C H, Child S Z and Carstensen E L 1995 Intestinal haemorrhage from exposure to pulsed ultrasound *Ultrasound Med. Biol.* **21** 1067–72

Duck F A 1989 Output data from European studies *Ultrasound Med. Biol.* **15** (Suppl. 1) 61–4

Duck F A and Martin K 1991 Trends in diagnostic ultrasound exposure *Phys. Med. Biol.* **36** 1423–32

Duck F A, Starrit H C, ter Haar G R and Lunt M J 1989 Surface heating of diagnostic ultrasound transducers *Br. J. Radiol.* **62** 1005–13

Dyson M 1991 The susceptibility of tissues to ultrasound *The Safe Use of Diagnostic Ultrasound* ed M F Docker and F A Duck (London: British Institute of Radiology) pp 24–9

Farmery M J and Whittingham T A 1978 A portable radiation-force balance for use with diagnostic ultrasonic equipment *Ultrasound Med. Biol.* **3** 373–9

FDA 1997 *FDA 510(k): Information for Manufacturers Seeking Market Clearance of Diagnostic Ultrasound Systems and Transducers* (Food and Drug Administration, Department of Health and Human Services, 1390 Piccard Drive, Rockville, MD 20850, USA)

Henderson J, Jago J, Willson K and Whittingham T A 1993 Towards a protocol for measurement of maximum spatial peak temporal average intensity from diagnostic B-mode ultrasound scanners in the field *Phys. Med. Biol.* **38** 1611–21

—— 1994a A survey of the acoustic outputs of diagnostic ultrasound equipment in current clinical use in the Northern Region *Ultrasound Med. Biol.* **21** 699–705

—— 1994b Development of protocols for measurement of maximum spatial peak temporal average intensity from scanners operating in pulsed Doppler and colour Doppler modes *Br. J. Radiol.* **67** 716

Hill C R 1969 Acoustic intensity measurement on ultrasound diagnostic devices *Ultrasonographia Medica, Vol 2, Proc. 1st World Congress on Ultrasonic Diagnostics in Medicine and SIDUO III* ed J Bock and K Ossoining (Vienna Academy of Medicine) p 21

IEC 1991 *IEC 61102: Measurement and Characterisation of Ultrasonic Fields Using Hydrophones in the Frequency Range 0.5 MHz to 15 MHz* (Geneva: International Electrotechnical Commission)

IEC 1992 *IEC 61157: Requirements for the Declaration of the Acoustic Output of Diagnostic Ultrasound Equipment* (Geneva: International Electrotechnical Commission)

IEC 1993 *IEC 61220: Guidance for the Measurement and Characterisation of Ultrasonic Fields Generated by Medical Ultrasonic Equipment Using Hydrophones in the Frequency Range 0.5 MHz to 15 MHz* (Geneva: International Electrotechnical Commission)

Jago J, Henderson J, Whittingham T A and Willson K 1995 How reliable are manufacturers' reported acoustic output data? *Ultrasound Med. Biol.* **21** 135–6

Jago J R, Mitchell G, Whittingham T A, Henderson J and Willson K 1994 Experimental measurement of the spatial distribution of temporal average intensity for complex beam shapes and scanning modes *British Medical Ultrasound Society Annual Scientific Meeting, Eastbourne, 7–9 December 1993* Abstract: *Br. J. Radiol.* **67** 716

Martin K 1986 Portable equipment and techniques for acoustic power output and intensity measurement *Physics in Medical Ultrasound* ed J A Evans (London: IPEM)

Perkins M A 1989 A versatile force balance for ultrasound power measurement *Phys. Med. Biol.* **34** 1645–51

Preston R C 1988 The NPL ultrasound beam calibrator *IEEE Trans. Ultrasonics, Ferroelectr. Freq. Contr.* **35** 122–38

Smith R A 1989 Are hydrophones of diameter 0.5 mm small enough to characterise diagnostic ultrasound equipment? *Phys. Med. Biol.* **34** 1593–607

Swindell W 1984 The temperature fields caused by acoustic standing waves in biological tissues *Br. J. Radiol.* **57** 167–8

ter Haar G, Duck F, Starrit H and Daniels S 1989 Biophysical characterisation of diagnostic ultrasound equipment—preliminary results *Phys. Med. Biol.* **34** 1533–42

Whittingham T A 1994 The safety of ultrasound *Imaging* **6** 33–51

Ziskin M C 1972 Survey of patient exposure to diagnostic ultrasound *Interaction of Ultrasound and Biological Tissues* eds J M Reid and M R Sikov (US Government Printing Office, Washington DC: DHEW Publication (FDA) 73-8008)

PART 3

ULTRASOUND HYPERTHERMIA AND SURGERY

CHAPTER 8

ULTRASOUND HYPERTHERMIA AND THE PREDICTION OF HEATING

Jeffrey W Hand

INTRODUCTION

The main purposes of this chapter are (i) to summarise the key features of ultrasound induced hyperthermia and associated therapies and (ii) to discuss thermal transport mechanisms in tissues and their modelling, a topic not only of relevance to thermal therapies but also to safety issues related to ultrasound induced temperature changes within tissues.

8.1. ULTRASOUND HYPERTHERMIA

8.1.1. Introduction

The benefit of adding hyperthermal treatment to radiotherapy in the treatment of tumours located either superficially or deep in the body has been demonstrated in the results of several randomised controlled clinical trials (Vernon *et al* 1996, Overgaard *et al* 1995, van der Zee and van Rhoon 1993). Ultrasound is an agent commonly used to induce hyperthermia and offers some marked advantages over microwave and radiofrequency techniques. These include:

• relatively large values for the ratio source dimensions/wavelength for transducers of practical size, enabling collimated and focused beams to be produced;
• relatively low attenuation in soft tissues (up to 1 dB cm^{-1} MHz^{-1}) for frequencies of interest (between 0.5 and 5 MHz).

The ability of ultrasound to produce small foci of high energy density opens up the possibility of high temperature, short duration hyperthermia and tissue ablation (sometimes called ultrasound surgery). These techniques for thermal therapies are attracting considerable research, clinical and commercial interest which are dealt with further in Chapter 9.

The fundamentals of ultrasound are discussed at length in other chapters in this book and in several excellent reviews. Those written specifically with hyperthermia in mind include Hunt (1990), Hynynen (1990) and Swindell (1986).

8.1.2.   Ultrasound intensity, attenuation and absorption

The acoustic intensity I is the rate of energy flow through unit area normal to the direction of wave propagation. For a plane continuous wave, the temporal average intensity is

$$I = \frac{p_0^2}{2Z} \tag{8.1}$$

where p_0 is the amplitude of the acoustic pressure and Z $(= \rho_0 c)$ is the acoustic impedance. c is the speed of sound in the tissue which has equilibrium density ρ_0. The SI unit of intensity is W m^{-2}. A useful practical unit for hyperthermia is W cm^{-2}.

When propagating through a medium such as tissue, ultrasound energy is attenuated. For a plane wave, the relationship between I_0, the intensity at the surface and I_z, the intensity at a depth z is

$$I_z = I_0 \, e^{-2\alpha z} \tag{8.2}$$

where α is the amplitude attenuation coefficient. The attenuation coefficient consists of two parts, one (α_a) due to absorption and the other (α_s) due to scattering:

$$\alpha = \alpha_a + \alpha_s. \tag{8.3}$$

The variation of intensity with depth for frequencies between 0.5 and 3 MHz, assuming attenuation is either 0.5 or 1 dB cm^{-1} MHz^{-1}, is shown in figure 8.1. These values for the attenuation coefficient represent the lower and upper ends of the range of values for soft tissues (for a deeper discussion of the relevant physical properties of tissue see Chapter 4).

Energy absorption leads to local heating. The scattered energy also contributes to heating caused by the ultrasound field. Although the relative contributions of α_a and α_s are dependent upon tissue type and frequency (Goss *et al* 1978, 1979, 1980 and Chapter 4), when considering most hyperthermia applications it is often assumed that almost all of the energy lost from the beam is absorbed (Hynynen 1990). For many tissues,

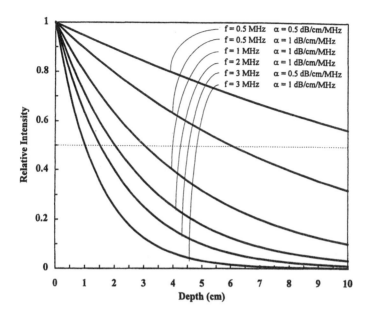

Figure 8.1. *Attenuation of plane wave ultrasound at 0.5, 1, 2 and 3 MHz in tissue with attenuation assumed to be either 0.5 or 1 dB cm^{-1} MHz^{-1}. The dotted line indicates the 50% relative intensity level.*

attenuation and absorption exhibit an approximately linear relationship with frequency over the range of frequencies used for hyperthermia (see Chapter 4).

The local heating capability associated with the propagation of plane wave ultrasound in tissue depends upon the depth into the tissue and the attenuation and absorption properties of the tissue. The absorbed power per unit volume, W_V, (equal to the specific absorption rate† (SAR) × density, ρ_0) is

$$W_V = 2\alpha_a I_0 \, e^{-(2\alpha z)} \tag{8.4}$$

and, if for simplicity we take $\alpha = \alpha_a$ and $\alpha_a = \alpha_{a0} f$ (α_{a0} is the absorption coefficient at 1 MHz and f is the frequency in MHz), then it follows that

$$\frac{W_V}{I_0} = 2\alpha_{a0} f \, e^{-(2\alpha_{a0} f z)} \tag{8.5}$$

† Specific absorption rate (SAR) is the time derivative of the incremental energy (dE) absorbed by an incremental mass (dm) in a volume element (dV) of a given density (ρ_0):

$$\mathrm{SAR} = \frac{\mathrm{d}}{\mathrm{d}t}\left(\frac{\mathrm{d}E}{\mathrm{d}m}\right) = \frac{\mathrm{d}}{\mathrm{d}t}\left(\frac{\mathrm{d}E}{\rho_0 \, \mathrm{d}V}\right).$$

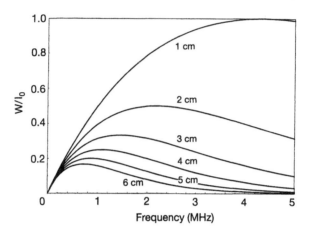

Figure 8.2. *Relative power deposition versus frequency at depths from 1 cm to 6 cm in soft tissue. W/I_0 is normalised to the maximum value calculated at a depth of 1 cm. The absorption coefficient at 1 MHz, α_{a0}, is taken to be 1 dB cm^{-1}. The frequency dependence of α is assumed to be linear.*

leading to an optimal frequency, f_{opt}, from the point of view of local deposition of power at depth z given by

$$f_{opt} = \frac{1}{2\alpha_{a0}z}. \tag{8.6}$$

The dependence of W_V/I_0 on frequency at various depths in tissue is shown in figure 8.2. The optimum frequency for focused ultrasound surgery, taking into account the frequency-dependent attenuation of tissue, has been discussed by Hill (1995).

8.1.3. Transducers for hyperthermia

8.1.3.1. Planar transducer systems. The simplest type of ultrasound device for hyperthermia consists of a plane circular disk transducer that is air backed and mounted in such a way that its front face is in contact with an integral bolus of degassed, and usually temperature controlled, water. The acoustic field associated with such a source has been discussed by several authors (e.g. Zemanek 1971, Harris 1981) and is described in Chapter 1.

In practice, the water column should be sufficiently long to contain the intensity maxima of the near field and is usually closed by a latex membrane. Even with the water bolus, tissues are usually exposed to part of the near field and so the drive is often frequency modulated to reduce local hot spots. This type of applicator, depicted schematically in figure 8.3, was used in early clinical investigations of hyperthermia, combined with radiotherapy or

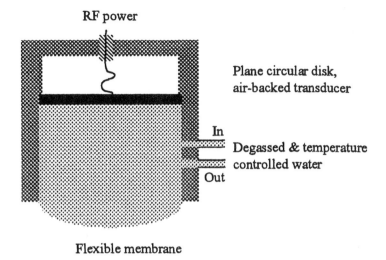

RF power

Plane circular disk,
air-backed transducer

In

Degassed & temperature
controlled water

Out

Flexible membrane

Figure 8.3.   *A schematic diagram of a plane-wave ultrasound hyperthermia applicator.*

chemotherapy, in the treatment of patients with superficial tumours (Marmor *et al* 1979, Marchal *et al* 1982). Frequencies used were typically in the range 1–3 MHz, for the reasons highlighted in figure 8.2 and transducer diameters were usually between 3 and 10 cm. With this type of device, a monotonically decreasing temperature versus depth profile is generally obtained; the only means of significantly changing the nature of this profile is by lowering the temperature of the most superficial tissues through aggressive cooling with a cold water bolus. Such simple systems probably approach a worst case scenario as far as ultrasound-induced hyperthermia is concerned. A useful appraisal of their clinical utility for superficial tumours in a range of anatomical sites is to be found in Corry *et al* (1987). The ability to dynamically adjust the specific absorption rate distribution beneath a plane wave ultrasound device offers greater flexibility in clinical use. Such features are to be found in devices described by Benkeser *et al* (1989) in which sub-sections of the applicator's aperture may be driven independently, offering spatial control of energy delivered to the tissues, and at either the fundamental (around 1 MHz) or a harmonic frequency (around 3 MHz), offering some control over energy deposition versus depth.

The advantages that may be gained by using a large array of plane transducers are the bases underlying a system designed for treating the intact breast (Lu *et al* 1996). A cylindrical applicator consisting of a stack of eight rings with an inner diameter of 25 cm is used. There are 48 transducers, each 1.5 cm × 1.5 cm with 0.24 cm inter-element spacing, in the four upper rings and 24 in the lower rings. In each ring, alternate transducers are driven at high (4.3–4.8 MHz) and low (1.8–2.8 MHz) frequency. Transducers

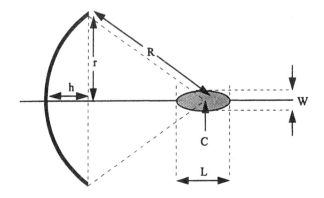

Figure 8.4. *Parameters describing a concave bowl transducer.*

with overlapping fields are driven incoherently to avoid interference effects. The patient lies in a prone position with the breast positioned within the water filled cylindrical applicator. The use of both high and low frequency transducers enables power deposition at the boundaries of the breast (breast surface and chest wall) and within the breast to be adjusted independently of each other, which is advantageous when attempting to produce uniform temperature distributions in regions with differing heat losses.

Montes and Hynynen (1995) describe the use of an array of planar circular transducers (1 cm diameter, operating frequency 2 or 4 MHz) incorporated into an applicator used for delivering intraoperative radiotherapy. A reflecting surface formed at the interface of degassed water and styrofoam is used to reflect the ultrasound beam through 90° so that the beam propagates in the direction parallel to the incident ionising radiation. Using this device, hyperthermia and radiotherapy may be delivered simultaneously, offering the possibility of an enhanced therapeutic effect.

8.1.3.2. Focused transducer systems. The greatest advantage that ultrasound offers over other means of inducing hyperthermia, such as microwaves, is the ability to produce focused fields. The simplest method of achieving a focused ultrasound field is to use a spherically curved shell (or concave bowl) transducer similar to that depicted in figure 8.4. The focus of such a transducer lies on the central axis, near the centre of curvature of the bowl (Kossoff 1979). The axial intensity distribution, $I(z)$, excluding the effect of attenuation, is given by

$$I(z) = I_0 \left[\frac{R}{R-z} \sin\left(\frac{\pi r_a^2 (R-z)}{2\lambda z R} \right) \right]^2 \qquad (8.7)$$

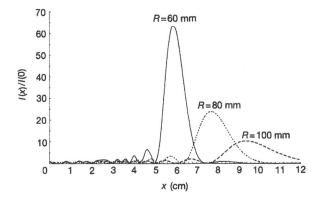

Figure 8.5. *Axial intensity distribution for three 1 MHz concave bowl transducers of diameter 60 mm and radius of curvature 60, 80 and 100 mm. The attenuation coefficient of the medium is taken to be 1 dB cm⁻¹.*

where I_0 is the average intensity at the surface of the transducer of diameter r_a. At the centre of curvature, the intensity is

$$I_R = I_0 \left[\frac{\pi r_a^2}{2R\lambda} \right]^2. \tag{8.8}$$

The ratio I_R/I_0 is known as the intensity gain factor. Figure 8.5 shows the axial intensity variation for three 1 MHz transducers with $R = 60$, 80 and 100 mm, respectively. The diameter of each transducer is 60 mm and the attenuation coefficient is taken to be 1 dB cm⁻¹. The shape of the focal region tends to be long and narrow, the actual dimensions being dependent upon the transducer's frequency, diameter and radius of curvature, R. The ratio R/r_a is often used to describe a transducer's geometrical focusing properties. The full width at half maximum of the intensity distribution in the focal plane (-3 dB level) is

$$d_3 = \frac{0.257\lambda R}{r_a} \tag{8.9}$$

while on axis, the depth range over which the intensity remains within 3 dB of the maximum is

$$l_3 = 0.443 \frac{R\lambda}{h} \tag{8.10}$$

where h is the depth of the spherical surface of the transducer, shown in figure 8.4. (The dimensions d_3 and l_3 are marked as W and L on this figure.) The ratio d_3/l_3 is $1.72r_a/h$ (Cline *et al* 1994). The focal volume of a focused ultrasound transducer is usually much smaller than the volume of the tumour to be treated. Several approaches to overcome this difficulty have been considered. Much pioneering work in the use of scanned focused ultrasound hyperthermia was carried out by Lele (e.g. Lele 1983). The intensity gain associated with focused transducers is due to the energy in the beam passing through a smaller area in the focal region compared with that near to the transducer surface, and so a mechanically scanned transducer should be moved over a large area at the surface (window) while being continuously directed at the target.

In other systems, several focused transducers are orientated so that their focal volumes are close but offset from each other. Nussbaum *et al* (1991) and Lindsley *et al* (1993) describe a large aperture array of 30 transducers with polystyrene lenses arranged in four concentric rings on a cylindrical surface. The transducers were driven at a frequency within the range 0.5 to 1 MHz and could be adjusted to form a ring focus of variable diameter (up to 11 cm) at a distance of approximately 32 cm from the array. Further control allowed for the complete array to be translated in three orthogonal directions and rotated about its central axis. The complete array was immersed in a water bath which was positioned below the patient. Pre-clinical studies using this system suggested that selective heating could be achieved in volumes with lateral dimensions between 4 and 8 cm at depths of 11 cm in soft tissue.

In another design, a non-coherent array of focused transducers, each positioned so that their focal volumes are coincident, is mounted on a gantry which provides translation in three orthogonal directions, rotated and tilted. The trajectory of the common focal volume may be moved through the target volume by controlling the movement of the gantry (Hynynen *et al* 1987, 1990, Hand *et al* 1992). These systems have shown that bulky tumours in the breast and some pelvic tumours can be treated when an adequate acoustic window is available. Anhalt *et al* (1992) added higher frequency transducers to the array and extended the application of the technique to the treatment of chest wall tumours of large area.

8.1.3.3. Phased array systems. An ultrasound phased array consists of relatively small transducers which are driven coherently. The distribution of phases and amplitudes of the signals applied to the elements can be chosen to produce either a focused beam which may be scanned through the treatment volume (at a greater rate than can be achieved using mechanical systems) or a customised power deposition pattern with specified local maxima and minima. The first approach, referred to as 'spot scanning' is the electronic analogue of the mechanical systems discussed in section 8.1.3.2. Electronic spot scanning may obviate the need for large scale movement of the array with respect to the patient, thus simplifying the coupling between the ultrasound transducers and the patient. In the second approach, SAR distributions consisting of multiple foci are synthesised directly and complex heating patterns can be produced by switching between two or more predetermined multi-foci patterns (mode scanning) (McGough *et al* 1994).

An advantage offered by this approach is that the time-averaged intensities that are necessary to induce the required increase in temperature over a volume of tissue can be produced without unacceptably high spatial-peak intensities that may be needed in the case of spot scanning.

Although phased arrays offer several advantages over single focused transducers (variable focal distance, rate of scanning, multi-foci patterns), they also have some disadvantages when compared to the simpler transducers. Single transducers may generate sound beams with very high intensities in the focal region and much smaller intensities within the intervening tissues. This is especially valuable for ultrasound surgery when it is necessary to ablate a target within tissue locally without damaging the surrounding tissues (see Chapter 9).

Another disadvantage of linear phased arrays arises from the presence of grating lobes and secondary local maxima. To avoid problems with grating lobes, the distance between centres of array elements ideally should be less than $\lambda/2$ (Steinberg 1976) but satisfying this condition while maintaining a sufficiently large aperture can result in a large number of elements and high costs. The common need to suppress grating lobes associated with practical phased array therapeutic systems has been investigated by several authors. The use of non-uniform inter-element spacing to suppress grating lobes in applicators for deep localised ultrasound hyperthermia was suggested by Ebbini and Cain (1991b). Hutchinson *et al* (1996) and Hutchinson and Hynynen (1996) adopted a similar approach for a linear phased array for transrectal use and produced an aperiodic array of 57 plane elements. Dupenloup *et al* (1996), using annular arrays, demonstrated a method for reducing grating lobes using a wide-band CW source. De-activation of a large subset of elements was reported by Hutchinson *et al* (1995) as being useful in reducing grating lobes when generating foci located at relatively large distances off an array's central axis, while Gavrilov *et al* (1997) investigated the efficacy of suppressing grating lobes using element de-activation to optimise the number of active elements in three linear arrays with different geometrical and acoustical characteristics and for various focusing conditions.

Ibbini and Cain (1990) investigated the focusing properties of a concentric ring array and showed that, in addition to single spot focusing on the central axis, both single and multiple concentric focal rings of variable width may be produced. However, unwanted secondary foci produced in front of and beyond the focal plane are also produced. A further limitation of this array is that only annular patterns may be produced due to its circular symmetry. The sector-vortex array (Umemura and Cain, 1989) comprises a disc transducer cut into M tracks, each of which is divided into N sectors of equal size. This two-dimensional array has additional geometric focusing provided by a lens. A prototype device, driven at 0.5 MHz and consisting of two tracks with 16 sectors per track, is described by Umemura *et al* (1992).

There have been several designs reported that attempt to reduce the number of elements and/or complexity associated with two-dimensional phased array systems. Ocheltree *et al* (1984) describe a stack of linear phased arrays in which electronic steering of a set of neighbouring arrays (e.g. the $(n-1)$th, nth and $(n+1)$th) is achieved in one plane while adjustment of the power deposition pattern in the orthogonal plane is achieved by switching to another set of arrays (e.g. the nth, $(n+1)$th and $(n+2)$th). In this way, the need to provide control and power circuitry is required only for the elements within a small number of linear arrays, rather than for all elements within the complete two-dimensional array. Another approach is that of Benkeser *et al* (1987) who describe arrays in which the elements have tapered thickness. The region along the element at which the thickness is resonant at the driving frequency will produce the greatest acoustical output. The focal region of such an array can be moved in two dimensions by controlling the phases of the signals applied to the elements while movement in the third direction can be achieved through appropriate shifts in frequency. This structure requires N elements compared to N^2 elements for an $N \times N$ two-dimensional array with the same centre–centre spacing.

Full two-dimensional arrays that have been investigated include $N \times N$ square element arrays with plane, cylindrical or spherical geometry (Ibbini and Cain 1990, Ibbini *et al* 1990, Ebbini and Cain 1991a, McGough *et al* 1996). Since there are $(2N_1 N_2 - 1)$ values to be allocated to the phases and amplitudes of driving signals for an array of $N_1 \times N_2$ elements, use of an optimisation algorithm is an important aid in achieving a desired acoustic field. Ebbini and Cain (1989) describe a method for direct synthesis of multiple focus field patterns. The method determines the phase and amplitude of driving signals to produce specified field levels at a set of 'control points' within the treatment volume. These control points can be the desired foci or points at which reduced field levels are required. The relationship between the complex pressure at the M control points (located at $r_1, r_2, \ldots, r_m, \ldots, r_M$) and the complex velocity at the surface of the N array elements can be written

$$Hu = p$$

where $u = [u_1, u_2, \ldots, u_n, \ldots, u_N]^t$ is the complex velocity at the surface of the elements ($[]^t$ denotes transpose), $p = [p(r_1), p(r_2), \ldots, p(r_m), \ldots, p(r_M)]^t$ is the complex pressure at the control points and H is the forward propagation operator with elements $h_{m,n}$ given by

$$h_{m,n} = \frac{i\rho_0 c k}{2\pi} \int_{S'_n} \frac{e^{-ik|r_m - r'_n|}}{|r_m - r'_n|} \, dS'_n \qquad (8.11)$$

where S'_n is the surface area of the nth element and r'_n represents points on that element. Given p, we need to determine u. In practice, $M < N$, i.e. the number of foci or points at which the field is forced to some reduced level is less than the number of elements in the array. Ebbini and Cain (1989) show that a solution is given by

$$u = H^{*t}(HH^{*t})^{-1}p \qquad (8.12)$$

where H^{*t} is the conjugate transpose of H and represents the backward operator (relating the control point space to the source space). In this solution for u, the amplitudes of the driving signals to each element are dependent upon the contribution made by that element to the pressure at the control point. Consequently there is a wide range of amplitudes leading to a decrease in total power radiated by the array with respect to the case when all elements are driven at full power and focusing is achieved by applying a suitable phase distribution. In some situations, for example for arrays with a small aperture, the latter may be the method of choice since delivery of adequate power will be the overriding requirement.

8.1.3.4. Ultrasound hyperthermia devices for specific applications.   In addition to the general types of hyperthermia device previously described, there has been development of devices for treatment of particular anatomical sites, such as the eye, and for endocavitary and interstitial applications.

Endocavitary devices.   Most endocavitary devices have been designed for transrectal application for delivering various forms of thermal therapy to the prostate, either for treatment of benign prostatic hyperplasia or prostatic carcinoma. Early transrectal multi-element devices for hyperthermia consisted of non-focused segmented transducers, with varying transducer geometry (cylindrical, semi-cylindrical and planar), operating at around 1–1.6 MHz (Diederich and Hynynen 1987, 1989, 1990, 1991, Prieur *et al* 1988). Subsequently, electrically focused arrays for endocavitary use were developed. Diederich and Hynynen (1991) used an array of 16 half-cylindrical elements, each 2.25 mm long and cut from a larger 0.5 MHz cylindrical transducer with outer diameter 15 mm, and Buchanan and Hynynen (1994) improved the design to include 64 elements (1.5 mm wide and 15 mm OD) in an array 110.5 mm long with 1.73 mm centre-to-centre spacing. The use of cylindrical elements tended to increase the volume over which energy was deposited and was useful for inducing hyperthermia in tumour masses surrounding a body cavity. Hand *et al* (1993) investigated designs for a linear array of planar elements that would reduce divergence of the ultrasound field and so achieve greater local heating. One of the aims of that approach was to investigate the use of planar arrays not only for hyperthermia but also for more highly localised ablation of prostate tissues. Practical linear phased arrays were later fabricated (figure 8.6) and investigated independently in several groups; designs incorporated either

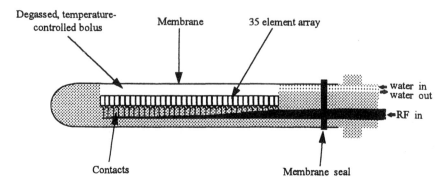

Figure 8.6. *A schematic diagram of an endocavitary device containing a 35 element linear phased array (Gavrilov and Hand, unpublished data).*

planar elements (Hutchinson *et al* 1995, 1996, Hutchinson and Hynynen 1996, Gavrilov and Hand 1997, Gavrilov *et al* 1997) or concave elements (Sheljaskov *et al* 1997).

Interstitial devices.   The combination of hyperthermia and brachytherapy is often used in the treatment of bulky or unresectable tumours. Although this form of therapy has been used during the last two decades (employing radiofrequency electrodes, microwave antennas, inductively heated ferromagnetic seeds, hot water sources, etc), the use of ultrasound interstitial devices is relatively new (Hynynen 1992, Jarosz 1991). Two approaches to ultrasound interstitial devices have been reported. In one (Jarosz 1991) an 18 to 24-G stainless steel needle is coupled to a 1 MHz planar disc transducer by a variable tapered conical velocity transformer. A plastic sleeve is placed over the proximal part of the needle such that the air gap encapsulated prevents transmission of ultrasound to the tissue except over the distal unclad 'active' length. The method permits the use of a relatively low frequency but does not offer control over the axial power deposition pattern. The second approach is based on cylindrical piezoceramic tubes, typically 1–2 mm outer diameter with wall thickness 0.25 mm and length 25 mm, which form transducers that are driven at a frequency in the range 4–12 MHz. These transducers are resonant across the wall thickness and deliver 2–3 W of acoustic power and induce therapeutic temperatures (42–45°C) over a radial distance of 15–17 mm, although some asymmetry of the radial penetration around these transducers, attributed to variations in wall thickness, is observed. The transducers are placed within a brachytherapy implant catheter and surrounded by water. As a consequence of their favourable radial penetration characteristics relative to most other interstitial hyperthermia techniques, ultrasound interstitial devices can be used in arrays with spacings of up to 25 mm between implants (Hynynen and Davis 1993).

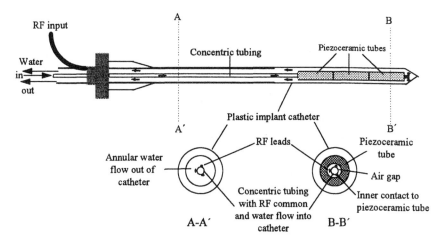

Figure 8.7. *A schematic diagram of an interstitial ultrasound applicator with an integrated cooling facility (after Diederich 1996).*

Diederich (1996) developed and evaluated multiple-element applicators combined with circulating temperature controlled degassed water to control the temperature of the catheter/tissue interface and to provide cooling of the transducers (figure 8.7). The transducers in these applicators were 1.5–1.6 mm outer diameter and could be inserted into 13 G brachytherapy catheters. The driving frequency, in the range 5–9 MHz had only a small influence on the radial penetration. Acoustic power of up to 5–6 watts per centimetre length of transducer were reported. In a subsequent report Diederich *et al* (1996) described multi-element devices which are compatible with the insertion of radiation sources to provide simultaneous, rather than sequential, thermo-brachytherapy. The applicators can be implanted directly within the target region, with the plastic coated transducers forming the outer wall of the 'implant catheter'. Two or more transducers, each 2.5 mm outer diameter, 0.33 mm wall thickness and 10 mm long and made from PZT piezoceramic formed the applicator. [125]I seeds or ^{192}Ir sources from remote afterloaders or iridium wire implants are placed in the lumen of the piezoceramic tube. Since, unlike applicators designed for sequential thermobrachytherapy, there is no water cooling, these devices have the characteristics of both hot source and acoustic heating devices.

8.1.4. High-intensity short-duration hyperthermia

The effect of hyperthermia is dependent upon both the elevated temperature and the time during which the elevated temperature is maintained. For hyperthermia to have an isoeffect on a tissue at different temperatures and durations, the following relationship must be satisfied

$$t_2 = t_1 B^{(T_1 - T_2)}$$

where t_1, t_2 are the durations at temperatures T_1, T_2, respectively. For temperatures greater than a transition temperature (and up to about 57°C), $B = 2$, that is the duration of heating must be decreased by a factor of two when the temperature is increased by 1°C. Below this transition temperature the value of B has been observed *in vivo* to be between 4 and 8. For most tissues, the transition temperature is in the range 42–43°C (Dewey 1994). This dependence is often described in terms of a thermal (isoeffect) dose which in practice has been calculated using

$$\text{Thermal dose} = \sum_{\substack{\text{duration of} \\ \text{treatment}}} \Delta t_{\text{eq}}(T) \qquad (8.13)$$

where $\Delta t_{\text{eq}}(T) = 2^{(T-43)}\Delta t$ if $T \geqslant 42.5$°C or $\Delta t_{\text{eq}}(T) = 2^{-0.5}6^{(T-42.5)}\Delta t$ if $T < 42.5$°C and Δt is the time (in minutes) between temperature measurements (Hand *et al* 1989). The units of thermal dose are 'equivalent minutes at 43°C'. Thus an alternative to conventional hyperthermia is to induce higher temperatures (up to approximately 55–60°C) for periods of the order of 10 seconds.

As alluded to earlier, scanned focused ultrasound is unique among non-invasive methods of inducing hyperthermia in that the spatial distribution of SAR may be controlled on a scale of the order of 1 cm or better due to the small focal volumes achievable. This high degree of control implies that differences in local tissue cooling due to heterogeneity in perfusion, variations in the density of discrete thermally significant vessels and even local cooling around single large vessels may be compensated for.

High-temperature short-duration ultrasound therapy reduces the dependence of the thermal dose delivered on blood flow (Hunt *et al* 1991, Dorr and Hynynen 1992) and so a relatively uniform thermal dose might be given without knowledge of perfusion or flow within large vessels. Lagendijk *et al* (1994) simulated thermal dose distributions in the presence of single vessels or a simulated vessel network due to sonicating the volume of interest with a series of 5 second pulses producing 17.5 W cm^{-3} in the focal volume. While the technique resulted in a significantly more uniform thermal dose distribution than was achievable when the volume was subject to a uniform SAR, the focus needed to be accurately positioned and stepped slowly and in a random manner through the target volume. In a rapid heating protocol, a large fraction of the thermal dose delivered is accrued after the ultrasound field is turned off and while the tissue is cooling. Kolios *et al* (1996) simulated lesion formation using thermal models (see sections 8.2.2 and 8.2.3) and showed that blood flow can influence the shape and size of the thermal dose distribution, even for exposures as short as approximately 8 s.

There are several practical problems to be overcome before this form of hyperthermia becomes clinically viable. These include accurate targeting of the focal volume, accounting for patient movement and limitations in terms of patient pain.

8.2. PREDICTION OF HEATING

Whenever there is a spatial dependence of temperature within a medium, energy in the form of heat flows from the region of higher to that of lower temperature; this is known as heat transfer. Three major types of heat transfer may be considered: conduction, convection and radiation. Heat transfer by convection results from a combination of conduction in a fluid and energy transport due to the motion of the fluid in the direction of energy flow. Convective effects may be encountered when determining boundary temperatures, such as the temperature of the skin in contact with a water bolus, and the thermal significance of blood vessels, as discussed later. Every body emits electromagnetic radiation in proportion to the fourth power of the absolute temperature of its surface. The balance between this energy loss and that gained by radiation from its environment may be a net loss or gain and in practice is dependent upon the emissivity and absorptance of the body's surface. The effects of conduction are discussed below. Reviews of heat transport within tissues are to be found in Chato (1990), Lagendijk (1990) and Arkin *et al* (1994).

8.2.1. *Thermal conduction*

According to Fourier's law, the conductive heat transfer in a particular direction is proportional to the cross-sectional area of the flow and to the temperature gradient. Thus

$$Q_x = -A_x \left(k_{xx} \frac{\partial T}{\partial x} + k_{xy} \frac{\partial T}{\partial y} + k_{xz} \frac{\partial T}{\partial z} \right) \qquad (8.14)$$

where Q_x is the 'heat flux' (in the x-direction) and has units W m^{-2}. The inclusion of the minus sign leads to heat flow in the direction of decreasing temperature. k is the coefficient of thermal conductivity and has units W m^{-1} K^{-1}. In general thermal conductivity is a tensor, as shown in the above expression for Q_x. However, in most cases, the heat flux in x, y and z can be considered to be decoupled (only k_{xx}, k_{yy} and $k_{zz} \neq 0$), each being driven by a single temperature gradient. If the material is isotropic, $k_{xx} = k_{yy} = k_{zz}$ and the thermal conductivity becomes a scalar constant.

According to the first law of thermodynamics, energy must be conserved. Thus energy balance can be applied to an elemental volume (dx·dy·dz) of the medium (figure 8.8) and, considering only conduction, the three-dimensional heat balance equation

$$\rho_0 s \frac{\partial T}{\partial t} = \frac{\partial}{\partial x} \left(k_x \frac{\partial T}{\partial x} \right) + \frac{\partial}{\partial y} \left(k_y \frac{\partial T}{\partial y} \right) + \frac{\partial}{\partial z} \left(k_z \frac{\partial T}{\partial z} \right) + Q_i \qquad (8.15)$$

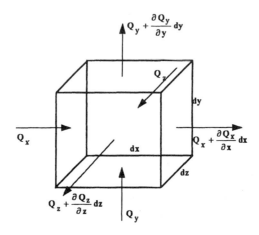

Figure 8.8. *Energy balance for an elemental volume.*

may be derived. Here, ρ_0 and s are the density and specific heat of the medium and any effects due to volumetric expansion have been neglected. In the case of tissues subjected to hyperthermia, the term Q_i contains distinct contributions, including the volumetric rate of metabolism Q_m, the power per unit volume absorbed from the ultrasound device Q_{us} and a term accounting for the effects of blood perfusion, Q_{BF}. The form adopted for Q_{BF} is dependent upon the particular requirements for the description of the resulting temperature field $T(x, y, z)$.

8.2.2. Pennes' bioheat transfer equation

Seminal work in the development of models describing heat transfer within tissues was that of Pennes (1948) who made a quantitative analysis of the relationship between arterial blood and tissue temperatures. Temperatures of human forearm tissues and brachial arterial blood were measured using thermocouple probes implanted radially to depths of 3–4 cm to evaluate the applicability of a heat flow theory to the forearm in terms of the local rate of tissue heat production and volumetric flow of blood. The effects of circulatory occlusion on the radial temperature distribution were also investigated. Pennes used the expression

$$Q_{BF} = w_b s_b K (T - T_{art})$$

to represent the cooling effect of perfusion, where w_b is the volumetric perfusion rate (kg m^{-3} s^{-1}), s_b is the specific heat of blood, T_{art} is the local arterial temperature (taken to be equal to body core temperature), T is the local tissue temperature and K is a constant ($0 \leqslant K \leqslant 1$) related

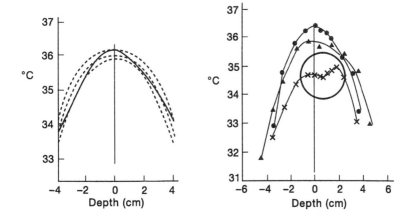

Figure 8.9. *Data from Pennes (1948). Left: Mean experimental data (solid line) and theoretical predictions based on equations (9) and (11) of the original paper for (dotted lines from top to bottom) volumetric blood flow through the tissue, $V = 3.0 \times 10^{-4}$, 2.5×10^{-4} and 2.0×10^{-4} g cm^{-3} s^{-1}, and rate of tissue heat production $h_m = 10^{-4}$ g cal cm^{-3} s^{-1}. Arterial blood temperature was $36.25°C$ (adapted from Pennes' figure 16). Right: Experimental temperature versus depth curves. Pennes attributed the form of the biphasic curve encircled to being 'undoubtedly caused by some local variation in vascular pattern' (adapted from Pennes' figure 15).*

to thermal equilibrium between capillary blood and local tissue. Figure 8.9 shows some of Pennes' original data. His model was in very good qualitative agreement and reasonable (to within about 0.5°C) quantitative agreement with measurements when considered over a length scale of a few centimetres or more. Pennes not only pointed to the assumed uniformity of his model as contributing to discrepancies between predictions and measurements but also stressed that it was 'possible that the venous return from the distal portion of the extremity modified the deep tissue temperature considerably'. Figure 8.9 also shows that Pennes observed temperature depth profiles that exhibited local minima and he attributed these to local effects of the vasculature.

Considering the underlying assumptions, Pennes' formulation of the bioheat transfer has been successful in fitting experimental temperature distributions, although the values of other parameters underlying these fits were not verified (Sekins and Emery 1982). In another case (Schreier *et al* 1990), agreement between predictions and in measurements *in vivo* of temperatures within a region of tissue (pig and rabbit thigh) implanted by hot water tubes was reasonable if perfusion was assumed not to exceed a value associated with resting muscle, which may be questionable in the presence of hyperthermia. However, some simulations in this study predicted temperatures that differed from those measured by up to 2°C. In a study of temperature and thermal dose changes associated with heating by focused ultrasound, Lagendijk *et al* (1994) highlighted the important effects large vessels and a high density of vessels have on local heat transfer within tissues.

The shortcomings of this approach have also been discussed widely in the literature and include:

• the assumption that blood enters the local tissue volume at arterial temperature and leaves at local tissue temperature. Anticipating a discussion regarding the thermal equilibrium length (X_{eq}) of vessels, this implies that X_{eq} is infinite for all vessels except capillaries for which $X_{eq} = 0$;
• heat transport related to mass transport of blood is neglected;
• the actual temperature of the blood entering the local tissue volume is neglected;
• individual cooling or heating by large vessels is neglected;
• the effects of the complete venous vessel network are neglected.

For these reasons and the need to describe the detailed temperature field with finer spatial resolution, other descriptions of heat transfer within tissues have been developed.

Nevertheless, the use of Pennes' bioheat transfer equation is widespread in the assessment of potential thermal hazards of ultrasound fields for diagnostic purposes (NCRP 1992, AIUM/NEMA 1992, Shaw 1994, Shaw *et al* 1996). While this may be adequate for semi-quantitative predictions of the resulting temperature increase (e.g. 'reasonable worst case' or 'typical' scenarios), a more detailed prediction of the local temperature field may require use of other descriptions of heat transport.

8.2.3.   *Other approaches to thermal modelling*

An important question to be answered when considering heat transfer between blood flow within a vessel and the vessel wall is at what distance from the entrance will the blood temperature equilibrate with that of the local tissue? According to Chen and Holmes (1980) the vessel length over which the temperature difference is reduced by a factor *e* is

$$X_{eq} = \Lambda \frac{\rho_b s_b v r^2}{k_b} \tag{8.16}$$

where r is the vessel radius and ρ_b, v, k_b are the density, flow velocity and thermal conductivity of blood, respectively. Λ is dependent upon the interaction of the vessel with the surrounding tissue. Assuming thermal conductivities of blood and tissue are similar and that the vessel thermally influences a cylindrical volume of tissue with radius approximately $10r$, then $\Lambda \approx 1.5$. Some examples of the value of X_{eq} related to vascular data obtained from a dog are given in table 8.1.

Table 8.1. *Properties of vascular compartments (after Chen and Holmes (1980) and Crezee (1993)).*

	Diameter (mm)	Length l_v (mm)	Flow (mm s^{-1})	X_{eq} (mm)	X_{eq}/l_v
Aorta	10	400	500	125000	310
Large arteries	3	200	130	2900	15
Main branches	1	100	80	200	2
Secondary branches	0.6	40	40	72	1.8
Tertiary branches	0.14	14	34	1.7	0.1
Terminal branches	0.05	1	20	0.13	0.1
Terminal arteries	0.03	1.5	4	0.009	0.006
Arterioles	0.02	2.2	3	0.003	0.002
Capillaries	0.008	1.1	0.7	0.0001	0.0001
Venules	0.03	2.2	0.7	0.0016	0.001
Terminal branches	0.07	1.5	0.7	0.009	0.006
Terminal veins	0.13	1	3	0.13	0.1
Tertiary veins	0.28	14	8	1.6	0.1
Secondary veins	1.5	40	13	73	1.8
Main veins	2.4	100	15	220	2.2
Large veins	6	200	36	3200	16
Vena cava	12.5	400	330	129000	320

A useful ratio in thermal modelling is that of the equilibrium length X_{eq} to the typical physical length of the vessel l_v. This is a measure of the degree to which the temperature of blood in the vessel has equilibrated with the local tissue temperature. For example if $X_{eq}/l_v \gg 1$ then the blood temperature will never reach that of the local tissue, while if $X_{eq}/l_v \ll 1$ then the blood temperature will equilibrate with that of the local tissue well within the length of that particular vessel. From table 8.1 it can be seen that for single vessels thermal equilibrium takes place ($X_{eq}/l_v \approx 1$) somewhere between the secondary and tertiary branches and not in the capillaries as assumed in the Pennes form of the bioheat transfer equation. However, most arteries and veins are arranged in counter-current pairs and this structure leads to faster thermal equilibration than suggested in table 8.1. Weinbaum *et al* (1984) investigated several vessel configurations, including a counter-current pair, and suggested that the value of X_{eq} is reduced by up to a factor of two due to counter-current heat exchange and so equilibration will occur at a larger generation of vessels. Following this new insight into the major heat exchange between vessels and tissue, several new models evolved.

Chen and Holmes (1980) considered that vessels were grouped into two categories: large vessels, each to be treated separately, and small vessels, to be treated as part of the tissue continuum, and that the boundary between these two groups occurred when $X_{eq} \approx l_v$. For the small vessels, three contributions to heat transfer were suggested (CH model). These were (i) a heat sink term similar to Pennes' perfusion term for the largest, thermally non-equilibrated vessels $Q_p = w^* \rho_b s_b (T_a^* - T)$ where w^* and T_a^* refer only to blood flow, and the temperature of the blood in the largest vessels, respectively, within a local elemental volume; (ii) a unidirectional flow term to account for convective heat transport associated with the equilibrated small vessels; (iii) a perfusion enhanced conductivity or 'effective conductivity' to describe small temperature fluctuations of nearly equilibrated blood along the temperature gradient.

Weinbaum *et al* (1984) concluded from anatomical observations in peripheral tissue that the main contribution to heat transfer from local blood perfusion is not at the level of the capillaries but is associated with the incomplete counter-current heat exchange between pairs of arteries and veins. A model (WJL) based on a pair of thermally significant vessels connected by capillaries was proposed but in practice was difficult to apply to most tissues, other than peripheral tissue, because of the detail of the vascular structure required. Weinbaum and Jiji (1985) subsequently proposed a simplified model (WJ) in which the mean tissue temperature is approximated by an average temperature of the adjacent counter-current pair of nearly equilibrated vessels and that most of the heat conducted through the wall of an arteriole enters the wall of its paired vein. These assumptions lead to an effective conductivity that represents both the capillary bleed-off and the heat exchange between tissue and adjacent vessels. Criticisms of the underlying assumptions in the WJ model were subsequently raised and alternate formulations suggested by Baish *et al* (1986a,b), Wissler (1987a,b) and Charny *et al* (1989, 1990).

In their review of the topic, Arkin *et al* (1994) summarise the Pennes, CH, WJ and WJL models and findings from experimental verifications of the models. Local temperature fluctuations existing close to large vessels should be accounted for by a convective term similar to that suggested in the CH model. The CH model can be viewed as a modification of Pennes' model by using a more precise description of perfusion and by including a convection term that is significant only close to large vessels. The CH and Pennes' models are applicable in the same tissue regions (vessel diameters <0.3 mm; $X_{eq}/l_v < 0.6$). In addition, Charny *et al* (1990) claim that Pennes' model describes the bleed off from large vessels to the tissue (for vessels 0.5–1.0 mm in diameter and with $X_{eq}/l_v \gg 1$). The WJ model appears to be valid only for smaller vessels (diameters <0.15–0.3 mm; $X_{eq}/l_v < 0.2$–0.3). The Pennes', CH and WJ models describe a continuum by a single equation in which perfusion effects are incorporated implicitly in the heat transfer equation. In contrast, the WJL model and variants thereof (Baish *et al* 1986b, Wissler 1987a, Charny *et al* 1989) consist of three equations coupling heat transfer of the venous blood, arterial blood and tissue and are applicable where most of the vessels are paired. Although the WJL model appears to be valid only in regions where vessels are small (diameters <0.3 mm; $X_{eq}/l_v < 0.3$), the variants work for vessels >0.01 mm diameter. However, their practical use requires detail of tissue and organ vascularity that is currently unavailable in general. Although none of the single equation models can be applied to all tissues, combinations may be valid. For example, Pennes can be used for large vessels and WJ for small vessels in regions such as muscle where most vessels are paired. However, the valid application of the various models to regions of the human body remains to be verified.

Crezee *et al* (1994) considered predictions associated with four models: a heat-sink model (Pennes'), a model using effective thermal conductivity (k_{eff}) (for vessels $\lesssim$ 0.5 mm diameter), a mixed heat-sink/k_{eff} model and a discrete vessel model and compared these with experimental findings in perfused bovine tongues. Both conduction and venous heat transfer mechanisms were found to be important, as predicted by the mixed models (k_{eff} for small vessels and discrete vessel/heat sink for larger vessels). In their effort to develop models for hyperthermia treatment planning, Mooibroek and Lagendijk (1991) described a model in which bending and branching of discrete vessels was accommodated and more recently Kotte *et al* (1996) used a parametric description of vessels in three dimensions. An analytical/numerical technique incorporating vascular trees has been used by Baish (1994) to determine steady state temperature distributions in tissue.

8.3. SUMMARY

From a physical point of view, the use of ultrasound to induce hyperthermia has some marked advantages over commonly used electromagnetic techniques, including relatively low attenuation in soft tissues (up to 1 dB cm^{-1} MHz^{-1}) and short wavelength (3–0.3 mm) in soft tissues for frequencies of interest ($\approx$0.5–5 MHz). Thus collimated and, more importantly, focused beams may be produced from transducers of convenient size. These characteristics enable a range of practical systems to be designed which offer control over the SAR distribution within tissues on a scale of the order of 1 cm, which is important for effective and safe treatment.

Clinical experience of ultrasound hyperthermia has been gained in the treatment of both superficial and deep seated tumours. In many cases, the systems used were either primitive or prototype focused devices and as such were sub-optimal. In some cases, patient tolerance to ultrasound treatments has been low, particularly when relatively low frequency non-focused devices were used. These problems are likely to be overcome as new systems involving electronically steered phased arrays become available. The recent successes of clinical hyperthermia combined with radiotherapy provide an incentive to investigate the full clinical potential of ultrasound hyperthermia.

There has been progress in understanding and modelling heat transport processes within tissues in recent years. Although Pennes' bioheat transfer equation continues to be used for many purposes, the strengths and weakness of other models based on more realistic descriptions of vascular effects on heat transport are now understood. Numerical algorithms for the practical application of such models to the prediction of spatially accurate temperature and thermal dose distributions are now available and will have significant impact on hyperthermia treatment planning.

ACKNOWLEDGMENTS

I am grateful to Dr L R Gavrilov and Dr J J W Lagendijk for their constructive comments.

REFERENCES

AIUM/NEMA 1992 Standard for real-time display of thermal and mechanical acoustic output indices on diagnostic ultrasound equipment *Publication UL-3* (Rockville, MD: American Institute of Ultrasound in Medicine)

Anhalt D, Hynynen K, Roemer R B, Nathanson S M, Stea B and Cassady J R 1992 Scanned ultrasound hyperthermia for treating superficial disease *Hyperthermic Oncology 1992* vol 2, ed E W Gerner (Tucson: Arizona Board of Regents) pp 191–2

Arkin H, Xu L X and Holmes K R 1994 Recent developments in modeling heat transfer in blood perfused tissues *IEEE Trans. Biomed. Eng.* **41** 97–107

Baish J W 1994 Formulation of a statistical model of heat transfer in perfused tissue *ASME J. Biomech. Eng.* **116** 521–7

Baish J W, Foster K R and Ayyaswamy P S 1986a Small scale temperature fluctuations in perfused tissue during local hyperthermia *ASME J. Biomech. Eng.* **108** 246–50

—— 1986b Heat transport mechanisms in vascular tissues: a model comparison *ASME J. Biomech. Eng.* **108** 324–31

Benkeser P J, Frizzell L A, Goss S A and Cain C A 1989 Analysis of a multielement ultrasound hyperthermia applicator *IEEE Trans. Ultrasonics Ferroelectr. Freq. Control* **36** 319–25

Benkeser P J, Frizzell L A, Ocheltree K B and Cain C A 1987 A tapered phased array ultrasound transducer for hyperthermia treatment *IEEE Trans. Ultrasonics Ferroelectr. Freq. Control* **34** 446–53

Buchanan M T and Hynynen K 1994 Design and experimental evaluation of an intracavitary ultrasound phased array system for hyperthermia *IEEE Trans. Biomed. Eng.* **41** 1178–87

Charny C K, Weinbaum S and Levin R L 1989 Bioheat transfer in a branching countercurrent network during hyperthermia *ASME J. Biomech. Eng.* **111** 263–70

—— 1990 An evaluation of the Weinbaum–Jiji bioheat transfer model for normal and hyperthermic conditions *ASME J. Biomech. Eng.* **112** 80–7

Chato J C 1990 Fundamentals of bioheat transfer *Thermal Dosimetry and Treatment Planning* ed M Gautherie (Berlin: Springer) pp 1–56

Chen M M and Holmes K R 1980 Microvascular contributions in tissue heat transfer *Ann. NY Acad. Sci.* **335** 137–50

Cline H E, Hynynen K, Hardy C J, Watkins R D, Schenk J F and Jolesz F A 1994 MR temperature mapping of focused ultrasound surgery *Magn. Reson. Med.* **31** 628–36

Corry P M, Jabboury K and Armour E P 1987 Clinical experience with plane-wave ultrasound systems *A Categorical Course in Radiation Therapy: Hyperthermia* ed R A Steeves and B R Paliwal (Oak Brook, IL: Radiological Society of North America) pp 151–8

Crezee J 1993 Experimental verification of thermal models *PhD Thesis* University of Utrecht

Crezee J, Mooibroek J, Lagendijk J J W and van Leeuwen G M J 1994 The theoretical and experimental evaluation of the heat balance in perfused tissue *Phys. Med. Biol.* **39** 813–32

Dewey W C 1994 Arrhenius relationships from the molecule and cell to the clinic *Int. J. Hypertherm.* **10** 457–83

Diederich C J 1996 Ultrasound applicators with integrated catheter-cooling for interstitial hyperthermia: theory and preliminary experiments *Int. J. Hypertherm.* **12** 279–97

Diederich C J and Hynynen K 1987 Induction of hyperthermia using an intracavitary ultrasonic applicator *Proc. IEEE Ultrasonics Symp.* vol 2, ed B R McAvoy (Piscataway, NJ: IEEE) pp 871–4

—— 1989 Induction of hyperthermia using an intracavitary multi-element ultrasonic applicator *IEEE Trans. Biomed. Eng.* **36** 432–8

—— 1990 The development of intracavitary ultrasonic applicators for hyperthermia: a design and experimental study *Med. Phys.* **17** 626–34

—— 1991 The feasibility of using electrically focused ultrasound arrays to induce deep hyperthermia via body cavities *IEEE Trans. Ultrasonics Ferroelectr. Freq. Control* **38** 207–19

Diederich C J, Khalil I S, Stauffer P R, Sneed P K and Phillips T L 1996 Direct-coupled interstitial ultrasound applicators for simultaneous thermobrachytherapy: a feasibility study *Int. J. Hypertherm.* **12** 401–19

Dorr L N and Hynynen K 1992 The effects of tissue heterogeneities and large blood vessels on the thermal exposure induced by short high-power ultrasound pulses *Int. J. Hypertherm.* **8** 45–59

Dopenloup F, Chapelon J Y, Cathignol D J and Sapozhnikov O A 1996 Reduction of the grating lobes of annular arrays used in focused ultrasound surgery *IEEE Trans. Ultrasonics Ferroelectr. Freq. Control* **43** 991–8

Ebbini E S and Cain C A 1989 Multiple-focus ultrasound phased-array pattern synthesis: optimal driving-signal conditions for hyperthermia *IEEE Trans. Ultrasonics Ferroelectr. Freq. Control* **36** 540–8

—— 1991a Experimental evaluation of a prototype cylindrical section ultrasound hyperthermia phased array applicator *IEEE Trans. Ultrasonics Ferroelectr. Freq. Control* **38** 510–20

—— 1991b A spherical-section ultrasound phased array applicator for deep localized hyperthermia *IEEE Trans. Biomed. Eng.* **38** 634–43

Gavrilov L R and Hand J W 1997 Development and investigation of ultrasound linear phased arrays for transrectal treatment of prostate *Ultrasonics Sonochemistry* in press

Gavrilov L, Hand J W, Abel P and Cain C A 1997 A method of reducing grating lobes associated with an ultrasound linear phased array intended for transrectal thermotherapy *IEEE Trans. Ultrasonics Ferroelectr. Freq. Control* **44** in press

Goss S A, Frizzell L A and Dunn F 1979 Ultrasonic absorption and attenuation in mammalian tissues *Ultrasound Med. Biol.* **5** 181–6

Goss S A, Johnston R L and Dunn F 1978 Comprehensive compilation of empirical ultrasonic properties of mammalian tissues *J. Acoust. Soc. Am.* **64** 423–57

—— 1980 Compilation of empirical ultrasonic properties of mammalian tissues: II *J. Acoust. Soc. Am.* **68** 93–108

Hand J W, Ebbini E, O'Keefe D, Israel D and Mohammadtaghi S 1993 An ultrasound linear array for use in intracavitary applicators for thermotherapy of prostatic diseases *Proc. 1993 IEEE Ultrasonics Symp.* ed M Levy and B R McAvoy (Piscataway, NJ: IEEE) pp 1225–8

Hand J W, Lagendijk J J W, Bach Andersen J and Bolomey J C 1989 Quality assurance guidelines for ESHO protocols *Int. J. Hypertherm.* **5** 421–8

Hand J W, Vernon C C and Prior M V 1992 Early experience of a commercial scanned focused ultrasound hyperthermia system *Int. J. Hypertherm.* **8** 587–607

Harris G R 1981 Review of transient field theory for a baffled planar piston *J. Acoust. Soc. Am.* **70** 10–20

Hill C R 1995 Optimum acoustic frequency for focused ultrasound surgery *Ultrasound Med. Biol.* **20** 271–7

Hunt J W 1990 Principles of ultrasound used for generating localized hyperthermia *Practical Aspects of Clinical Hyperthermia* ed S B Field and J W Hand (London: Taylor and Francis) pp 371–422

Hunt J W, Lalonde R, Ginsberg H, Urchuk S and Worthington A 1991 Rapid heating: critical theoretical assessment of thermal gradients found in hyperthermia treatments *Int. J. Hypertherm.* **7** 703–18

Hutchinson E B, Buchanan M T and Hynynen K 1995 Evaluation of an aperiodic ultrasound phased array for prostate thermal therapies *Proc. 1995 IEEE Ultrasonics Symp.* ed M Levy, S C Schneider and B R McAvoy (Piscataway, NJ: IEEE) pp 1601–4

—— 1996 Design and optimization of an aperiodic ultrasound phased array for intracavitary prostate thermal therapies *Med. Phys.* **23** 767–76

Hutchinson E B and Hynynen K 1996 Intracavitary ultrasound phased array for noninvasive prostate surgery *IEEE Trans. Ultrasonics Ferroelectr. Freq. Control* **43** 1032–42

Hynynen K 1990 Biophysics and technology of ultrasound hyperthermia *Methods of External Hyperthermic Heating* ed M Gautherie (Berlin: Springer) pp 61–115

—— 1992 The feasibility of interstitial ultrasound hyperthermia *Med. Phys.* **19** 979–87

Hynynen K and Davis L S 1993 Small cylindrical ultrasound sources for induction of hyperthermia via body cavities or interstitial implants *Int. J. Hypertherm.* **9** 263–74

Hynynen K, Roemer R, Anhalt D, Johnson C, Xu Z K, Swindell W and Cetas T C 1987 A scanned focused multiple transducer ultrasonic system for localised hyperthermia treatments *Int. J. Hypertherm.* **3** 21–35

Hynynen K, Shimm D, Anhalt D, Stea B, Sykes H, Cassady J R and Roemer R B 1990 Temperature distributions during clinical scanned, focused ultrasound hyperthermia treatments *Int. J. Hypertherm.* **6** 891–908

Ibbini M S and Cain C A 1990 The concentric ring array for ultrasound hyperthermia: combined mechanical and electrical scanning *Int. J. Hypertherm.* **6** 401–19

Ibbini M S, Ebbini E S and Cain C A 1990 $N \times N$ square-element ultrasound phased array applicator: simulated temperature distributions associated with directly synthesized heating patterns *IEEE Trans. Ultrasonics Ferrorelectr. Freq. Control* **37** 491–500

Jarosz B J 1991 Temperature distribution in interstitial ultrasonic hyperthermia *Proc. Ann. Int. Conf. IEEE Engineering in Medicine and Biology Society* ed J H Nagel and W M Smith (Piscataway, NJ: IEEE) pp 179–80

Kolios M C, Sherar M D and Hunt J W 1996 Blood flow cooling and ultrasonic lesion formation *Med. Phys.* **23** 1287–98

Kossoff G 1979 Analysis of focusing action of spherically curved transducers *Ultrasound Med. Biol.* **5** 359–65

Kotte A, van Leeuwen G, de Bree J, van der Koijk J, Crezee H and Lagendijk J J W 1996 A description of discrete vessel segments in thermal modelling of tissues *Phys. Med. Biol.* **41** 865–84

Lagendijk J J W 1990 Thermal models: principles and instrumentation *An Introduction to the Practical Aspects of Clinical Hyperthermia* ed S B Field and J W Hand (London: Taylor and Francis) pp 478–512

Lagendijk J J W, Crezee J and Hand J W 1994 Dose uniformity in scanned focused ultrasound hyperthermia *Int. J. Hypertherm.* **10** 775–84

Lele P P 1983 Physical aspects and clinical studies with ultrasonic hyperthermia *Hyperthermia in Cancer Therapy* ed F K Storm (Boston: Hall) pp 333–67

Lindsley K, Stauffer P R, Sneed P, Chin R, Phillips T L, Seppi E, Shapiro E and Henderson S 1993 Heating patterns of the Helios ultrasound hyperthermia system *Int. J. Hypertherm.* **9** 675–84

Lu X-Q, Burdette E C, Bornstein B A, Hansen J L and Svensson G K 1996 Design of an ultrasonic therapy system for breast cancer treatment *Int. J. Hypertherm.* **12** 375–99

Marchal C, Bey P, Metz R, Gaulard M L and Robert J 1982 Treatment of superficial human cancerous nodules by local ultrasound hyperthermia *Br. J. Cancer* **45** (suppl V) 243–5

Marmor J B, Pounds D, Postic T B and Hahn G M 1979 Treatment of superficial human neoplasms by local hyperthermia induced by ultrasound *Cancer* **43** 188–97

McGough R J, Kessler M L, Ebbini E S and Cain C A 1996 Treatment planning for hyperthermia with ultrasound arrays *IEEE Trans. Ultrasonics Ferroelectr. Freq. Control* **43** 1074–84

McGough R J, Wang H, Ebbini E S and Cain C A 1994 Mode scanning: heating pattern synthesis with ultrasound phased arrays *Int. J. Hypertherm.* **10** 433–42

Montes H and Hynynen K 1995 A system for the simultaneous delivery of intraoperative radiation and ultrasound hyperthermia *Int. J. Hypertherm.* **11** 109–19

Mooibroek J and Lagendijk J J W 1991 A fast and simple algorithm for the calculation of convective heat transfer by large vessels in three-dimensional inhomogeneous tissues *IEEE Trans. Biomed. Eng.* **38** 490–501

NCRP 1992 Exposure criteria based on thermal mechanisms *Report No 113* (Bethesda, MD: National Council on Radiation Protection and Measurements)

Nussbaum G H, Straube W L, Drag M D *et al* 1991 Potential for localized, adjustable deep heating in soft tissue environments with a 30 beam ultrasonic hyperthermia system *Int. J. Hypertherm.* **7** 279–99

Ocheltree K B, Benkeser P J, Frizzell L A and Cain C A 1984 An ultrasonic phased array applicator for hyperthermia *IEEE Trans. Sonics Ultrasonics* **31** 526–31

Overgaard J, González González D, Hulshof M C C M, Arcangeli G, Dahl O, Mella O and Bentzen S M 1995 Randomised trial of hyperthermia as adjuvant to radiotherapy for recurrent or metastatic malignant melanoma *Lancet* **345** 540–3

Pennes H H 1948 Analysis of tissue and arterial blood temperatures in the resting human forearm *J. Appl. Physiol.* **1** 93–122

Prieur G, Nadi M, Marchal C, Chitnallah A and Bey P 1988 Development of a new intracavitary ultrasound applicator *Proc. 10th Ann. Conf. IEEE Engineering in Medicine and Biology Society* ed G Harris and C Walker (Piscataway, NJ: IEEE) pp 932–3

Schreier K, Budhina M, Lesnicar H, Handl-Zeller L, Hand J W, Prior M V, Clegg S T and Brezovich I A 1990 Preliminary studies of interstitial hyperthermia using hot water *Int. J. Hypertherm.* **6** 431–44

Sekins K M and Emery A 1982 Thermal science for physical medicine *Therapeutic Heat and Cold* ed J Lehmann (Baltimore: Williams and Wilkins) pp 70–132

Shaw A 1994 Prediction of temperature rise in layered media from measured ultrasonic intensity data *Phys. Med. Biol.* **39** 1203–18

Shaw A, Preston R C and Bacon D R 1996 Perfusion corrections for ultrasonic heating in nonperfused media *Ultrasound Med. Biol.* **22** 203–16

Sheljaskov T, Lerch R L, Bechtold M, Newerla K and Schätzle U 1997 A phased array for simultaneous HIFU therapy and sonography *Proc. 1996 IEEE Ultrasonics Symp.* (Piscataway, NJ: IEEE) pp 1527–30

Steinberg B D 1976 *Principles of Aperture and Array System Design* (New York: Wiley)

Swindell W 1986 Ultrasonic hyperthermia *Physical Techniques in Clinical Hyperthermia* ed J W Hand and J R James (Letchworth: Research Studies Press) pp 288–326

Umemura S and Cain C A 1989 The sector-vortex phased array: acoustic field synthesis for hyperthermia *IEEE Trans. Ultrasonics Ferroelectr. Freq. Control* **39** 32–8

Umemura S-I, Holmes K R, Frizzell L A and Cain C A 1992 Insonation of fixed porcine kidney by a prototype sector-vortex-phased array applicator *Int. J. Hypertherm.* **8** 831–42

van der Zee J and van Rhoon G C 1993 The value of loco-regional deep hyperthermia in addition to a standard series of radiotherapy for the treatment of large, inoperable pelvic tumours *Final Report of a Study within the Frame of Investigative Medicine* (a program of the Dutch Health Fund Insurance Council)

Vernon C C, Hand J W, Field S B, Machin D, Whaley J B, van der Zee J, van Putten W L J, van Rhoon G C, van Dijk J D P and González González D 1996 Radiotherapy with or without hyperthermia in the treatment of superficial localized breast cancer: results from five randomized controlled trials *Int. J. Radiat. Oncol. Biol. Phys.* **35** 731–44

Weinbaum S and Jiji L M 1985 A new simplified bioheat transfer equation for the effect of blood flow on local average tissue temperatures *ASME J. Biomech. Eng.* **107** 131–9

Weinbaum S, Jiji L M and Lemons D E 1984 Theory and experiment for the effect of vascular microstructure on surface tissue heat transfer—part 1: anatomical foundation and model conceptualization *ASME J. Biomech. Eng.* **106** 321–30

Wissler E H 1987a Comments on the new bioheat transfer equation proposed by Weinbaum and Jiji *ASME J. Biomech. Eng.* **109** 226–33

—— 1987b Comments of Weinbaum and Jiji's discussion of their proposed bioheat transfer equation *ASME J. Biomech. Eng.* **109** 355–6

Zemanek J 1971 Beam behaviour within the nearfield of a vibrating piston *J. Acoust. Soc. Am.* **49** 181–91

CHAPTER 9

FOCUSED ULTRASOUND SURGERY

Gail R ter Haar

INTRODUCTION

Most medical ultrasound operates at frequencies in the range 0.5–10 MHz, with associated wavelengths of 0.15–3 mm in tissue. These short wavelengths mean that an ultrasonic beam may in principle be brought to a tight focus of the order of a millimetre, a few centimetres away from its source. We thus have the possibility that a focused beam may be used to produce heating solely within the focal region, and not elsewhere in the beam provided high focal gains are used. This property is used in focused ultrasound surgery where focal intensities of 1–4 kW cm^{-2} are used to produce temperatures in excess of 60°C with the aim of ablating tissue in the well circumscribed focal region, while sparing tissue elsewhere [1]. The volume of ablated tissue is termed a 'lesion'. It has been shown that the demarcation between dead cells and live cells is very sharp, the boundary being only about six cells wide.

The principle of the technique is shown schematically in figure 9.1. A source of ultrasound is placed at a distance from the target tissue volume. The source may be extracorporeal, or intracavitary (situated, for example, in the rectum). The focus of the beam is set to lie within the tissue volume to be treated. The source, which may lie several centimetres from the skin, is coupled to the patient by an enclosed volume of degassed water. Exposure times of 1–2 seconds are used to ensure that heat delivery may be independent of perfusion. By using high acoustic powers for short times, ablative temperatures are achieved rapidly, and heat dissipation by blood perfusion and thermal conduction become insignificant, thus overcoming one of the disadvantages of conventional hyperthermia techniques. A typical lesion in *ex vivo* liver is shown in plate 1(top). Plate 1(bottom) shows the sharp boundary between the inside of the lesion and the viable cells outside.

177

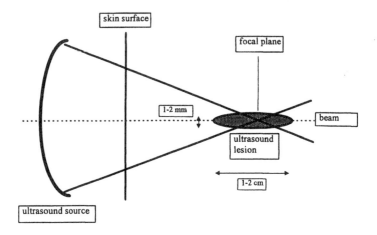

Figure 9.1. *Schematic diagram showing the principle involved in focused ultrasound surgery. The field from a high power ultrasound source is brought to a focus at depth within the tissue, thus producing a 'trackless' lesion.*

9.1. MECHANISMS OF LESION PRODUCTION

There are two main mechanisms by which ablative damage is achieved. Firstly absorption of the ultrasonic energy leads to tissue heating; and secondly interaction between the ultrasonic field and gas contained in the tissue may lead to acoustic cavitation. These two mechanisms will be dealt with separately here, but in reality cannot always be separated.

9.1.1. Thermal effects

A full discussion of the heating of tissue by ultrasound is given in Chapter 8. For the purposes of this chapter, an approximate estimate of the rate of temperature rise that may be experienced when a beam of intensity, I, passes through a tissue with intensity absorption coefficient, μ_a, may be made. The amount of heat produced, Q, is given by

$$Q = \mu_a I. \tag{9.1}$$

In the absence of cooling by blood, for a tissue of density ρ_t, and specific heat capacity s_t, the rate of temperature rise dT/dt is therefore given by

$$dT/dt = \mu_a I / \rho_t s_t. \tag{9.2}$$

For typical soft tissues at 1.7 MHz, $\mu_a = 0.5$ dB cm^{-1} (see Chapter 4), $\rho_t = 10^3$ kg m^{-3}, $s_t = 4.18 \times 10^3$ J kg^{-1} K^{-1}. In focused ultrasound surgery, spatial peak intensities of around 1000 W cm^{-2} are used, and this leads to a

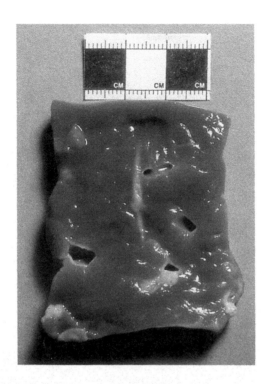

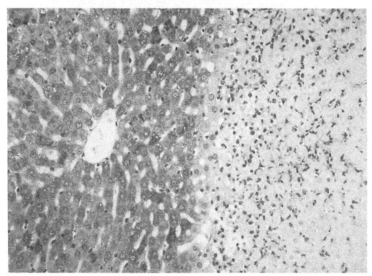

Plate 1.

(Top) Single focused ultrasound lesion formed in a piece of bovine liver. The focal peak was placed 1 cm below the liver capsule.

(Bottom) Histological section showing (left) undamaged 'normal' cells in rat liver, and (right) dead cells inside the lesion.

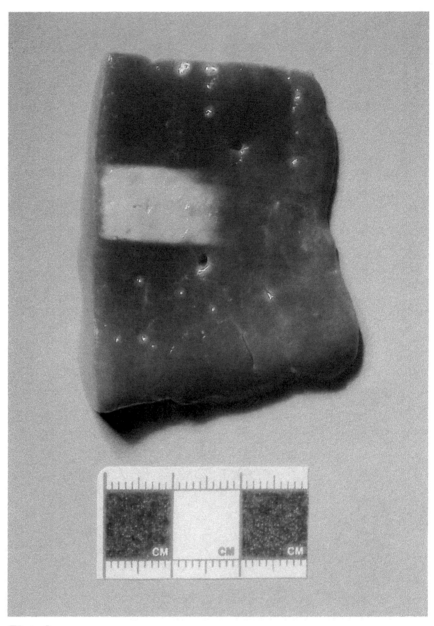

Plate 2.
Photograph showing a volume of tissue 'painted' out using a moving transducer.

rate of temperature rise of 49°C s^{-1}. It is known that the temperature required to produce a lesion is about 56°C. Thus, exposure times are typically 1–2 s. It has been shown [2, 3] that if exposure times of less than 3 s are used, the temperature rise achieved in tissue is essentially independent of perfusion, and the ellipsoidal shape of the ablative lesion follows closely that of the 6 dB contour of the ultrasound field.

9.1.2. Cavitation

The high spatial peak intensities used in focused ultrasound surgery lie close to the cavitation threshold. Both inertial and non-inertial cavitation may occur (see Chapter 11). While cavitation bubbles aid visualisation of a lesioned volume, there is evidence that the occurrence of cavitation renders the lesion shape and position unpredictable, although some authors suggest that it is an essential requirement for successful lesioning [4].

9.2. LESION SHAPE AND POSITION

In order to capitalise on the sharp definition of tissue damage that is produced in a focal lesion, it is important that both the position and shape of a lesion can be accurately predicted.

It is expected from a linear consideration of lesion formation that the damage should form more or less symmetrically across the focal plane. Figure 9.2 shows a contour plot for a focused therapy beam. Under linear conditions, the shape and size of the lesion follows the 6 dB contour closely. It has been demonstrated [5] that as the intensity is increased, the lesion migrates forwards towards the transducer. In addition, the lesions may become asymmetric, with a 'bulbous' end facing the transducer. A number of factors may be instrumental in producing these effects. Firstly, cavitation bubbles formed at the 'front' of the lesion can act as scatterers, enhancing locally absorbed energy, and screening the rear of the focal region. Secondly, the increase of attenuation coefficient with temperature will lead to more absorption at the front of the lesion, and a reduction in the amount of energy reaching the rear. The third factor that may play a rôle is nonlinear propagation (see Chapter 2). The high harmonics will be absorbed more rapidly (ahead of the focus) than the fundamental, thus also enhancing the energy deposition at the lesion front.

9.3. SOURCES OF ULTRASOUND

The optimum source for focused ultrasound surgery exhibits a high peak in the focal region, while having low prefocal peaks on axis, and no side

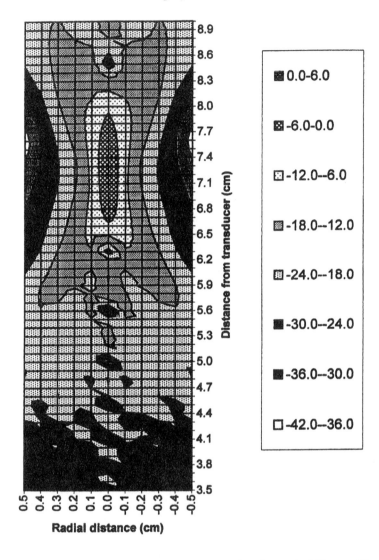

Figure 9.2. *Beam plot showing the pressure distribution in 6 dB steps. The lesion closely follows the 6 dB contour.*

lobes. The importance of the prefocal intensity peaks derives from the high acoustic absorption of skin. At 1.7 MHz, the absorption coefficient of skin has been measured as up to six times that of soft tissues such as liver. Thus, there is the possibility of producing a skin burn if a high prefocal intensity peak occurs at the skin. This problem may be compounded by the fact that when the ultrasonic beam is scanned to 'paint' out clinically relevant tissue volumes, there is overlapping of the ultrasonic beams at the skin surface.

The majority of transducers used for focused ultrasound surgery are formed from air-backed, high power piezoelectric ceramic such as PZT4. In the early days of focused ultrasound surgery quartz was used [6], but piezoelectric materials have now replaced it. Piezoelectric composite materials are now being developed. These have the advantage that they can readily be shaped, and can be driven as phased arrays to give electronic beam steering (see Chapter 5).

Two types of focused ultrasound surgery device are currently available commercially: trans-rectal devices designed primarily for urological treatments (especially of the prostate), and extracorporeal systems designed primarily for treatment of pathology in the abdomen and breast. The transrectal devices incorporate an imaging and a therapy transducer into one head. One applicator uses a 4 MHz crystal alternately for imaging and ablation [7], and has a focal length of 25 mm. When a focal intensity of 1680 W cm^{-2} is used, the lesion is 7 mm in length and 0.6 mm in diameter. A second device uses a retractable 7.5 MHz imaging transducer, and a 2.25 MHz therapy transducer (focal length 35 mm, focal intensities 700–1000 W cm^{-2}) [8]. This gives lesions that are 8 mm long, and 1.5 mm in diameter.

There are currently three extracorporeal devices being used for clinical treatments. The modified lithotripter of Vallancien *et al* [9] uses multiple, confocal 1 MHz transducer elements. It has a focal length of 320 mm and gives lesions that are 10 mm long and 2 mm in diameter. Intensities in excess of 10 000 W cm^{-2} are quoted. Imaging is achieved with a central 3.5 MHz transducer. The patient lies on a waterproof membrane that is incorporated into a bed. The ultrasound source is positioned below the patient. A similar device has been incorporated into a magnetic resonance scanner [10]. In this case, a single 10 cm diameter 1.5 MHz transducer with radius of curvature 8 cm is used as the source [11].

A third extracorporeal device is that used at the Royal Marsden Hospital. This uses a 1.7 MHz, 10 cm diameter spherical bowl transducer, with a focal length of 14 cm [12]. It is mounted above the patient, and is capable of being positioned with five degrees of freedom. Acoustic contact with the skin is achieved through a small bag of degassed water. An *in situ* focal peak intensity of 1490 W cm^{-2} gives an ellipsoidal lesion that is approximately 1.5 cm along the beam axis, and 1.5 mm in diameter in a 1 s exposure.

Clearly, for many applications, the tissue volume ablated by a single focused ultrasound lesion is not clinically useful. Larger volumes are 'painted out' by mechanical translation of the transducer, or by electronic steering of the ultrasonic beam, or by a combination of these methods. Since the acoustic properties of tissue are changed upon 'lesioning', ablation is started at the point in the target volume that lies furthest from the ultrasound source. Plate 2 shows tissue ablation achieved by moving the transducer while excited.

9.4. IMAGING OF FOCUSED ULTRASOUND SURGERY TREATMENTS

9.4.1. *Ultrasound techniques*

The initial hope of focused ultrasound surgery (FUS) was that diagnostic ultrasound could be used to image the target volume, focused ultrasound used to treat it, and then ultrasound imaging used again to monitor the extent of damage. It now seems apparent that conventional B-mode imaging only displays a lesion if it contains gas. Other methods such as those based on changes in tissue attenuation or elasticity are more likely to yield useful visualisation of damaged tissue. Ultrasound temperature monitoring techniques may prove useful for placing the focal peak [13].

9.4.2. *Magnetic resonance imaging*

Magnetic resonance (MR) imaging is being used in combination with MR thermometry to determine the treatment volume, and to target the focused ultrasound beam. A low power pulse may be used to site the focus under MR guidance, and then high power treatments can be carried out [14]. Damaged tissue, which has a lower water content than normal tissue, can be visualised on a T1-weighted MR scan [11].

9.5. CLINICAL APPLICATIONS

While focused ultrasound surgery was first proposed by Lynn *et al* in 1942 [15] as a tool for neurosurgery, it has to date only found significant application in ophthalmology and urology. Current developments are expanding possible uses into oncology and vascular occlusion.

9.5.1. *Neurology*

The first studies of the use of focused ultrasound in the brain were those published by Lynn *et al* [15] and Lynn and Putnam [16]. They attempted to put lesions in the brain through the skull. Acoustic absorption and mode conversion in the bone resulted in profound damage to the scalp with only small lesions being seen in the brain, mostly at the beam entry site.

Several studies have been concerned with placing lesions in cat brain. Fry *et al* [17] showed that the white matter was more susceptible to damage than grey matter which has a lower absorption coefficient.

Subsequently Fry treated 50 patients with Parkinson's disease with focused ultrasound surgery [18, 19]. Following craniotomy, the beam was directed, under local anaesthetic, through the exposed dura to the *substantia negra* and *ansa lenticilaris*. The treatment took up to 14 hours. It was claimed that symptoms of Parkinsonism were eliminated. Despite these good results, the treatment was not taken up more widely, probably because of the introduction of L-dopa therapy at this time.

Ballantine *et al* [20] used focused ultrasound surgery for the treatment of painful neuromata in ten sites in seven patients. Despite reported success in seven out of the ten sites, no further reports of this treatment have been published.

9.5.2. Ophthalmology

Lavine *et al* [21] reported the formation of cataracts when ultrasound is focused in the eye. He therefore claimed this to be a method for selective destruction of ophthalmological targets. Purnell *et al* [22] were able to demonstrate choroidal bleaching, chorioretinal adhesions and localised destruction of the ciliary body. Rosenberg and Purnell [23] showed that high intensity ultrasound could be used to decrease intra-ocular pressure.

Focused ultrasound surgery has been used successfully for the treatment of glaucoma. The technique was first demonstrated in rabbits [24] and in pigs [25] and was then carried into humans [26]. In a trial of 1117 treatments (880 patients) 79.3% had a sustained lowering of intra-ocular pressure after one year. This technique has also been used for sealing traumatic capsular tears [27], for treating intra-ocular tumours [28, 29], for retinal detachment [30] and for the treatment of retinal haemorrhage [31]. The technique has not been widely taken up in ophthalmology, probably because of the introduction of laser surgery.

9.5.3. Urology

Trans-rectal ultrasound is usually the imaging modality of choice for urologists who wish to investigate the prostate. It is therefore an obvious step to incorporate high intensity focused ultrasound therapy into such an examination. In the urological field, therefore, FUS has been most widely used for the treatment of benign prostate hyperplasia (BPH). The most popular current treatment, transurethral resection (TURP), requires general anaesthesia, and carries with it significant morbidity and mortality rates. There is therefore considerable interest in thermal therapies of the prostate, of which FUS is an obvious favoured contender. Human trials have shown that transrectal FUS can lead to necrosis of prostatic tissue in the target volume [8, 32, 33], and an improvement in symptoms and urine flow rates have been demonstrated a year after treatment [34, 35]. Extensive long term follow-up studies are required before this will become the treatment of choice, but early results are encouraging.

It has been shown that extracorporeal FUS can be used to produce selective damage in bladder and kidney, in large animal models, and in man [36–40].

Vallancien *et al* [39] showed that 33% of patients with superficial primary bladder tumours had no recurrence of their disease.

9.5.4. Oncology

Ultrasound was first used to treat tumours in 1933 when Szent-Gyorgi [41] used plane wave ultrasound to treat neoplastic tissue. FUS has a significant advantage over conventional hyperthermia treatments (see Chapter 8) in that the energy is deposited sufficiently quickly that the final temperature achieved is independent of blood perfusion in the target volume. In addition, the temperatures used (in excess of 60°C) are cytotoxic, and so no adjuvant therapy is required.

Good local control has been achieved in 50% of prostate cancer patients in one study [42] and in 30% in another [43]. A Phase 1 trial of soft tissue tumours of the abdomen has shown that FUS treatment of prostate, kidney and liver can be tolerated on an outpatient basis without the need for anaesthesia or other form of sedation [44].

9.5.5. Other applications

An intriguing possible application lies in its ability to occlude blood flow when lesions are placed across blood vessels. This has been demonstrated in tumours [45] and in normal tissue [46–48]. Potential applications of this facility lie in fetal medicine for the occlusion of shunt vessels involved in feto–fetal transfusion syndrome [48] and in oncology in the treatment of renal tumours. The possibility of using FUS for the treatment of cardiac arrhythmias has been discussed [49].

9.6. CONCLUSION

Focused ultrasound surgery provides an alternative non-invasive method of the selective destruction of tissue at depth. Regions of ablated tissue can be imaged using ultrasound, magnetic resonance and X-ray CT imaging. This technique is likely to find application where precise tissue destruction is required, and where tolerance to other therapies (such as radio- or chemotherapy) has already been reached. There is no evidence for a build-up of toxicity with this technique, and thus repeat exposures in adjacent tissue volumes are a possibility. However, FUS has no role to play in organs that are overlain by gas or bone.

REFERENCES

[1]   ter Haar G R 1995 Ultrasound focal beam surgery *Ultrasound Med. Biol.* **21** 1089–100

[2] Billard B E, Hynynen K and Roemer R B 1990 Effects of physical parameters on high temperature ultrasound hyperthermia *Ultrasound Med. Biol.* **16** 409–20

[3] Chen L, ter Haar G R, Hill C R, Dworkin M, Carnochan P, Young H and Bensted J P M 1993 Effect of blood perfusion on the ablation of liver parenchyma with high-intensity focused ultrasound *Phys. Med. Biol.* **38** 1661–73

[4] Chapelon J Y, Prat F, Sibille A, Abou El Fadil F, Henry L, Theilliere Y and Cathignol D 1992 Extracorporeal, selective focused destruction of hepatic tumours by high intensity ultrasound in rabbits bearing VX-2 carcinoma *Min. Inv. Ther.* **1** 287–93

[5] Watkin N A, Rivens I and ter Haar G R 1996 The intensity dependence of the site of maximal energy deposition in focused ultrasound surgery *Ultrasound Med. Biol.* **22**(4) 483–91

[6] Fry W J 1958 Precision high intensity focused ultrasonic machines for surgery *Am. J. Phys. Med.* **37** 152–6

[7] Foster R S, Bihrle R, Sanghvi N T, Fry F J and Donohue J P 1993 High-intensity focused ultrasound in the treatment of prostatic disease *Eur. Urol.* **23** Suppl. 1 29–33

[8] Gelet A, Chapelon J Y, Margonari J, Theilliere Y, Gorry F, Souchon R and Bouvier R 1993 High-intensity focused ultrasound experimentation on human benign prostatic hypertrophy *Eur. Urol.* **23** Suppl. 1 44–7

[9] Vallancien G, Chartier-Kastler E, Bataille N, Chopin D, Harouni M and Bougaran J 1993 Focused extracorporeal pyrotherapy *Eur. Urol.* **23** Suppl. 1 48–52

[10] Cline H E, Schenck J F, Hynynen K, Watkins R D, Souza S P and Jolesz F A 1992 MR guided focused ultrasound surgery *J. Comput. Assist. Tomogr.* **16** 956–65

[11] Hynynen K, Freund W R, Cline H E, Chung A H, Watkins R D, Vetro J P and Jolesz F A 1996 A clinical, noninvasive, MR-imaging monitored ultrasound surgery method *Radiographics* **16** 185–95

[12] ter Haar G R, Rivens I, Chen L and Riddler S 1991 High intensity focused ultrasound in the treatment of rat tumours *Phys. Med. Biol.* **36** 1495–501

[13] Seip R, VanBaren P, Cain C and Ebbini E 1996 Non-invasive real-time multipoint temperature control for ultrasound phased array treatments *IEEE Trans. Ultrasonics Ferroelectr. Freq. Control* **43** 1063–73

[14] Hynynen K, Damianou C A, Colucci V, Unger E, Cline H E and Jolesz F A 1995 MR monitoring of focused ultrasonic surgery of renal cortex—experimental and simulations studies *J. Magn. Res. Imag.* **5** 259–66

[15] Lynn J G, Zwemer R L, Chick A J and Miller A E 1942 A new method for the generation and use of focused ultrasound in experimental biology *J. Gen. Physiol.* **26** 179–92

[16] Lynn J G and Putnam T J 1944 Histological and cerebral lesions produced by focused ultrasound *Am. J. Pathol.* **20** 637–49

[17] Fry W J, Barnard J W, Fry F J, Krumins R F and Brennan J F 1955 Ultrasonic lesions in the mammalian central nervous system *Science* **122** 517–8

[18] Fry W J and Fry F J 1960 Fundamental neurological research and human neurosurgery using intense ultrasound *IRE Trans. Med. Electron.* **ME-7** 166–81

[19] Meyers R, Fry F J, Fry W J, Eggleton R C and Schultz D F 1960 Determination of topological human brain representations and modifications of signs and symptoms of some neurological disorders by the use of high level ultrasound *Neurology* **10** 271–7

[20] Ballantine H T, Bell E and Manlapaz J 1960 Progress and problems in the neurological applications of focused ultrasound *J. Neurosurg.* **17** 858–76

[21]   Lavine O, Langenstrass K, Bowyer C, Fox F, Griffing V and Thaler W 1952 Effects of ultrasonic waves on the refractive media of the eye *Arch. Ophthalmol.* **47** 204–9

[22]   Purnell E, Sokollu A, Torchia R and Taner N 1964 Focal chorioretinitis produced by ultrasound *Invest. Ophthal.* **3** 657–64

[23]   Rosenberg R S and Purnell E 1967 Effects of ultrasonic radiation on the ciliary body *Am. J. Ophthalmol.* **63** 403–9

[24]   Coleman D J, Lizzi F L, Driller J, Rosado A L, Burgess S E P, Torpey J H, Smith M E, Silverman R H, Yablonski M E, Chang S and Rondeau M J 1985 Therapeutic ultrasound in the treatment of glaucoma—II clinical applications *Ophthalmology* **92** 347–53

[25]   Coleman D J, Lizzi F L, Driller J, Rosado A L, Chang S, Iwamoto T and Rosenthal D 1985 Therapeutic ultrasound in the treatment of glaucoma—I experimental model *Ophthalmology* **92** 339–46

[26]   Silverman R H, Vogelsang B, Rondeau M J and Coleman D J 1991 Therapeutic ultrasound for the treatment of glaucoma *Am. J. Ophthalmol.* **111** 327–37

[27]   Coleman D J, Lizzi F L, Torpey J H, Burgess S E P, Driller J, Rosado A L and Nguyen H T 1985 Treatment of experimental lens capsular tears with intense focused ultrasound *Br. J. Ophthalmol.* **69** 645–9

[28]   Burgess S E P, Silverman R H, Coleman D J, Yablonski M E, Lizzi F L, Driller J, Rosado A L and Dennis P H 1986 Treatment of glaucoma with high intensity focused ultrasound *Ophthalmol.* **93** 831–8

[29]   Lizzi F L, Coleman D J, Driller J, Ostromogilsky M, Chang S and Grenall P 1984 Ultrasonic hyperthermia for ophthalmic therapy *IEEE Trans. Sonics Ultrasonics* **SU-31** 473–81

[30]   Rosecan L R, Iwamoto T, Rosado A L, Lizzi F L and Coleman D J 1985 Therapeutic ultrasound in the treatment of retinal detachment: clinical observation and light and electron microscopy *Retina* **5** 115–22

[31]   Coleman D J, Lizzi F L, El-Mofty A A M, Driller J and Franzen L A 1980 Ultrasonically accelerated absorption of vitreous membranes *Am. J. Ophthalmol.* **89** 490–9

[32]   Susani M, Madersbacher S, Kratzik C, Vingers L and Marberger M 1993 Morphology of tissue destruction induced by focused ultrasound *Eur. Urol.* **23** Suppl. 1 34–8

[33]   Madersbacher S, Kratzik C, Szabo N, Susani M, Vingers L and Marberger M 1993 Tissue ablation in benign prostatic hyperplasia with high intensity focused ultrasound *Eur. Urol.* **23** Suppl. 1 39–43

[34]   Mulligan E D, Lynch T H, Mulvin D, Greene D, Smith J M and Fitzpatrick J M 1997 High-intensity focused ultrasound in the treatment of benign prostatic hyperplasia *Br. J. Urol.* **79** 177–80

[35]   Madersbacher S, Kratzik C, Susani M and Marberger M 1994 Tissue ablation in benign prostatic hyperplasia with high intensity focused ultrasound *J. Urol.* **152** 1956–61

[36]   Vaughan M G, ter Haar G R, Hill C R, Clarke R L and Hopewell J W 1994 Minimally invasive cancer surgery using focused ultrasound: a pre-clinical, normal tissue study *Br. J. Radiol.* **67** 267–74

[37]   Watkin N A, Morris S B, Rivens I, Woodhouse C R J and ter Haar G R 1996 A feasibility study of the non-invasive treatement of superficial bladder tumours with focused ultrasound *Br. J. Urol.* **78** 715–21

[38]   Watkin N A, Morris S B, Rivens I H and ter Haar G R 1997 High intensity focused ultrasound ablation of the kidney in a large animal model *J. Endourol.* **11** 191–6

[39] Vallancien G, Harouni M, Guillonneau B, Veillon B and Bougaran J 1996 Ablation of superficial bladder tumours with focused extracorporeal pyrotherapy *Urology* **47** 204–7

[40] Chapelon J Y, Gelet A, Margonari J, Lebars E and Cathignol D 1991 Tissue ablation with focused ultrasound (TAFU) *J. Endourol.* **5** Suppl. 1 S50

[41] Szent-Gyorgi A 1933 Chemical and biological effects of ultrasonic radiation *Nature* **131** 278–80

[42] Gelet A, Chapelon J Y, Bouvier R, Souchon R, Pangaud C, Abdelrahim A F, Cathignol D and Dubernard J M 1996 Treatment of prostate cancer with transrectal focused ultrasound: early clinical experience *Eur. Urol.* **29** 174–83

[43] Madersbacher S, Pedevilla M, Vingers L, Susani M and Marberger M 1995 Effect of high intensity focused ultrasound on human prostate cancer in-vivo *Cancer Res.* **55** 3346–51

[44] ter Haar G R, Rivens I H, Moskovic E, Huddart R and Visioli A 1998 Phase 1 clinical trial of the use of focused ultrasound surgery for the treatment of soft tissue tumours *Surgical Applications of Energy* ed T P Ryan *Proc. SPIE* **3249** 270–6

[45] Rivens I 1992 Quantitative studies of biological damage induced using high intensity focused ultrasound *PhD Thesis* University of London

[46] Delon-Martin C, Vogt C, Chignier E, Guers C, Chapelon J Y and Cathignol D 1995 Venous thrombosis generation by means of high intensity focused ultrasound *Ultrasound Med. Biol.* **21** 113–9

[47] Hynynen K, Chung A H, Colucci V and Jolesz F A 1996 Potential adverse effects of high-intensity focused ultrasound exposure on blood vessels in vivo *Ultrasound Med. Biol.* **22** 193–201

[48] Rivens I H, Rowland I J, Denbow M, Fisk N, Leach M O and ter Haar G R 1998 Focused ultrasound surgery induced vascular occlusion in fetal medicine *Surgical Applications of Energy* ed T P Ryan *Proc. SPIE* **3249** 270–6

[49] Zimmer J E, Hynynen K, He D S and Marcus F 1995 The feasibilty of using ultrasound for cardiac ablation *IEEE Trans. Biomed. Eng.* **42** 891–7

CHAPTER 10

ACOUSTIC WAVE LITHOTRIPSY

Michael Halliwell

INTRODUCTION

Renal and ureteric stones are very common in many countries and in Western Europe the incidence is estimated at 3% to 4% of the population. To this figure must also be added the frequent problem of gallstones. Conventional removal of renal, most ureteric and gall bladder stones involves a traumatic surgical incision, which is associated with a hospital stay of one to three weeks and a prolonged convalescent period.

Since about 1978 two new methods, percutaneous ultrasonic lithotresis and extracorporeal lithotripsy, have grown in importance and a large number of operations now involve one or the other of these methods. Both reduce or eliminate invasive surgery and greatly reduce hospitalisation and post-operational convalescence. Both procedures employ high intensity acoustic waves to produce disintegration of the stones.

Early ultrasonic equipment for this application was based on continuous wave sources and direct contact applicators. By contrast, stone destruction using extracorporeal methods is usually carried out using the application of a series of pressure pulses. The use of laser techniques for percutaneous lithotripsy has become important as another example of minimally invasive surgery.

10.1.   PERCUTANEOUS CONTINUOUS-WAVE SYSTEMS

The concept of the use of continuous-wave ultrasound for stone disintegration was reported as early as 1953 and practical work was carried out in the early 1970s. However, the slowness of stone erosion has prevented wide adoption of the method, although equipment is still produced. It is not proposed to discuss percutaneous or semi-invasive systems, including laser lithotripsy, although the latter may act through a combination of localised plasma and shock wave action. Only extracorporeally induced pressure pulse methods and instruments will be considered.

10.2.   EXTRACORPOREALLY INDUCED LITHOTRIPSY

Extracorporeal lithotripsy is rapidly growing as a primary method for kidney stone destruction despite the high capital cost of equipment. Pioneered in Germany the process has become popular both with patients and hospitals due to the non-invasive nature of the procedure and the short treatment and recovery time. Several different forms of the equipment are now available from a number of manufacturers.

An important factor in the procedure is the accurate determination of the stone location and the orientation of the transducer to position the focal volume at the stone. This is carried out using X-ray or diagnostic ultrasound scanning in three dimensions. With an ultrasonic scanning system an estimation of treatment progress can be undertaken in real time.

The treatment time is very variable and depends critically on the lithotripter system as well as on the form, location and size of the stone. With kidney stones the disintegrated fragments are passed through the ureter, bladder and urethra by normal excretory processes during the next few days. Gall stones are somewhat more difficult to deal with because of their anatomical location and in some cases their relative softness.

10.2.1.   Types of pressure wave transducer

Four techniques are at present employed for the generation of the required pressure pulse, the operating principles being: spark discharge; piezoelectric excitation; electromagnetic induction; chemical explosion. Future equipment may use magnetostrictive elements, based on the new massive magnetostriction materials. In the three electrical systems, the short-duration high-peak energy is supplied by the electrical discharge of a bank of capacitors into an electromechanical transducer. The chemical system employs small explosive charges detonated within a focusing reflector.

10.2.1.1.   Spark discharge.
An ellipsoidal reflector is used with a spark gap mounted at one of the two foci. During use the other focus should coincide with the location of the stone. The spark gap may be a self-contained unit that is easily demountable for servicing or replacement. In most cases the life of the discharge tips is relatively short and ease of replacement forms an important part of the design. The gap can be in the form of a cartridge or fitted as two separate electrodes. The reflector may be closed

by a flexible diaphragm and filled with degassed water or other acoustically efficient liquid with low transmission loss. Application to the patient may be from above or below and many different versions are available.

10.2.1.2. Piezoelectric. The transducer design can be one of three types. The most common form is composed of multiple piezoelectric ceramic elements mounted in a mosaic pattern on a spherical dish. Each element is synchronised to ensure simultaneous pulse transmission and the large aperture energy source provides accurate focusing of the pressure wave. Unlike a spark gap, each individual element has a relatively low emission energy, and high intensity occurs only at the focus. The low stress lengthens the maintenance-free period and ensures high reliability. Malfunction of a few isolated elements will not materially affect the overall performance and the construction enables these to be replaced. Another form uses a ring or tubular piezoceramic transducer mounted at the focus of a parabolic reflector. Alternatively, plane wave sources can also be used and focused using acoustic lenses manufactured from plastics or metals.

10.2.1.3. Electromagnetic. One type of electromagnetic transducer employs a spirally-wound 'pancake' coil to move a metal diaphragm in a liquid-filled cylinder. The plane wave front is focused with an acoustic lens to provide a convenient focal point within the patient. The front of the cylinder is in contact with the patient and is closed by a flexible diaphragm. Another type uses an ellipsoidal reflector with the electromagnetic transducer mounted at the internal focus.

10.2.1.4. Magnetostrictive. The development of high efficiency magnetostrictive materials has stimulated development of single pulse transducers operating with low voltage drives, focusing being achieved with either parabolic mirrors or plastic lenses. Commercial exploitation has not yet commenced.

10.2.1.5. Chemical. Small explosive charges mounted on an 'ammunition belt' are located one at a time at the internal focus of an ellipsoidal bowl. The discharge produces a rapidly expanding gas bubble which generates the acoustic pressure pulse.

10.2.2. Positioning systems

The accurate external positioning of the pressure pulse transmitter is of great importance if damage to the surrounding tissue is to be avoided. As the position is defined within three-dimensional space, the detection method must be capable of defining all axes. In some equipment a high resolution ultrasonic scanner is used with the probe positioned within the treatment head or mounted on a location arm. The position of the focus relative to the probe is fixed and a continuous image of the stone can be displayed during the operating procedure. Microprocessor control of the movement of the transmitter relative to the patient can be used to ensure accurate setting irrespective of patient movement.

The alternative system is X-ray fluoroscopy. However, in contrast to ultrasound imaging, X-ray use is limited by considerations of radiation dose, and therefore fluoroscopic screening cannot be used continuously to monitor the progress of treatment. Furthermore, when lithotripsy treatment of gallstones is being carried out, there may be poor X-ray contrast between the stone material and the surrounding soft tissue, making it difficult to evaluate the stone destruction.

10.2.3. Field measurement

10.2.3.1. Measurement probes and hydrophones.   A number of measuring devices have been used for determining the characteristics of pulsed lithotripter fields. Conventional hydrophones and pressure transducers are generally inadequate for these measurements due to the transient nature of the pulse, the short rise-time (tens of nanoseconds) and the relatively short time of the occurrence. The high pressures at the focus (up to 100 MPa for the peak positive pressure) enforce requirements on the type of detector which are not normally found in other applications. The broad frequency response required, from 0.5 to 100 MHz also imposes limits on the design of robust elements (Coleman *et al* 1989).

Detectors are required to perform two functions: the amplitude of the shock wave at the focus must be measured; the shape of the pressure envelope must be traced.

Piezoelectric polymer (PVDF) membrane hydrophones are widely employed but have the disadvantages of high cost and short operating life (De Reggie *et al* 1981, Preston *et al* 1983, Harris 1988). Miniature piezoelectric ceramic tubular elements offer an alternative with some advantages of robustness, lower cost and minimal disturbance of the acoustic field (Lewin and Schafer 1991). Some needle hydrophones with a rigid backing to the polymer element do not reproduce the rarefactional portion of shock waves demonstrated with membrane type hydrophones (Granz 1989, Ide and Ohdaira 1981, Platte 1985). Further associated discussion on the performance and use of membrane hydrophones in diagnostic ultrasound fields, when the waveforms have undergone significant distortion from nonlinear propagation effects, may be found in Chapters 2 and 7. Whichever type is used care must be taken to ensure that the impedance of the measuring device does not significantly load the hydrophone, and that the calibration takes account of any small loading which may exist. Often the hydrophone is supplied with a head amplifier to match the impedances appropriately.

Table 10.1. *Possible hydrophones for focus and field measurements.*

Description	Use	Remarks	Literature (examples)
PVDF single sheet spot poled membrane, less than 25 μm thick	Focus hydrophone	Life restricted to few shocks	De Reggie *et al* 1981 Preston *et al* 1983 Harris 1988
PVDF needle type	Field hydrophone	Widely used for lithotripter measurements	Granz 1989 Ide and Ohdaira 1981 Platte 1985
Laser optic fibre	Focus and field type	Easy repair and recalibration following stress failure (not commercially available but well described)	Staudenraus and Eisenmenger 1993 Koch *et al* 1997 Chan *et al* 1989

Table 10.2. *Measurement techniques and probes for quality assurance purposes.*

Probe	Features	Parameter measured	Literature (examples)
Capacitive coupling: PVDF spot poled membrane	Large sensitive area	Pressure waveform	Filipczynski and Etienne 1990
Steel ball electromechanical	Very robust	Energy per pulse	Pye *et al* 1991
Model stones	Mimics clinical application	Destructive force per pulse	Delius *et al* 1994
Pressure sensitive paper	Robust, qualitative field parameters	Spatial pressure distribution, semi-quantitative peak pressure measurement	Oyanagi *et al* 1994

Capacitance hydrophones and optical techniques involving interferometry are available but require relatively complicated and difficult handling procedures (Filipczynski and Etienne 1990). Acousto-optic fibre hydrophones have been developed. Quartz-glass fibres seem capable of reproducing the rarefactional acoustic pressure more faithfully than membrane hydrophones (Staudenraus and Eisenmenger 1993, Koch *et al* 1997, Chan *et al* 1989). They are reported to be more sensitive to the presence of cavitation bubbles and the fibre tip has a limited lifetime. Their repair and recalibration is described as uncomplicated. An electromagnetic probe has been developed which is based on the pressure-wave-stimulated movement of a metal ball

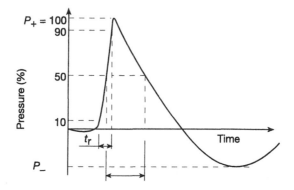

Figure 10.1. *Typical pressure pulse waveform at the focus.*

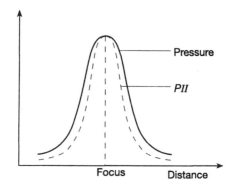

Figure 10.2. *Typical spatial pressure distribution at the focus. The pulse intensity integral (energy density) is shown as a dashed line.*

coupled to a coil held within a magnetic field. This extremely robust device is more useful for the indication of the total energy of the pressure pulse rather than its shape (Pye *et al* 1991). Pressure sensitive paper is also available for use at the pressures found in the focus of a lithotripter although its value for quantitative measurements is not clear (Oyanagi *et al* 1994).

Model stones may be used as a test of overall system efficacy. Stones are available with sufficiently well controlled construction that this method is valuable for routine checking (Delius *et al* 1994).

The limitations of lithotripter detectors prevent continuous day-to-day monitoring of performance in the operating theatre, restricting measurements to maintenance periods. This is a disadvantage and improved equipment is required.

10.2.3.2. Test chamber. The measurement of the pressure wave (figure 10.1) and the mapping of the focal zone (figure 10.2) can be carried out in a chamber filled with degassed water. In one type of test chamber a transducer positioning system is mounted at the top of the chamber with remotely controlled stepping motors used to drive the hydrophone carrier in three separate axes. The pressure wave generator is placed beneath the acoustic diaphragm and pressed firmly against it, after having been coated with a coupling gel to assist in energy transfer. In some lithotripters the pressure wave unit may be inclined and it is necessary to construct the chamber base at a matching angle. For measurements of pressure pulse waveform and peak-compressional acoustic pressure. it is essential that the hydrophone is accurately positioned at the focus.

10.2.3.3. Acoustic pulse energy. The energy per pulse can be calculated by integrating the field over the -6 dB surface around the focus. The energy in a lithotripsy pulse at the focus can be approximated by the expression

$$E_p = \frac{1}{Z} \iint_S p(x, y, z, t)^2 \, dt \, dS \tag{10.1}$$

where $p(x, y, z, t)$ is the acoustic pressure function; S is the focal surface, in the x–y plane containing the focus at depth z; t is time; and Z is the characteristic acoustic impedance of water ($= 1.5 \times 10^6$ kg m^{-2} s^{-1}).

For a beam with circular symmetry

$$E_p = 2\pi \int_0^{X_6} \text{PII}(r_z) r_z \, dr_z \tag{10.2}$$

where $\text{PII}(r)$ is the pulse intensity integral (energy density) at the radius r_z, at a depth z, given by $\sqrt{(x_z^2 + y_z^2)}$, and the integration extends to X_6, the -6 dB radius based on a measurement of the peak-positive acoustic pressure profile. At each radial position,

$$\text{PII}(r) = \frac{1}{Z} \int p(r_z, t)^2 \, dt. \tag{10.3}$$

One numerical solution of equation (10.2) is

$$E = 0.5\pi \sum_{i=1}^{N} (\text{PII}_i + \text{PII}_{i-1})(R_i^2 - R_{i-1}^2). \tag{10.4}$$

Here, it is assumed that a measurement of the beam has been made radially at N points from $r = 0$ to $r = R$ and that the pulse intensity integral at point $r = R_i$ is PII_i. Also $R_0 = 0$ and $R_N = R$.

REFERENCES

Chan H W L, Chiang K S, Price D C, Gardner J L and Brinch J 1989 Use of a fibre-optic hydrophone in measuring acoustic parameters of high power hyperthermia transducers *Phys. Med. Biol.* **34** 1609–22

Coleman A, Saunders J, Preston R C *et al* 1989 A survey of the acoustic output of commercial extracorporeal shock wave lithotripters *Ultrasound Med. Biol.* **15** 213–37

De Reggie A S, Roth S C, Kenney J M and Edelman S 1981 Piezoelectric polymer probe for ultrasonic applications *J. Acoust. Soc. Am.* **69** 853–9

Delius M, Ueberle F and Gambihler S 1994 Destruction of gallstones and model stones by extracorporeal shock waves *Ultrasound Med. Biol.* **20** 251–8

Filipczynski L and Etienne J 1990 Capacitance hydrophones for pressure determination in lithotripsy *Ultrasound Med. Biol.* **16** 157–65

Granz B 1989 PVDF hydrophone for the measurement of shock waves *IEEE Trans. Electr. Instrum.* **24** 499–502

Harris G R 1988 Hydrophone measurements in diagnostic ultrasound fields *IEEE Trans. Ultrasonics, Ferroelectr. Freq. Contr.* **35** 87–106

Ide M and Ohdaira E 1981 Wide frequency range miniature hydrophone for the measurement of pulse ultrasonic field *Japan. J. Appl. Phys.* **20** Suppl. 20-3 205–8

Koch C, Molkenstruck W and Reibold R 1997 Shock wave measurement using a calibrated interferometric fiber-tip sensor *Ultrasound Med. Biol.* **23** 1259–66

Lewin P A and Schafer M E 1991 Shock wave sensors: I. Requirements and design *J. Litho. Stone Disease* **3** 3–17

Oyanagi M, Kudo N, Yanagida Y, Iwama N and Okazaki K 1994 Measurement of shock wave pressure distribution of extracorporeal shock wave lithotripter using pressure sensitive paper: practical usefulness and limitations *Japan. J. Appl. Phys.* **33** 3155–8

Platte M 1985 A polyvinylidene fluoride needle hydrophone for ultrasonic applications *Ultrasonics* **24** 113–8

Preston R C, Bacon D R, Livett A J and Rajendram K 1983 PVDF membrane hydrophone performance properties and their relevance to the measurement of the acoustic output of medical ultrasonic equipment *J. Phys. E: Sci. Instrum.* **16** 786–96

Pye S D, Parr N J, Monro E G, Anderson T and McDicken W N 1991 Robust electromagnetic probe for the monitoring of lithotripter output *Ultrasound Med. Biol.* **17** 931–9

Staudenraus J and Eisenmenger W 1993 Fibre-optic probe for ultrasonic and shock wave measurements in water *Ultrasonics* **31** 267–73

PART 4

ULTRASOUND AND BUBBLES

CHAPTER 11

AN INTRODUCTION TO ACOUSTIC CAVITATION

Timothy G Leighton

11.1. THE ACOUSTIC PROPERTIES OF THE BUBBLE

Perhaps the simplest of the interactions between a gas pocket in a liquid, and an acoustic wave in that liquid, is that of geometrical scattering. This occurs when the wavelength of the acoustic wave is much less than the bubble radius. Sound is scattered because of the strong acoustic impedance mismatch between the gas and the liquid, and the bubble casts an acoustic shadow. In practice the highest frequencies employed in liquids tend to be 30 MHz (giving a wavelength in water of around 50 μm), and the wavelength even at 1 MHz is 1.5 mm. As such, true geometrical scattering would occur only for bubbles which are so large as to have an important implication: bubbles of a smaller size are likely to be more common in the population. While the scattering of sound from bodies which are not much larger than the wavelength differs from that described above (Morse and Ingard 1986), an even more important factor is this: that smaller bubbles tend to behave less like the inert bodies described above, whose walls act as simple reflective boundaries for the sound, and more as oscillators which can be driven by the sound field to pulsate. Such pulsation enhances the coupling between the sound field and the bubbles. This not only dramatically changes the scattering of sound but also causes a range of other effects which broadly characterise acoustic cavitation.

11.1.1. Stiffness and inertia

An oscillator requires two key elements: stiffness and inertia. In the bubble system, the first of these is provided by the gas. If compressed, it exerts a force which resists that compression and would tend to make the bubble expand and vice versa. Potential energy is stored in the gas as the bubble volume changes. When the bubble wall moves, the surrounding liquid must also move. If the system has spherical symmetry, then the velocity of an incompressible liquid falls off as an inverse square law away from the bubble wall (Minnaert 1933). Therefore there is a kinetic energy associated with bubble pulsations which is invested in the moving matter. Since the liquid is so much denser than the gas, this is primarily invested in the liquid (though motion of the gas contributes to a much smaller extent; Leighton *et al* 1995a). Comparison of the potential and kinetic energies (which is a comparison of the gas stiffness with the liquid inertia) allows the formulation of the natural frequency f_r of the oscillator. A simple calculation, based on the linear pulsations of a spherical air bubble of radius R_0 in water which is assumed to be incompressible and inviscid, gives

$$f_r R_0 \approx 3 \text{ Hz m} \qquad (R_0 \gtrsim 10 \ \mu\text{m}) \qquad (11.1)$$

an approximation which neglects surface tension, making this formulation less valid for smaller bubbles.

11.1.2. Resonance

When a bubble pulsates in a sound field, it loses energy through viscous and thermal mechanisms, and through the radiation of acoustic waves into the liquid (Devin 1959). The existence of a natural frequency and damping such as this means that when driven by a sound field, the bubble can exhibit a resonance at a frequency similar to that given in equation (11.1). This resonance possesses properties in common with others. Bubbles just larger than the size which is resonant with the sound field will pulsate in antiphase to those just smaller than resonance size; the amplitude of pulsation tends to be largest close to the resonance condition (Leighton 1994a, §4.1). However the bubble possesses a complex range of other properties associated with the resonance, which will be detailed in sections 11.2 and 11.3. In brief, as the wall pulsation amplitude increases near resonance, the nonlinear character of the oscillation becomes more pronounced, and the resultant bubble activity can be divided into two classes. In the first category are those effects, such as the emission of harmonics of the driving frequency, which increase continuously with increasing pulsation amplitude. Therefore if the bubble is resonant at a driving (or 'pump') acoustic frequency of f_p, then the emission of signals at $2f_p$, $3f_p$, etc may be taken to indicate that the bubbles are close to resonance size (Miller *et al* 1984, Leighton 1994b). This can be used as a method of sizing bubbles. However it has drawbacks in that phenomena other than resonant bubbles can give rise to harmonic emissions, and for such sizing purposes, a two-frequency technique may be superior (Leighton *et al* 1997, Phelps *et al* 1997). A second class of phenomenon occurs as a threshold with the increasing pulsation amplitude, such as the stimulation of surface waves on the bubble wall† (Phelps and Leighton 1996, 1997). While the two examples (harmonic emission and surface wave activity) reflect the dynamics of single bubbles, phenomena which have strong implications for the bubble population as a whole can be found in these two classes (respective examples of radiation forces and rectified diffusion are discussed in section 11.3). The most important of the threshold phenomena, historically studied in populations but with recent important findings through observation of single bubbles, is inertial cavitation, described in the following section.

11.1.3. Inertial cavitation

In 1917 Lord Rayleigh provided a theoretical analysis for the phenomenon we now call inertial cavitation. The impetus for this study had come from the problem of the erosion of ship propellers, and consideration was given to the possibility that gas bubbles, excited by the pressure fluctuations generated close to the propeller, were responsible for the 'pitting' seen on the blades. Inertial collapse of a spherical gas bubble in a liquid is characterised by a relatively slow (i.e. timescales of the order of half an acoustic cycle), approximately isothermal growth of an initial bubble nucleus to many times its original size. This is followed by a rapid collapse, the initial stages of which are dominated by the inertia of the spherically converging liquid. During the collapse, the gas temperature rises as it is compressed, and shocks can propagate within the gas. Both high temperatures (Neppiras 1980) and shocks (Bradley 1968) can potentially generate free radicals, the subsequent radiative recombination of which can emit light. Light emission from collapsing bubbles has been observed, and is called 'sonoluminescence'. The production of this emission is sometimes taken as indicative of (but not necessarily requisite for) the occurrence of inertial cavitation, and it is perhaps in this that the main practical exploitation of the phenomenon of sonoluminescence is currently realised. This is because many, but by no means all, of the physical and biological changes which ultrasonic cavitation can induce occur as a result of inertial cavitation (section 11.4).

However, a second, and some would say more important, reason for considering sonoluminescence is its ability to provide experimental evidence against which to test theories of bubble dynamics. In recent years there has been a significant increase in the number of publications supporting one mechanism or another for the generation of sonoluminescence, and the topic remains controversial (Leighton 1994a, §4.2.1). Throughout the 60 years of investigation to date into sonoluminescence, the proposed mechanisms by

† With associated implications for nucleation and microstreaming, which will be discussed in section 11.3.

which the emission is produced have been debated. A range of mechanisms has been proposed, and although there have been several periods where the broad consensus of opinion favours one, these have not lasted. Many people in the 1980s favoured a thermochemical mechanism (Walton and Reynolds 1984), proposed by Griffing (1950, 1952). In this, oxidising agents such as hydrogen peroxide are formed by the high temperatures (~5000 K; Suslick *et al* 1986) within the compressed gas, and give rise to chemiluminescent reaction. However, there has been support for over 50 years for mechanisms based on electrical discharge (Frenzel and Schultes 1934, Frenkel 1940, Bresler 1940), though details of the mechanism have changed (Margulis 1985, Lepoint *et al* 1994). Jarman (1960) proposed gas shocks as the source of the emission, and variations of this theory too have survived to present (Vaughan and Leeman 1989, Wu and Roberts 1993, Moss *et al* 1994). However other mechanisms proposed in the early days currently have very few supporters (Chambers 1936, Weyl and Marboe 1949, Gunther *et al* 1959). Replacing these in recent years have been publications introducing into the generation of sonoluminescence the phenomena of Casimir energies (Schwinger 1992); rectified diffusion (Crum and Cordry 1994); molecular collisions (Frommhold and Atchley 1994); quantum radiation (Eberlein 1996); and jetting (Prosperetti 1997).

Perhaps the main reason for the increase in recent years in publications on the proposed mechanisms for the emission was the discovery that it was possible to produce a sonoluminescing bubble which behaved quite differently to those envisaged previously by most people. After reaching minimum size, the bubble rebounds, emitting a pressure pulse into a liquid. Until recently, the bubble was then thought to fragment, and as such the phenomenon we now call 'inertial cavitation' was until recently termed 'transient cavitation'. However, Gaitan and Crum (1990) discovered sonoluminescence over measurement intervals of thousands of acoustic cycles from repeated collapses of a single bubble which apparently† did not break up. The discovery that the sonoluminescent flash in such circumstances is of less than 50 ps duration (Barber *et al* 1992) throws into question the mechanism by which light is generated in this 'single-bubble sonoluminescence' (SBSL). However in practical applications involving the cavitation of many bubbles, the breakdown of spatial symmetry might suggest that each bubble performs such inertial collapses once or only a few times and then fragments on rebound. In such 'cavitation-field sonoluminescence' (CFSL) or 'multibubble sonoluminescence' (MBSL) it may be that the same or another of the various mechanisms for generating light from bubble collapse dominates, depending on the details of the collapse. Currently the most widely accepted theory for CFSL is based on

† Barber *et al* (1992) did not observe fragmentation and re-coalescence occurring during each acoustic cycle, though it is possible that this might occur over timescales too rapid to measure.

the recombination of thermally generated free radicals (Kamath *et al* 1993). There are, however, differences between the predictions of this theory and recent measurements of CFSL emissions (Matula *et al* 1997). The authors suggest an upper limit on the pulse width during CFSL of 1.1 ns.

As mentioned at the close of section 11.1.2, the occurrence of inertial cavitation is a threshold phenomenon. Whether it occurs or not, for a free-floating spherical bubble nucleus of known gas content in a given liquid, depends primarily on the acoustic pressure amplitude, the acoustic frequency, and the size of the nucleus. If the nucleus is too small, then surface tension forces prevent the initial sudden growth, and inertial cavitation does not occur. If it is too large, then the bubble may grow, but be too 'sluggish' to concentrate the energy sufficiently on collapse to generate free radicals etc†. There is therefore a critical size range in which, for a given sound field, the initial size of the bubble must fall if it is to nucleate inertial cavitation (Flynn 1964, Flynn and Church 1984, Holland and Apfel 1989, Leighton 1994a). The lower the frequency, the wider this range.

This is clearly shown in figure 11.1, where the transition threshold between inertial and non-inertial cavitation is plotted, based upon calculations by Apfel and Holland (1991). They assumed that, in response to a single cycle of ultrasound, a bubble which is spherical at all times, would grow; and upon subsequent adiabatic collapse, the gas within the bubble should attain a temperature of at least 5000 K if the collapse is to be 'inertial' (Apfel 1981a,b, Apfel and Holland 1991). Though there are clear approximations and the choice of such a criterion for defining inertial cavitation is not fundamental, this is nevertheless an extremely useful calculation. It illustrates that the acoustic pressure amplitude required to cause a bubble to undergo inertial cavitation is dependent upon the initial radius of that bubble. Since in most applications the frequency is the easiest to control of the three parameters shown in figure 11.1, followed by the acoustic pressure amplitude *at the bubble*, with the radii of the nuclei present being the least accessible, then the graph can be interpreted in another manner. At a fixed frequency, say 10 MHz, an ultrasonic cycle with a peak negative pressure of 1.5 MPa (assumed to be constant throughout the field) will only generate inertial cavitation within a water sample if, according to this model, it contains bubbles having radii between 0.03 and 0.77 μm. As the pressure amplitude decreases, so does the range of bubble sizes which can nucleate inertial

---

† A simple catapult analogy is informative. Bubbles undergoing inertial cavitation tend to grow, prior to collapse, to a roughly similar size (Leighton 1994a, §4.2.3, §4.3.1b(iii)). This is like trying to achieve the same length of draw (i.e. about one arm's length) on a number of catapults. If the elastic on the catapult is too short, then it is not possible to draw a full arm's length (if the full draw were only a few centimetres, the throw of the catapult would be very short). This is similar to the case of the very small bubble, where surface tension prevents bubble growth. If the elastic in the catapult were, at equilibrium, very long, it would be easy to draw a full arm's length, but the elastic would not be taut, and again the throw would be short. This is akin to the case of the large bubble.

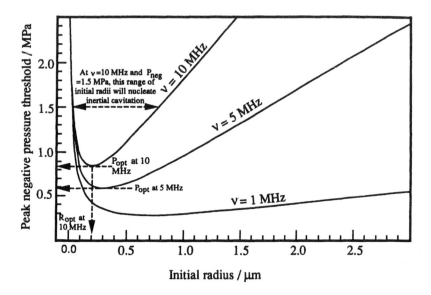

Figure 11.1. *The threshold for inertial cavitation, as predicted by the theory of Apfel and Holland. For each frequency a line can be plotted: if the conditions of peak negative pressure and of the initial bubble radius are such that the point of interest on the graph lies below the line, non-inertial cavitation will occur. If the point of interest is above the line, inertial cavitation will occur. Reprinted by permission of Elsevier Science from 'Gauging the likelihood of cavitation from short-pulse, low-duty cycle diagnostic ultrasound' by R E Apfel and C K Holland, Ultrasound in Medicine and Biology, vol 18, pp 267–81, copyright 1991 by World Federation of Ultrasound in Medicine and Biology.*

cavitation. The very lowest peak negative pressure which could give rise to inertial cavitation, according to this model, is p_{opt} (around 0.84 MPa at 10 MHz), and at this pressure only bubbles of a radius R_{opt} (0.2 μm at 10 MHz) could possibly nucleate inertial cavitation.

The theory of Apfel and Holland has given rise to a measure called the 'mechanical index', which may be used to indicate the likelihood of exceeding the threshold required to nucleate cavitation. As described for figure 11.1, the model demonstrates that for a given acoustic frequency there is a minimum in the curve. If the whole range of bubble size classes are present, then there is a specific initial bubble size for which the threshold acoustic pressure required to nucleate inertial cavitation is a minimum. As the insonation frequency increases, the bubble radius which requires minimum pressure to nucleate inertial cavitation decreases. This is because inertial and viscous forces increase with increasing frequency, and there is insufficient time to bring about the required amount of bubble growth. For the same reason the acoustic pressure required to nucleate inertial cavitation in all but the smallest bubbles increases with increasing frequency. Surface tension dominates the response of the smallest bubbles.

If one is interested in a worst case assessment of the likelihood that inertial cavitation will occur when a liquid is insonated, clearly one must assume that the bubble population contains bubbles at the radius corresponding to the minimum in the threshold curve. At a given frequency, it is bubbles of this radius, R_{opt}, which the analysis predicts will require the smallest peak negative pressure, p_{opt}, to undergo prompt inertial cavitation in response to a single acoustic cycle. Apfel and Holland (1991) generated a plot of p_{opt} against frequency for water and whole blood, using pure fluid bulk property values for the surface tension, density and viscosity relevant to the two fluids. The liquids are assumed to contain the relevant nuclei at size R_{opt}. Apfel and Holland employed a two parameter least-squares fit to these plots in order to obtain a relationship between p_{opt} and insonation frequency $f = \omega/2\pi$. They found a least-squares fit of:

$$\frac{p_{opt}^a}{f} = b \tag{11.2}$$

where if p_{opt} is measured in MPa and f in MHz, the constant a takes values of 2.10 for water and 1.67 for blood, and b has values 0.06 for water and 0.13 for blood. For a given sound field with a maximum negative pressure of p_r, then by taking a value of $a \approx 2$ to approximate the appropriate physiologically relevant liquid, a mechanical index MI can be defined for the sound in that liquid:

$$\text{MI} = \frac{p_r[\text{MPa}]}{\sqrt{f[\text{MHz}]}}. \tag{11.3}$$

The mechanical index MI for prompt cavitation represents an approximate measure of the worst case likelihood of nucleating inertial cavitation. As such it can be used to estimate the potential for nucleating inertial cavitation resulting from insonation by diagnostic ultrasound. Clearly it would be less appropriate to apply this index to tone-burst or continuous-wave insonations, during which a range of complicated processes, including rectified diffusion and enhancement of the cavitation by ultrasonic pulsing, can occur (section 11.3; Leighton 1994a, §4.4.3, §5.3). Holland and Apfel recommend that the pulse length should not exceed 10 cycles, nor the duty cycle 1:100. The peak negative pressure output from the device, as measured in water, must be derated to give the appropriate peak negative pressure that would be attained *in vivo* at the location of the maximum pulse intensity integral (AIUM/NEMA 1998). The centre frequency is used for f which, for accuracy, is expected to be of the order of MHz. Apfel and Holland (1991) suggest that a mechanical index value below $\sqrt{0.5} \approx 0.7$ would indicate that, even in the presence of a broad size distribution of nuclei, the conditions are not sufficient to allow significant bubble expansion. If MI $\geqslant \sqrt{0.5}$, Apfel and Holland suggest that 'the user should be advised of the potential for bubble activity'. The AIUM, NEMA and FDA have adopted the mechanical index as a real-time output display to estimate the potential for cavitation *in vivo* during diagnostic ultrasound scanning (AIUM/NEMA 1998). The use of MI related to practical exposure measurement procedures is discussed in Chapter 7.

Key points should be noted. First, the model for the index is based on the assumption of a free-floating spherical nucleus of optimum size. In certain circumstances it may be that the nucleus is of a different type (section 11.3.3). Second, the mechanical index gauges the likelihood of prompt cavitation, and nothing more: the effect of interest (e.g. a bio-effect) may be related to some other mechanism (section 11.4). Third, the effects of nonlinear propagation (Leighton 1994a, §1.2.3) are not included, and hence the index might underestimate conditions *in situ*. Fourth, when applied to diagnostic ultrasound instruments, the mechanical index describes conditions only at the focus, which is not necessarily the point of interest. Fifth, the mechanical index has arisen from a theory which gives smooth curves of the form shown in figure 11.1; such curves may in fact show peaks when other effects are incorporated (Roy 1996, Allen *et al* 1997).

Lastly, the underlying theory is applied to derive the mechanical index in a way intended to elucidate the conditions which attain the threshold for nucleating inertial cavitation. The amount by which the mechanical index is exceeded is therefore only a guide to the degree of cavitational activity, and by no means an exact predictor. Consider the MHz range illustrated in figure 11.1, where the range of nuclei size which can seed inertial cavitation is relatively narrow. In such a sound field of fixed frequency and increasing acoustic pressure, as one exceeds the threshold in figure 11.1, the range of nuclei which may nucleate inertial cavitation increases. In a field containing a broad range of bubble sizes, with a uniform number of bubbles in each size class, the total number of nucleated events would be expected to increase. However if there is only a narrow distribution of bubble sizes, exceeding the threshold by increasing amounts in this manner would, to first order, have little effect on the *number* of inertial events which are nucleated. What might be expected to increase is the energy associated with each collapse. Similarly, as the acoustic frequency changes, it is no simple matter to predict how the mechanical index might correlate with the amount of cavitational 'activity' observed (Leighton 1997).

11.2.   TYPES OF CAVITATION

In section 11.1 an isolated spherical bubble was discussed. Such a bubble can cast a geometric acoustic shadow in an ultrasonic field of sufficiently high frequency. This could, for example, monitor the slow dissolution one might expect of such a bubble (figure 11.2(*a*)); or the slow growth which might occur under decompression or heating (figure 11.2(*b*)). Since the bubble undergoes no oscillation in these two cases, it has no acoustic effect beyond that of an impedance mismatch. However, as a mechanical oscillator possessing stiffness and inertia, the bubble has a natural frequency (figure 11.2(*c*)), which corresponds to the note emitted when such a bubble is entrained (for example in a waterfall; Leighton and Walton 1987). Further coupling with an oscillatory pressure field occurs when a bubble is driven into pulsation by an incident sound field. The bubble might be driven such that the amplitude of oscillation of the bubble wall is low (for example, the incident sound field has a low acoustic pressure amplitude, or the bubble has a size which differs greatly from that required for resonance with the driving frequency). In such a case, energy is lost from the beam through acoustic re-radiation and, as discussed in section 11.2, through conversion to heat by viscous and thermal damping mechanisms associated with the bubble motion (Devin 1959, Eller 1970). Figure 11.2(*c*) shows just such an oscillation. Despite damping, the illustrated bubble pulsations are shown not to decrease in amplitude, suggesting a continuous-wave or tone-burst insonation, rather than a short pulse (which is the case illustrated in figure 11.2(*d*)).

Inertial cavitation, typified by the sudden expansion and then rapid collapse of the bubble, is shown in figure 11.2(*e*). The bubble may fragment, or repeat the growth/collapse cycle a number of times (Apfel 1981a, b); or, in specialised conditions, can pulsate for thousands of cycles, emitting a sonoluminescent flash at each collapse, as discussed in section 11.1.3. The generic oscillation shown in figure 11.2(*f*) is a high amplitude pulsation of a spherical bubble. Depending on the amplitude, such oscillation may be inertial, or non-inertial but of high amplitude. If non-inertial, there are a number of interpretations of how this situation, which differs in wall oscillation amplitude from that shown in figure 11.2(*c*), arises. It might, for example, occur if figures 11.2(*c*) and (*f*) show the same bubble, but in figure 11.2(*f*) the acoustic pressure amplitude of the driving field is greater. Alternatively the bubbles in figures 11.2(*c*) and (*f*) might be in the same sound field, but with the bubbles in figure 11.2(*f*) being closer to resonance. However as the bubble is more closely driven to resonance, other effects begin to occur. Most notable of these are surface waves, which visually cause a 'shimmer' to appear on the surface of the bubble (Neppiras 1980), and acoustically may be detected with a combination-frequency technique (Phelps and Leighton 1997). Such surface waves are illustrated in figure 11.2(*g*). Surface waves will be stimulated if the amplitude of the bubble wall displacement exceeds a certain value, which it will tend to do as the amplitude of the sound field increases (providing no other effect, such as inertial cavitation or fragmentation, occurs); or, more commonly, the closer the bubble is to the resonance condition. In general, the greater the degree by which the amplitude of the driving field exceeds the threshold condition to excite surface waves on resonant bubbles, the broader the range

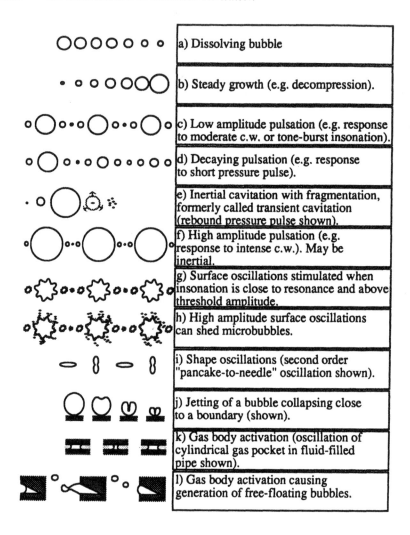

Figure 11.2. *A schematic illustration of the range of bubble behaviour. The behaviours are described in the text. The expansion ratios drawn for these bubbles are exaggerated to illustrate the pulsations more clearly.*

of bubble sizes, centred about the resonance, on which surface waves will be stimulated (Phelps and Leighton 1996). Surface waves can be associated with an erratic 'dancing' translational motion (Crum and Eller 1969) and, at high amplitudes, microbubbles might break off from the tips of the surface waves (figure 11.2(h); Leighton 1995). Surface wave activity is also associated with microstreaming, local circulation currents around a vibrating body which can influence mass transport in the liquid (section 11.3.1) and generate shear forces (section 11.4) close to a bubble.

Other departures from spherical symmetry include shape oscillations (figure 11.2(*i*)), found particularly in larger bubbles, where the tendency of surface tension to promote sphericity is weaker (Strasberg 1956, Longuet-Higgins 1992a, Leighton 1994a, §3.6). If extreme, such shape oscillations can break a bubble up, usually generating a small number of fragments of roughly similar size, in contrast to the shedding of many microbubbles from a larger parent bubble, shown in figure 11.2(*h*) (Longuet-Higgins 1992b, Leighton 1995, Leighton *et al* 1995a, 1998). Shape oscillations are encouraged by anisotropies in the local environment. Common causes for these are the presence of other bubbles, particles or walls. If such are present, then during the subsequent collapse the bubble may involute, one wall passing through the bubble to form a high-speed liquid jet (Lauterborn and Bolle 1975, Plesset and Prosperetti 1977). This is shown in figure 11.2(*j*). Though the jetting of a bubble may be quite complicated, with for example the formation of counterjets, the net effect is that the bubble will usually fragment (Neppiras 1980).

Such behaviour necessitates the adoption of models having a greater degree of sophistication than that of the isolated spherical bubble in an infinite fluid employed at the outset of this chapter. Departures from spherical symmetry in both the bubble wall and the environment must be incorporated. The concept of the 'stabilised gas body' goes further than the interactions described above, assimilating solid structures into the wall bounding the gas pocket. The pocket is partially bounded by liquid, and partially by solid structures, which can stabilise the gas body against dissolution. They might comprise, for example, approximately cylindrical pockets of gas contained within tubular vessels, such as are found in biological structures. Examples include structures in plants (Miller 1979), insect tracheae (Child *et al* 1981), and, speculatively, mammalian blood vessels and ear canals (Leighton *et al* 1995b) (figure 11.2(*k*)). The structures may be rigid, in which case the curved interfaces between the solid and gas would not move. However the gas/liquid interfaces which comprise the end-walls of the cylinder might oscillate in a piston-like or a membrane-like manner (Miller and Nyborg 1983, Miller 1984, Miller and Neppiras 1985). Oscillations of such gas pockets may cause the nucleation of free-floating bubbles into the liquid (figure 11.2(*l*); section 11.5.2).

Figure 11.2 gives rise to three questions. What determines which of the types of behaviour a given bubble will undergo? What are the implications of the occurrence of one type for the occurrence of another? What are the implications of the occurrence of one type of cavitation for causing change to the medium? The first of these three questions has already been discussed, in terms of the resonance and the thresholds for inertial cavitation, surface wave activity, etc. The second and third questions will be answered in sections 11.3 and 11.4 respectively.

11.3.   THE IMPLICATIONS OF THE OCCURRENCE OF ONE TYPE OF CAVITATION FOR THE OCCURRENCE OF ANOTHER

Given a fixed acoustic frequency and in the absence of inhomogeneities in the medium, the type of bubble activity present depends on the resonance condition (which in turn depends on the bubble size), and the amplitude and homogeneity of the acoustic field *at the bubble*. In sections 11.3.1 and 11.3.2, two examples are given to show how these can change, and to allow illustration of their interdependence. In a bubble population, such interdependence and feedback can work through a complicated series of relationships to set up a steady-state cavitating condition (Leighton 1995).

11.3.1.   Alteration of the bubble size by rectified diffusion

By *rectified* diffusion a bubble can, in an appropriate sound field, increase its equilibrium radius through a net flux of previously dissolved gas out of solution and into the bubble. This flux is therefore counter to the direction that would be towards equilibrium were the sound field not present, since a gas bubble in a liquid will tend to dissolve owing to the excess internal gas pressure required to balance the pressure $2\sigma/R_0$ due to surface tension σ (Epstein and Plesset 1950). Thus, in the absence of a sound field or any stabilising mechanism†, bubbles gradually dissolve (Gupta and Kumar 1983). In the presence of a sound field, the situation is quite different. During non-inertial cavitation, since evaporation and condensation take place so much more rapidly than the bubble dynamics, it is commonly assumed that, to first order, the vapour pressure within the bubble remains constant at the equilibrium value. However this is not so for the gas content of the bubble, a gas which will also be dissolved in the liquid. Harvey *et al* (1944), studying the formation of bubbles in animals, suggested a mechanism by which bubbles undergoing stable cavitation in a sound field can experience a steady increase in their equilibrium radius, R_0. This inwardly directed 'rectified diffusion' comes about through the active pumping of gas, initially dissolved in the liquid, into the bubble, using the energy of the sound field. There are two contributory elements to a full description of the processes, an 'area effect' and a 'shell effect' (Leighton 1994a, §4.4.3(a)).

The area effect arises through the correlation between the direction of mass flux and the area of the bubble wall. While the bubble radius is less than equilibrium, the gas inside is at a greater pressure than the equilibrium value, and thus diffuses out into the liquid. Conversely, when the bubble radius is significantly greater than R_0, the internal gas pressure is less than the equilibrium value, and so gas diffuses from the liquid into the bubble interior. The net flow rate of the gas, however, is not equal during the

† See section 11.3.3.

compressed and expanded phases of the bubble motion, because the area of the bubble wall (the transfer interface) is greater in the latter case than in the former. Therefore, over a period of time, there will be a net influx of gas to the bubble interior.

The shell effect occurs because the diffusion rate of a gas in a liquid is proportional to the concentration gradient of the dissolved gas. As the bubble pulsates, a spherical shell of liquid surrounding the bubble will change volume, and so the concentration gradient will change. When the bubble is expanded, each liquid shell contracts. The concentration of dissolved gas in the liquid adjacent to the bubble wall is less than the equilibrium value (Henry's law), but the shell is thinner than when the bubble is at equilibrium radius, so that the gradient across the shell is higher. Therefore the rate of diffusion of gas towards and into the bubble is high. When the bubble is contracted, the liquid shells surrounding the bubble are expanded. Though the concentration of gas near the bubble wall is higher than when the bubble is expanded (Henry's law), the increased thickness of the shell means that the concentration gradient is not as great as when the bubble is expanded. The two factors (gas concentration at the bubble wall, and shell thickness) work together when the bubble is expanded, but against one another when the bubble is contracted. On expansion there is a large concentration gradient driving gas a short distance, and in the second case a smaller gradient drives the gas a longer distance. The former effect is dominant.

The result is that both the bubble wall surface area and the dissolved gas diffusion rate are asymmetrical with respect to expansion and contraction: the 'area' and 'shell' effects reinforce one another. The combined effect means that during nonlinear cavitation in acoustic fields which are sufficiently intense, the equilibrium radius about which the bubble pulsates will tend to increase (Eller and Flynn 1965, Eller 1972, Crum 1984, Church 1988a,b).

The threshold acoustic pressure for growth is dependent on the bubble size and the acoustic frequency. Local minima in the acoustic pressure threshold are seen at harmonics and subharmonics of the resonance (Church 1988a). The growth rate once the threshold has been exceeded is strongly influenced by the presence of surface-active agents, and by microstreaming (introduced in 11.2), which affects the transport of dissolved gas beyond the bubble wall (Church 1988b). Microstreaming was not incorporated into the above simple discussion, but its effects are qualitatively simple. As a bubble grows by rectified diffusion, the dissolved gas is taken from the liquid near the bubble. If there is no flow, then the rate at which the deficit is met depends on the rate at which dissolved gas can diffuse from regions further out from the bubble. Since this is in general a slow process, the liquid outside the bubble wall will become depleted of dissolved gas. The resulting change in concentration gradient reduces the rate of further growth. However, microstreaming flows will tend to bring liquid from further out close to the bubble wall. The convection of dissolved gas reduces the depletion and increases the growth rate. Microstreaming will continually refresh the liquid at the bubble wall, giving it a dissolved gas concentration close to that found far from the bubble. The converse process is of course valid: if a bubble is dissolving, microstreaming will tend to remove from the region outside the bubble wall the excess dissolved gas concentration, so increasing the rate of dissolution (Church 1988b).

Rectified diffusion and dissolution allow bubbles to steadily change their equilibrium size, and therefore modify the cavitational activity they undertake. Among the wide range of possibilities are the following. A bubble may grow to become closer to the resonance condition (where, for example, surface waves and microstreaming require lower acoustic pressure amplitudes to be stimulated). If a bubble fragments following an inertial collapse, and the resulting small bubbles are smaller than the size required to nucleate inertial cavitation, but larger than that required to grow by rectified diffusion, such growth can make inertial cavitation a self-nucleating process. Church (1988a) outlines some of these possible scenarios, and assesses their likelihood.

11.3.2. Alteration of the acoustic pressure field at the bubble by radiation forces

In an accelerating liquid, a gas bubble, being less dense than the surrounding fluid, will accelerate in the same direction as the surrounding liquid, but to a greater degree (Leighton 1994a, §3.3.2(d)). That degree is governed by the magnitude of the difference in density between gas and liquid or, equivalently†, the instantaneous bubble volume. As an acoustic field passes through a liquid, the liquid particle acceleration oscillates, reversing direction periodically such that it is aligned with the direction of propagation for half the acoustic cycle, and contrary to it for the remainder. The bubble, being less dense than the liquid, will follow suit, accelerating in the same direction as the continually alternating liquid particle acceleration‡. However if the bubble is pulsating with the same periodicity as the driving field (which a linear oscillator will do), then the phase relation is such that the bubble will be in the expansion phase (i.e. with its radius greater than equilibrium) as it travels in one direction, and in the compression phase as it travels in the other direction (Leighton *et al* 1990). The bubble accelerates more during the expansion phase than the compression, since then its volume is clearly greater, and its density less. Therefore such a bubble in a sound field will experience forces which reverse direction twice each acoustic cycle. However the net effect will be that it travels in the direction taken by the liquid acceleration when the bubble is in the expansion half-cycle. If the

† Assuming the bubble contains a fixed mass of gas.
‡ Assuming that inertial effects on the response times are small.

bubble is assumed to be an oscillator with a single degree of freedom, bubbles of less than resonance size will pulsate in antiphase to those of greater than resonance size. Therefore if one type of bubble is accelerated in one direction by the sound field, the other type will be forced in the opposite direction.

This behaviour is most readily observed in a standing wave field, where the radiation forces are commonly called 'primary Bjerknes forces'. They cause bubbles of less than resonance size to travel up pressure gradients to collect at acoustic pressure antinodes, while bubbles of larger than resonance size migrate down the gradients to the nodes. Similar comments apply in focused acoustic fields, where bubbles of less than resonance size migrate up pressure gradients towards the focus. Such migrations change the acoustic pressure amplitude at the bubble, and therefore affect the type of cavitation. However there can be important indirect effects, such as the acoustic shielding which such aggregations of bubbles produce, which also affect the local acoustic field (Leighton 1995).

Formulations also exist for the radiation force on a bubble in travelling-wave conditions, where the force is greatest on resonant bubbles; and for the force on a particle suspended in a liquid close to a bubble (Coakley and Nyborg 1978), an example of the latter being the aggregation of platelets in blood (Miller *et al* 1979). The so-called 'secondary' or 'mutual' Bjerknes force is exerted between two pulsating bubbles (Bjerknes 1906, 1909, Prandtl 1954, Batchelor 1967, Leighton 1994a, §4.4.1). There is a general rule that two bubbles which are both less than, or both greater than, resonance size attract; but that if one bubble is greater than, and the other less than, resonance size, they repel. However this is a simplification, and the results may be more complicated if the bubble population density is high or the incident sound field is strong.

Radiation forces therefore can cause mutual repulsion or attraction between bubbles. The latter can cause the net bubble size to increase by causing coalescence; or decrease, because the proximity of one bubble can induce shape oscillations in another (figure 11.2(i)). If sufficiently pronounced, such oscillations can lead to bubble fragmentation (Leighton 1995, Leighton *et al* 1995a, 1998). As discussed with respect to figure 11.2, the equilibrium size of a bubble (along with the acoustic pressure and frequency) strongly influences the type of cavitation it will undergo. Inertial cavitation can be nucleated by bubbles of an appropriate size (section 11.1.3). Details of the nucleation will be discussed in section 11.3.3; and section 11.3.4 will return to the theme of how radiation forces, nucleation and the acoustic field, as well as gas diffusion, determine the cavitation type and effects which will occur when ultrasound is passed through a liquid.

The radiation forces associated with bubbles which have been described in this section can be considered as particular extensions of the more general radiation forces acting at interfaces which have been discussed in Chapter 3.

11.3.3. Nucleation

At the start of section 11.1.3, a free-floating bubble is considered to be an appropriate nucleus for inertial cavitation. The theory outlined there assumes nucleation of inertial cavitation within the first acoustic cycle, so-called 'prompt cavitation', from a free-floating spherical bubble nucleus. Clearly effects relating to longer insonation periods, such as growth by rectified diffusion, are not covered. In practice of course it is not a requirement that such nuclei be present before the start of insonation. Not only may bubbles which are initially too large to nucleate inertial cavitation enter the critical range (through, for example, dissolution, or fragmentation through a shape oscillation or microbubble shedding, as discussed above). Bubbles too small to nucleate cavitation may also enter it through rectified diffusion or coalescence. Radiation forces can affect both size increases and size reductions, by relocating bubbles to regions of greater or lesser acoustic pressure amplitude, or to the presence of neighbours. Radiation forces in focused fields can even convect suitable nuclei into the focus to nucleate inertial cavitation there (Madanshetty *et al* 1991, Madanshetty 1995).

The model employed to produce figure 11.1 is based upon the dynamics of isolated, spherical bubbles, which *de facto* must be free floating. Such bodies are not stable with respect to loss from the liquid as a result of buoyant forces, and indeed will tend to dissolve. Why therefore any suitable nuclei can be found in a sample of liquid which has been left standing for hours is an interesting question (Leighton 1994a, §2.1). Hydrophobic impurities, if present, can collect on a bubble wall over time, and hinder further reduction in size (Akulichev 1966, Sirotyuk 1970, Yount 1979, 1982, Yount *et al* 1984). Other possible nuclei can be found naturally as gas pockets, stabilised against dissolution in crevices and cracks in the container wall or within free-floating particles within the liquid (Harvey *et al* 1944, Trevena 1987, Atchley and Prosperetti 1989, Leighton 1994a). The process by which these bring about nucleation is illustrated in figure 11.2(*l*). High amplitude ultrasonic waves cause the gas pockets to either expand out of their crevice, or conceivably shed microbubbles through surface waves, to generate free-floating nuclei for cavitation. If inertial cavitation is undesirable, then such particles can be removed by filtering. However even this will not completely remove all suitable nuclei for inertial cavitation, which may be generated, for example, by the passage of cosmic rays through the sample (Greenspan and Tschiegg 1967).

11.3.4. Population effects

For a given liquid (including its gas and solid content), it was shown in section 11.1.3 that, to first order, whether a bubble undergoes inertial or non-inertial cavitation depends on three parameters: the acoustic frequency, the acoustic pressure amplitude *at the bubble*, and the equilibrium bubble size. In section 11.2 this idea was extended to show that, of the wide range of types of cavitation that exist, the ones that occur in a given situation will depend on the above parameters plus others (such as the proximity of the bubble to inhomogeneities such as other bubbles, particles or walls). Section 11.3 has outlined how such key parameters as bubble location and size can be altered during insonation (by radiation forces, rectified diffusion, coalescence and fragmentation etc). It is usually simple to control the acoustic frequency to which a sample is subjected. It is, understandably, often much more difficult to control the bubble size and location.

The type of cavitational behaviour a bubble undertakes depends on the relation between its size and the other critical sizes (figure 11.2). These include: the radius which is resonant with the sound field (governing, for example, radiation force effects, surface wave activity and microbubble shedding etc); the upper and lower limits of the radius range for the nucleation of inertial cavitation; and the threshold for rectified diffusion. How such changes affect the bubble size distribution with respect to the critical sizes mentioned above determines the bubble activity seen. The issues involved may be complex, as the following scenarios suggest. Dissolution and fragmentation provide mechanisms by which bubble size reductions in the population can occur. Coalescence and rectified diffusion provide the ways to produce larger bubbles. To produce a bubble which is larger than resonance, it must either pre-exist (and continue to persist during insonation despite buoyancy and possible fragmentation); form through coalescence; or grow to a size larger than resonance by rectified diffusion. Acting against the latter scenario is the fact that it is on reaching resonance size that the bubble is most likely to lose gas through microbubble emission from surface waves (figure 11.2(h)). Also, if the bubble does pass through resonance intact, then rectified diffusion becomes far less efficient once a bubble exceeds resonance size, the pressure threshold increasing and the growth rate decreasing. There is also the general trend that the larger the bubble, the more easily it is fragmented. Therefore in many fields, it is the aggregation of the smaller bubbles at pressure antinodes or the focus which is more commonly observed (figure 11.3). Bubble aggregations such as the one mentioned above are acoustically active. They may shield, channel, or scatter the acoustic field. This leads to the surprising observation that, of the three parameters mentioned at the start of this section, the acoustic pressure amplitude *at the bubble* may also sometimes be difficult to control.

The types of cavitation (and there may be many) which occur when a liquid is insonated therefore depend on a large number of interacting parameters, introduced above and summarised in figure 11.4 (Leighton 1995). It is possible to control such interactions to produce surprising effects, such as 'pulse enhancement'. This occurs when the magnitude of an effect produced when ultrasound is used in pulsed mode exceeds that observed when the same amount of ultrasonic energy is delivered in continuous wave.

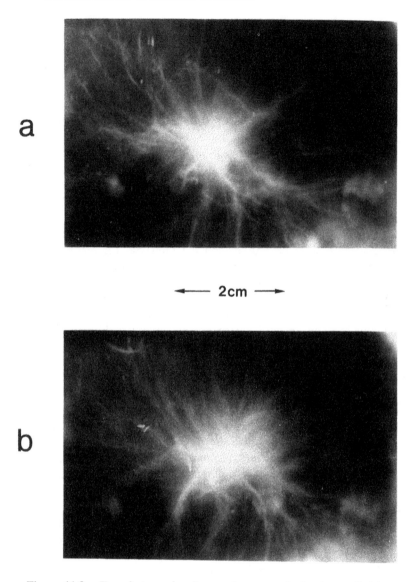

Figure 11.3. *Two photographs of aerated water cavitating in a cylindrically focused 10 kHz continuous-wave sound field, taken approximately 3 s apart. The view is along the line of the axial focus, where the acoustic pressure amplitude is 0.24 MPa. Exposure time, 1/30th second. Streamers are clearly visible, comprising bubbles moving rapidly towards the focus, driven by radiation forces. Comparison of (a) with (b) illustrates that, although the general form is constant, the details change.*

There are a number of mechanisms by which this can be brought about, involving complex interactions of the parameters described above. Detailed

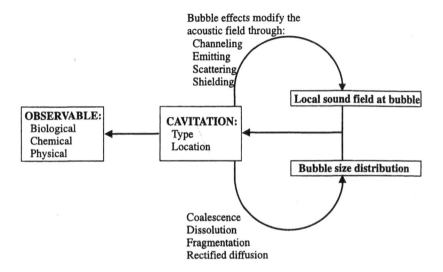

Figure 11.4. *A schematic illustration of how the various factors interact to determine the types of cavitation which occur, and hence the effect observed. The latter may be chemical, physical or biological. It depends both on the type of cavitation (e.g. inertial, non-inertial, jetting, fragmentary etc) and its location. Both these factors depend strongly on the local sound field at the bubble, and on the sizes of bubble present in the population. These two together, for example, characterise the inertial cavitation threshold, and also where bubbles will migrate and accumulate under radiation forces. Such accumulation will in turn affect the local sound field, through the processes of channelling, scattering, and shielding; and will affect the bubble size distribution through its influence on the processes of coalescence, fragmentation, and rectified diffusion. In summary therefore, the observed effect depends on the characteristics of the cavitation, which is determined by the local sound field and the bubble size distribution. However there is feedback from the cavitation which influences these two key parameters.*

discussion of such mechanisms (Leighton 1994a, §5.3) is beyond the scope of this chapter. In the final section, the observables which can be produced by cavitation, the 'end result' in figure 11.4, will be discussed.

11.4. THE IMPLICATIONS OF THE OCCURRENCE OF ONE TYPE OF CAVITATION FOR CAUSING CHANGE TO THE MEDIUM

As shown in figure 11.4, cavitation can produce physical, chemical and biological effects. Given the complexity of interactions shown in that figure and, further, the range of behaviour a single bubble can undergo in an acoustic field (figure 11.2), it is not surprising that the production of a given effect may not be uniquely the result of one form of bubble activity. The exploitation of the chemical and mechanical effects resulting from the acoustically induced cavitation of a cloud of bubbles in many common applications (such as cell killing, ultrasonic cleaning, cavitation erosion and sonochemistry) relies in action upon the thermal, sonochemical, gas and liquid shock, and jetting processes described above. The following discussion illustrates the ability of the types of cavitation to produce an effect on the medium through the example of, primarily, erosion, with special emphasis on damage to biological material. For further details see Leighton (1994a, §5).

That bubble activity may cause erosion has been known since the start of the century (Rayleigh 1917). The most dramatic erosive effects are produced by the impact of a liquid jet when a bubble involutes on collapse (figure 11.2(j)) and by the rebound pressure pulses emitted when a bubble undergoes an inertial collapse (figure 11.2(e)) (Noltingk and Neppiras 1950, Neppiras and Noltingk 1951, Plesset and Prosperetti 1977). While the pressure pulse from an individual bubble is rapidly attenuated with distance such that only surfaces within about one bubble radius from the centre of collapse may be damaged by the rebound pressure emission of a single bubble, clouds of bubbles may collapse co-operatively, enhancing the damage (Vyas and Preece 1976). It is likely that cavitation erosion resulting from jetting and inertial collapses plays a part in lithotripsy (Coleman *et al* 1992, 1993).

If non-inertial pulsations are of sufficient amplitude, bubbles may cause damage by a number of mechanisms. For example, cell disruption may be brought about because the bubbles can travel rapidly through the liquid under the influence of acoustic radiation forces, generating hydrodynamic stresses which have been shown to produce haemolysis (Miller and Williams 1989). If the acoustic field amplitude exceeds a certain threshold, which is lower the closer the bubbles are to resonance (Phelps and Leighton 1996), surface waves are stimulated (figure 11.2(g)). Associated with these are the microstreaming currents discussed in sections 11.2 and 11.3.1, which can cause an erosive effect (for example, with dental ultrasonics; Ahmad *et al* 1987). Non-inertial cavitation may indirectly influence erosion through affecting the damage which results from inertial cavitation. This might occur through shielding of the sound field, or through the production of nuclei, through surface wave activity (figure 11.2(h)) or rectified diffusion (section 11.3.1).

The ability of non-inertial cavitation to directly bring about physical damage is usually of greatest importance in the absence of inertial cavitation (which would otherwise dominate the erosion, masking the contribution from stable bubbles). Bio-effect from non-inertial cavitation is therefore more readily identifiable when low intensity fields are applied to sensitive chemical or biological systems. Demonstrations include microstreaming-induced cell death associated with gas body activation (Miller 1985, Vivino *et al* 1985) and the use of photometry to detect ATP release from *in vitro* human erythrocytes (Williams and Miller 1980), a technique which has probably produced effects at the lowest continuous-wave intensity recorded for a detectable bio-effect, 4 mW cm^{-2} (Williams 1983). The mechanism was probably rupture or a change in permeability of the cell membrane brought about through microstreaming stresses at the cell wall.

11.5. CONCLUSION

There are many types of bubble activity which can occur in an acoustic field. The details of how an isolated bubble would behave will depend on key parameters, including the acoustic frequency; the relationship between the bubble size and the thresholds for inertial cavitation, rectified diffusion, surface wave activity; and the acoustic pressure amplitude *at the bubble*. However in most practical situations the bubble is not isolated, and inhomogeneities in the fluid and sound field may be important. That said it is often the effects of cavitation, rather than the type of bubble activity *per se*, which are important, and these will depend on the interactions between the bubble population, the sound field, and the medium.

REFERENCES

Ahmad M, Pitt Ford T R and Crum L A 1987 Ultrasonic debridement of root canals: acoustic streaming and its possible role *J. Endodont.* **13** 490–9

Akulichev V A 1966 Hydration of ions and the cavitation resistance of water *Sov. Phys.–Acoust.* **12** 144–9

Allen J S, Roy R A and Church C C 1997 On the role of shear viscosity in mediating inertial cavitation from short-pulse, megahertz-frequency ultrasound *IEEE Trans. Ultrasonics, Ferroelectr. Freq. Control* **44** 743–51

AIUM/NEMA 1998 *Standard for Real-time Display of Thermal and Mechanical Acoustic Output Indices on Diagnostic Ultrasound Equipment* 2nd edn (Rockville, MD: American Institute of Ultrasound in Medicine)

Apfel R E 1981a Acoustic cavitation *Methods in Experimental Physics* vol 19 ed P D Edmonds (New York: Academic Press) pp 355–413

—— 1981b Acoustic cavitation prediction *J. Acoust. Soc. Am.* **69** 1624–33

Apfel R E and Holland C K 1991 Gauging the likelihood of cavitation from short-pulse, low-duty cycle diagnostic ultrasound *Ultrasound Med. Biol.* **17** 179–85

Atchley A A and Prosperetti A 1989 The crevice model of bubble nucleation *J. Acoust. Soc. Am.* **86** 1065–84

Barber B P, Hiller R, Arisaka K, Fetterman H and Putterman S 1992 Resolving the picosecond characteristics of synchronous sonoluminescence *J. Acoust. Soc. Am.* **91** 3061–3

Batchelor G K 1967 *An Introduction to Fluid Dynamics* (Cambridge: Cambridge University Press) pp 452–5

Bjerknes V F J 1906 *Fields of Force* (New York: Columbia University Press)

—— 1909 *Die Kraftfelder* (Brunswick) no 28 of the series *Die Wissenschaft*

Bradley J N 1968 *Shock Waves in Chemistry and Physics* (London: Methuen) pp 246–63

Bresler S 1940 *Acta Physiochemica URSS* **12** 323 (in Russian)

Chambers L A 1936 The emission of visible light from pure liquids during acoustic excitation *Phys. Rev.* **49** 881

Child S Z, Carstensen E L, and Lam S K 1981 Effects of ultrasound on *Drosophila*—III. Exposure of larvae to low-temporal-average-intensity, pulsed irradiation *Ultrasound Med. Biol.* **7** 167–73

Church C C 1988a Prediction of rectified diffusion during nonlinear bubble pulsations at biomedical frequencies *J. Acoust. Soc. Am.* **83** 2210–7

—— 1988b A method to account for acoustic microstreaming when predicting bubble growth rates produced by rectified diffusion *J. Acoust. Soc. Am.* **84** 1758–64

Coakley W T and Nyborg W L 1978 Chapter II: Cavitation; dynamics of gas bubbles; applications *Ultrasound: Its Applications in Medicine and Biology* ed F Fry (New York: Elsevier) Part 1, pp 77–159

Coleman A J, Choi M J, Saunders J E and Leighton T G 1992 Acoustic emission and sonoluminescence due to cavitation at the beam focus of an electrohydraulic shock wave lithotripter *Ultrasound Med. Biol.* **18** 267–81

Coleman A J, Whitlock M, Leighton T G and Saunders J E 1993 The spatial distribution of cavitation induced acoustic emission, sonoluminescence and cell lysis in the field of a shock wave lithotripter *Phys. Med. Biol.* **38** 1545–60

Crum L A 1984 Rectified diffusion *Ultrasonics* **22L** 215–23

Crum L A and Cordry S 1994 Single-bubble sonoluminescence *Bubble Dynamics and Interface Phenomena (Proc. IUTAM Symposium, Birmingham, UK, 1993)* ed J R Blake *et al* (Dordrecht: Kluwer Academic) pp 287–97

Crum L A and Eller A I 1969 Motion of bubbles in a stationary sound field *J. Acoust. Soc. Am.* **48** 181–9

Devin C Jr 1959 Survey of thermal, radiation, and viscous damping of pulsating air bubbles in water *J. Acoust. Soc. Am.* **31** 1654

Eberlein C 1996 Sonoluminescence as quantum vacuum radiation *Phys. Rev. Lett.* **76** 3842

Eller A I 1970 Damping constants of pulsating bubbles *J. Acoust. Soc. Am.* **47** 1469–70

—— 1972 Bubble growth by diffusion in an 11-kHz sound field *J. Acoust. Soc. Am.* **52** 1447–9

Eller A I and Flynn H G 1965 Rectified diffusion through nonlinear pulsations of cavitation bubbles *J. Acoust. Soc. Am.* **37** 493–503

Epstein P S and Plesset M S 1950 On the stability of gas bubbles in liquid–gas solutions *J. Chem. Phys.* **18** 1505–9

Flynn H G 1964 Physics of acoustic cavitation in liquids *Physical Acoustics* vol 1, part B ed W P Mason (New York: Academic Press) pp 57–172

Flynn H G and Church C C 1984 A mechanism for the generation of cavitation maxima by pulsed ultrasound *J. Acoust. Soc. Am.* **76** 505–12

Frenkel Ya I 1940 *Acta Phisiochemica URSS* **12** 317–23 (in Russian)

Frenzel J and Schultes H 1934 Luminescenz im ultraschall beschickten wasser. Kurze mitteilung *Z. Phys. Chem.* B **27** 421–4

Frommhold L and Atchley A A 1994 Is sonoluminescence due to collision-induced emission? *Phys. Rev. Lett.* **73** 2883–6

Gaitan D F and Crum L A 1990 Observation of sonoluminescence from a single cavitation bubble in a water/glycerine mixture *Frontiers of Nonlinear Acoustics, 12th ISNA* ed M F Hamilton and D T Blackstock (New York: Elsevier) p 459

Greenspan M and Tschiegg C E 1967 Radiation-induced acoustic cavitation; apparatus and some results *J. Res. Natl Bur. Stand.* C **71** 299–311

Griffing V 1950 Theoretical explanation of the chemical effects of ultrasonics *J. Chem. Phys.* **18** 997–8

—— 1952 The chemical effects of ultrasonics *J. Chem. Phys.* **20** 939–42

Gunther P, Heim E, Eichkorn G 1959 Phasenkorrelation von schallwechseldruck und sonolumineszenz *Z. Angew. Phys.* **11** 274–7 (in German)

Gupta R S and Kumar D 1983 Variable time step methods for the dissolution of a gas bubble in a liquid *Comput. Fluids* **11** 341–9

Harvey E N, Barnes D K, McElroy W D, Whiteley A H, Pease D C and Cooper K W 1944 Bubble formation in animals *J. Cell Comp. Physiol.* **24** 1–22

Holland C K and Apfel R E 1989 An improved theory for the prediction of microcavitation thresholds *IEEE Trans. Ultrasonics, Ferroelectr. Freq. Control* **36** 204–8

Jarman P D 1960 Sonoluminescence: a discussion *J. Acoust. Soc. Am.* **32** 1459–63

Kamath V, Prosperetti A and Egolfopoulos F N 1993 A theoretical study of sonoluminescence *J. Acoust. Soc. Am.* **94** 248–60

Lauterborn W and Bolle H 1975 Experimental investigations of cavitation-bubble collapse in the neighbourhood of a solid boundary *J. Fluid Mech.* **72** 391–9

Leighton T G 1994a *The Acoustic Bubble* (London: Academic Press)

—— 1994b Acoustic bubble detection. I. The detection of stable gas bodies *Environmental Eng.* **7** 9–16

—— 1995 Bubble population phenomena in acoustic cavitation *Ultrasonics Sonochemistry* **2** S123–S136

—— 1997 A strategy for the development and standardisation of measurement methods for high power/cavitating ultrasonic fields: review of cavitation monitoring techniques *ISVR Technical Report No. 263* University of Southampton

Leighton T G, Ramble D G and Phelps A D 1997 The detection of tethered and rising bubbles using multiple acoustic techniques *J. Acoust. Soc. Am.* **101** 2626–35

Leighton T G, Schneider M F and White P R 1995a Study of bubble fragmentation using optical and acoustic techniques *Sea Surface Sound '94 (Proc. 3rd Meeting on Natural Physical Processes Related to Sea Surface Sound)* ed M J Buckingham and J R Potter (Singapore: World Scientific) pp 414–28

Leighton T G and Walton A J 1987 An experimental study of the sound emitted from gas bubbles in a liquid *Eur. J. Phys.* **8** 98–104

Leighton T G, Walton A J and Pickworth M J W 1990 Primary Bjerknes forces *Eur. J. Phys.* **11** 47–50

Leighton T G, White P R and Marsden M A 1995b Applications of one-dimensional bubbles to lithotripsy, and to diver response to low frequency sound *Acta Acustica* **3** 517–29

Leighton T G, White P R and Schneider M F 1998 The detection and dimensions of bubble entrainment and comminution *J. Acoust. Soc. Am.* **103** 1825–35

Lepoint T, Voglet N, Faille L and Mullie F 1994 Bubbles deformation and interface distortion as a source of sonochemical and sonoluminescent activity *Bubble Dynamics and Interface Phenomena (Proc. IUTAM Symposium, Birmingham, UK, 1993)* ed J R Blake *et al* (Dordrecht: Kluwer Academic) pp 321–33

Longuet-Higgins M S 1992a Nonlinear damping of bubble oscillations by resonant interaction *J. Acoust. Soc. Am.* **91** 1414–22

—— 1992b The crushing of air cavities in a liquid *Proc. R. Soc. Lond.* **439** 611–26

Madanshetty S I 1995 A conceptual model for acoustic microcavitation *J. Acoust. Soc. Am.* **98** 2681–9

Madanshetty S I, Roy R A and Apfel R E 1991 Acoustic microcavitation: its active and passive detection *J. Acoust. Soc. Am.* **90** 1515–26

Margulis M A 1985 Sonoluminescence and sonochemical reactions in cavitation fields. A review *Ultrasonics* **23** 157–69

Matula T J, Roy R A and Mourad P D 1997 Optical pulse width measurements of sonoluminescence in cavitation-bubble fields *J. Acoust. Soc. Am.* **101** 1994–2002

Miller D L 1979 A cylindrical bubble model for the response of plant-tissue gas-bodies to ultrasound *J. Acoust. Soc. Am.* **65** 1313–21

—— 1984 Gas body activation *Ultrasonics* **22** 261–9

—— 1985 Microsteaming as a mechanism of cell death in *Elodea* leaves exposed to ultrasound *Ultrasound Med. Biol.* **11** 285–92

Miller D L and Neppiras E A 1985 On the oscillation mode of gas-filled micropores *J. Acoust. Soc. Am.* **77** 946–53

Miller D L and Nyborg W L 1983 Theoretical explanation of the response of gas-filled micropores and cavitation nuclei to ultrasound *J. Acoust. Soc. Am.* **73** 1537–44

Miller D L, Nyborg W L and Whitcomb C C 1979 Platelet aggregation induced by ultrasound under specialized condition in vitro *Science* **205** 505

Miller D L and Williams A R 1989 Bubble cycling as the explanation of the promotion of ultrasonic cavitation in a rotating tube exposure system *Ultrasound Med. Biol.* **15** 641–8

Miller D L, Williams A R and Gross D R 1984 Characterisation of cavitation in a flow-through exposure chamber by means of a resonant bubble detector *Ultrasonics* **22** 224–30

Minnaert M 1933 On musical air-bubbles and sounds of running water *Phil. Mag.* **16** 235–48

Morse P M and Ingard K U 1986 *Theoretical Acoustics* (Princeton, NJ: Princeton University Press)

Moss W C, Clarke D B, White J W and Young D A 1994 Hydrodynamic simulations of bubble collapse and picosecond sonoluminescence *Phys. Fluids* **6** 2979–85

Neppiras E A 1980 Acoustic cavitation *Phys. Rep.* **61** 159–251

Neppiras E A and Noltingk B E 1951 Cavitation produced by ultrasonics: theoretical conditions for the onset of cavitation *Proc. Phys. Soc.* B **64** 1032–8

Noltingk B E and Neppiras E A 1950 Cavitation produced by ultrasonics *Proc. Phys. Soc.* B **63** 674–85

Phelps A D and Leighton T G 1996 High resolution bubble sizing through detection of the subharmonic response with a two frequency excitation technique *J. Acoust. Soc. Am.* **99** 1985–92

—— 1997 The subharmonic oscillations and combination frequency emissions from a resonant bubble *Acta Acustica* **83** 59–66

Phelps A D, Ramble D G and Leighton T G 1997 The use of a combination frequency technique to measure the surf zone bubble population *J. Acoust. Soc. Am.* **101** 1981–9

Plesset M S and Prosperetti A 1977 Bubble dynamics and cavitation *Ann. Rev. Fluid Mech.* **9** 145–85

Prandtl L 1954 *Essentials of Fluid Dynamics* (Glasgow: Blackie) pp 180, 342–5

Prosperetti A 1997 A new mechanism for sonoluminescence *J. Acoust. Soc. Am.* **101** 2003–7

Rayleigh Lord 1917 On the pressure developed in a liquid during the collapse of a spherical cavity *Phil. Mag.* **34** 94–8

Roy R A 1996 The demographics of cavitation produced by medical ultrasound *J. Acoust. Soc. Am.* **99** 2485 (abstract)

Schwinger J 1992 Casimir energy for dielectrics: spherical geometry *Proc. Natl Acad. Sci. USA* **89** 11 118–20

Sirotyuk M G 1970 Stabilisation of gas bubbles in water *Sov. Phys.–Acoust.* **16** 237–40

Strasberg M 1956 Gas bubbles as sources of sound in liquids *J. Acoust. Soc. Am.* **28** 20–6

Suslick K S, Hammerton D A and Cline R E Jr 1986 The sonochemical hot-spot *J. Am. Chem. Soc.* **108** 5641–2

Trevena D H 1987 *Cavitation and Tension in Liquids* (Bristol: Adam Hilger)

Vaughan P W and Leeman S 1989 Acoustic cavitation revisited *Acustica* **69** 109–19

Vivino A A, Boraker D K, Miller D and Nyborg W 1985 Stable cavitation at low ultrasonic intensities induces cell death and inhibits ^{3}H-TdR incorporation by con-a-stimulated murine lymphocytes in vitro *Ultrasound Med. Biol.* **11** 751–9

Vyas B and Preece C M 1976 Stress produced in a solid by cavitation *J. Appl. Phys.* **47** 5133–8

Walton A J and Reynolds G T 1984 Sonoluminescence *Adv. Phys.* **33** 595–660

Weyl W A and Marboe E C 1949 *Research* **2** 19

Williams A R 1983 *Ultrasound: Biological Effects and Potential Hazards* (New York: Academic Press)

Williams A R and Miller D L 1980 Photometric detection of ATP release from human erythrocytes exposed to ultrasonically activated gas-filled pores *Ultrasound Med. Biol.* **6** 251–6

Wu C C and Roberts P H 1993 Shock-wave propagation in a sonoluminescing gas bubble *Phys. Rev. Lett.* **70** 3424–7

Yount D E 1979 Skins of varying permeability: a stabilisation mechanism for gas cavitation nuclei *J. Acoust. Soc. Am.* **65** 1429–39

—— 1982 On the evolution, generation, and regeneration of gas cavitation nuclei *J. Acoust. Soc. Am.* **71** 1473–81

Yount D E, Gillary E W and Hoffman D C 1984 A microscopic investigation of bubble formation nuclei *J. Acoust. Soc. Am.* **76** 1511–21

CHAPTER 12

ECHO-ENHANCING (ULTRASOUND CONTRAST) AGENTS

David O Cosgrove

INTRODUCTION

Unlike all other imaging technologies, ultrasound has lacked agents that can be administered to patients to improve or enhance the diagnostic information available. The natural contrast due to differences in backscatter (see Chapter 4) is insufficient in many circumstances to differentiate one tissue from its neighbour. With the recent introduction of safe and effective agents that enhance the ultrasonic information, this has now changed [1–3]. Most important by far are microbubbles but some other attempts to improve imaging by other means are of interest. Commonly called 'contrast agents', the implied analogy with X-ray and MR agents may be unhelpful because in many situations the ultrasound agents actually reduce the target-to-background contrast (e.g. when an intravenous microbubble agent is used in a large vascular space such as the heart or major blood vessels) while the purpose of administering non-vascular agents is often to improve visualisation of deeper-lying structures (e.g. gastric agents to aid imaging the retroperitoneum) rather than to outline the gut itself. The neutral term 'echo-enhancing agents' is preferred here.

12.1.   NON-BUBBLE APPROACHES

The simple approaches to improving image quality in the abdomen and pelvis by filling the gut and the bladder with fluid, which acted by displacing bowel gas, has been improved upon with the development of a chopped cellulose and simethicone mix (SonoR$_x$, Ima R$_x$ Inc., Tucson, Arizona) for oral use in amounts of up to 500 ml. It is claimed not only to displace gas but also to act as an adsorber. Reports of its effectiveness, including comparisons with degassed water, have shown it to be clinically useful for the upper retroperitoneum [4].

Instillation of simple fluids into body spaces can also be useful. An example is saline sono-hysterography where an echo free fluid outlines the uterine cavity and provides excellent images of polyps and developmental anomalies. The similar technique of sono-salpingography requires instillation of an echogenic fluid and two microbubble agents have been used to good effect (Echovist (Schering AG, Berlin) [5] and Albunex (Malinkrodt, St Louis, Missouri)). The microbubbles tend to obscure uterine detail and so the sequential combination (saline first for the uterus, microbubbles afterwards for the tubes) provides the most complete information. Sono-salpingography, though not providing as detailed anatomical information as a conventional X-ray salpingogram, is sufficient for establishing tubal patency and thus can be used as a screening test to avoid unnecessary ionising radiation.

Isotonic fluid instilled into the colon following colonoscopy improves delineation of the bowel wall with exquisite detail. The entire colon can be visualised in most patients and, just as in other parts of the gut, the use of high resolution probes allows the five-layered structure within the 5–6 mm wall thickness to be discerned. Polyps form echogenic projections into the lumen, the structure of the wall at the point of attachment being preserved. Carcinomas may have the same appearance but invasion into the wall and surrounding tissues destroys the layers, replacing them with a mass generating echo of lower amplitude. The advantage over colonoscopy of revealing submucosal changes is clear. Accuracies of 83 to 100% in the detection and staging of colonic carcinoma have been reported [6–9] and the technique seems to be useful in children with bleeding from familial polyposis [10]. Changes in the wall allow Crohn's disease to be distinguished from ulcerative colitis, in which thickening rarely extends deeper than the inner (mucosal) layer [11]. However, in pseudopolyposis the ultrasonic distinction becomes difficult or impossible. In most cases of Crohn's colitis there is effacement of some or all of the layers of the wall together with thickening of up to 10 mm: cases where the layers are normal have only superficial ulcers at colposcopy.

Hydrostatic reduction of an intestinal intussusception can be monitored by ultrasound [12].

12.2. MICROBUBBLE AGENTS

12.2.1. History

The behaviour of gas-filled bubbles in an acoustic field has been studied for many years, and is reviewed fully in Chapter 11. Nevertheless it was a chance observation by a cardiologist, Dr Charles Joiner, as long ago as the late 1960s that led to the development of effective microbubble echo-enhancing agents: he was performing a transthoracic M-mode echocardiogram while injections of indocyanine green were being given through a left ventricular catheter to measure cardiac output using the Fick dye-dilution principle [13]. Joiner noticed transient increases in echo strength in the outflow tract of the heart following each injection. Much further research showed that these signals were not specific to the particular fluid injected and the proposal that the echoes were due to microbubbles was supported by *in vitro* observations that an increase in ambient pressure obliterated the enhancement [14, 15]. Injection of hand-made bubbles formed by squirting physiological saline between two syringes via a three-way tap in has continued to be widely used in echocardiography.

Though effective enhancers, the transience and poor reproducibility of these bubbles limits their clinical use and so ways to improve them were studied. It turned out that mixing some of the patient's own blood with the saline improved the stability of the bubbles. Feinstein pursued this line and found that serum albumin was the critical blood component and that sonication produced more reliable microbubbles of controlled size [16]. Eventually this approach led to the production of Albunex, the first pharmaceutical echo-enhancer (Molecular Biosystems Inc., San Diego, California) [17].

Subsequently many pharmaceutical houses set up development projects using a variety of techniques to form stable microbubbles of the right size range. Of these, Levovist (Schering AG, Berlin) is the most widely studied and is now commercially available in many parts of the world, including Europe [18]. It is a modification of Echovist and both are made of galactose, structured as microcrystals that act as nucleation sites for air in the ampoule. When mixed for injection, a solution–suspension is formed with microbubbles of air attached to minute irregularities on the surface of those microcrystals that remain in the solid state. On injection, the remaining crystals dissolve, releasing the microbubbles into the blood. Echovist was of limited value for the heart because the microbubbles dissolved before it crossed the lung bed, though, as mentioned, it has found an important application in salpingography (HyCoSy, *h*ystero *c*ontrast *s*alpingograph*y*) [5]. In an important development, the survival of the microbubbles was improved by adding a surfactant (palmitic acid) in trace amounts and, in this form, a bolus of Levovist lasts for 1 to 4 min in the blood (this can be prolonged *ad lib* if the agent is infused rather than given as a bolus [19]). It provides some 10–20 dB increase in echo signals that can be detected on grey scale imaging in large vessels and in the heart but requires the higher sensitivity of Doppler for its detection in smaller vessels, especially in the microvascular circulation [20].

EchoGen (developed by Sonus, Bothel, Washington, licensed to Abbott, Chicago, Illinois) is another example of an agent where the microbubbles are formed *in vivo*, but the mechanism of formation of its microbubbles is entirely different [21]. Essentially it consists of a perfluoro compound (perfluoropentane) which is liquid at room temperature but is a gas at body temperature. It is prepared as an aqueous emulsion with surfactants and, on injection, the droplets change to the gas phase, expanding to form microbubbles in the 2–8 μm range. In practice, the phase transition is greatly improved if the colloidal suspension is activated by subjecting it to a vacuum before injection. This is achieved by withdrawing the plunger of the closed-off syringe and then quickly releasing the tension: this 'hypobaric activation' method is popularly referred to as 'popping'.

Albunex, on the other hand, is typical of the preformed microbubble or, more precisely, microcapsule type, where the entire structure is present in the ampoule, and only needs suspension of the bubbles before injection. Most newer microbubbles fall into this category but both the capsule material and the gas they contain vary widely. Some also use biological materials, such as the phospholipids of Sonovue (BR1, Bracco, Milan) [22] and Aerosomes (Du Pont Jerk, N Billerica, Massachusetts) [23], while others use biodegradable synthetic capsules, an example being Sonovist (Schering AG, Berlin) [24]. In this interesting agent the shell is cyanoacrylate ('tissue glue'), a material that is so stable that the injected microcapsules are treated as foreign particles and disposed of by phagocytosis into the macrophage system in the body. They are taken up by the liver's Kupffer cells where they persist for many hours before eventually being degraded. This property underlies the use of Sonovist to delineate normal liver tissue (see 'sonoscintillation' below).

While air is the gas used in the early agents and also in some of the newer ones, there are advantages in using inert gases of high molecular weight, chiefly because they diffuse more slowly and so confer a longer life in the blood. Apart from their use in EchoGen, perfluoro gases are used in FS 069 [25] (MBI), a derivative of Albunex [16], as well as in Aerosomes and in Imagent (Alliance Pharmaceuticals, San Diego, California) [26]. Sulphur hexafluoride is used in Sonovue (Bracco, Milan); this inert gas was formerly used to measure gas transfer in the lungs [22] and so is known to be well tolerated. The choice of gas also affects the echogenicity of the agent in complex ways which are not yet well understood. When the partial pressure of the microbubble's gas is lower than that of blood (effectively when any gas other than air is used) blood nitrogen may diffuse into the bubbles. This adds to the rectified diffusion that can cause them to enlarge at some phases of their life and so their survival behaviour may be quite complex.

Thus the development of clinically useful echo-enhancing agents has been a long trail of exploration during which many ingenious methods have been devised to make them. Their physical structure is at least as important as their chemical constituents and the field is perhaps best regarded as an example of engineering nanotechnology rather than as conventional pharmacology.

12.2.2. *Safety of contrast agents*

An important clinical feature of microbubbles is their excellent safety profile: no significant adverse events attributable to the agents have been reported over many thousand injections in several thousand patients and volunteers. However, their safety has to be considered more widely than that of other diagnostic agents because the interaction of the insonating sound beam with the bubbles is an additional topic of possible concern over and above the usual issues of the safety of the constituents and of the microcapsules themselves.

The constituents have all been chosen for their known safety, biological and biocompatible materials being selected for the capsule and for the surfactants, together with air or inert gases for the contents. Minor residual concerns regarding allergic responses [27] and the difficulty of sterilising them seem to be without foundation.

Turning to the effects of injecting particles, the most obvious worry is the possibility of embolism but there are theoretical and practical reasons for this to be a minor concern. Since the microbubbles are designed to be small enough to cross capillary beds, they should not embolise unless they enlarge or aggregate into a foam. Both are possible but unlikely to be important considerations, especially in view of the small amounts of gas involved. In Levovist each clinical dose contains less than 200 μl of air, much less than is often injected inadvertently, for example, when an intravenous infusion is set up. In comparison with emboli from other sources, e.g. platelet emboli from an unstable atheromatous plaque, the short life of microbubbles in the blood (some 5 minutes) acts as a protecting factor while much larger particles deliberately designed to embolise are used in perfusion lung scans with no untoward effects. Extensive animal and human tests looking for biochemical products of central nervous system damage, as well as neurological testing, have been consistently negative [28].

Similar considerations apply to other possible effects of injecting particles; while triggering the release of vaso-active and bronchoconstrictor compounds in the lung is a known problem with some agents in some species (especially in the pig), these effects do not seem to occur in man.

The possibility of cavitation (see Chapter 11) produced by the interaction of the sound beam with microbubbles has been given prominence by the description of haemorrhagic damage following lithotripsy sonification *in vivo*, but these effects require much higher power deposition than is achieved with diagnostic scanners. They seem to have very sharp lower power cut-off levels and have not been demonstrated in situations similar to those that obtain *in vivo* with diagnostic scanner output levels [29]. No biochemical markers of free radical production (e.g. peroxides) have been found in experiments designed to test this point.

Thus far, therefore, these agents seem to be very safe, much more so, in fact, than the iodinated agents that continue to be used for X-rays.

Nevertheless, prudent use and pre- and post-marketing surveillance for adverse events should be continued [30].

12.2.3. Basic principles

The essential mechanisms whereby microbubbles act as echo-enhancers are the same as pertain to scatterers in general: the echo intensity is proportional to the change in acoustical impedance as the sound beam crosses from the blood to the gas in the bubbles. The impedance mismatch at such an interface is very high so that essentially all of the incident sound is scattered (though not all will travel back to the transducer). However, though the reflection is near-complete, by itself this would not produce a very effective enhancing agent because the microbubbles are small and sparse; reflectivity is proportional to the sixth power of the particle diameter and directly to their concentration [31]. The scattering of sound is discussed in Chapter 4.

The extraordinary echogenicity of microbubbles results from the fact that they resonate when insonated (see Chapter 11) and this makes them behave as though they have a very much larger cross-section than a rigid bubble of the same diameter by a factor of 10^{14} [32] (figure 12.1). Designing microbubbles to exploit this resonance is central to pharmaceutical research in the field. Obviously they must be made small enough to cross the capillary beds ($\leqslant 8$ μm) and, like any mechanical resonance system, the critical frequency depends on their diameter. The fortunate coincidence that microbubbles of 1–7 μm diameter happen to have their resonance frequencies in the 2–15 MHz range of ultrasound that is used for clinical diagnosis is the fortuitous basis for the entire subject.

An important difference from the ionic agents used for X-ray and MRI needs to be borne in mind: since microbubbles do not diffuse across the capillary endothelium, there is no interstitial phase of enhancement. Thus they are essentially markers for the blood pool (or for any other body space into which they have been placed) and their distribution is similar to red cells.

12.2.4. Clinical applications

The enhancement of ultrasound echo amplitude given by microbubbles can be exploited in a number of different ways, of which the most obviously useful is to enable studies that were otherwise technically difficult or unreasonably protracted (table 12.1) [33]. This 'rescue' role of microbubbles is currently the most important clinical advantage they provide and is used widely, both in B-mode and for Doppler. However, the effect also makes possible some new types of application that cannot be performed without microbubbles, especially demonstrating flow in the microcirculation, e.g. in the myocardium and in tumours, and providing the unique opportunity to

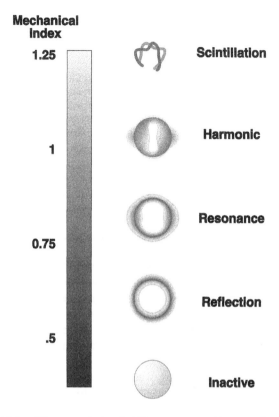

**Mechanical
Index**

1.25 — Scintillation

1 — Harmonic

0.75 — Resonance

.5 — Reflection

Inactive

Figure 12.1. *Diagram of microbubble resonance with transmitted power. Simple reflection occurs at low insonating intensities but, as the transmitted power is increased, microbubbles begin to resonate, at first in a linear mode and then nonlinearly. Harmonic and transient phenomena such as stimulated emission (sonoscintillation) result at higher powers. All these effects occur within the permitted output levels.*

perform functional studies by tracking the transit of a microbubble bolus through a region of interest. In the future many other applications are expected to emerge with the development of new types of microbubble agent and promising new ways to use them (see below).

Agents that are currently licensed give sufficient enhancement on a grey scale to be useful in the heart and great vessels. They fill ('opacify') the cardiac chambers and thus improve delineation of the left ventricular endocardial border. This improves detection of wall motion abnormalities (especially important during stress echocardiography) and makes it possible to perform these measurements in 'difficult to image patients' (plate 3) [34]. They are also useful in demonstrating shunts and valvular disease and have been used *in utero* to demonstrate twin–twin transfusion [35].

However, in the lower concentrations obtained in the myocardium, special

Table 12.1. *Uses of microbubbles.*

Use	Comments
1. Doppler rescue	Of poor quality study because of attenuation or weak signals (e.g. portal vein in cirrhosis) Of impractically protracted study (e.g. renal arteries for stenosis)
2. Extend the applications of Doppler	Microvasculature (e.g. in tumours)
3. Functional studies	Transit timing for dynamic information Functional diagnostic indices Functional images
4. New fields	Targeted agents: diagnostic drug delivery

techniques such as intermittent imaging and use of the harmonic mode are needed to reveal their presence. Doppler is more sensitive than B-mode, but the tissue motion of the myocardium swamps the Doppler signals unless they are suppressed by imaging in the harmonic mode.

This bulk tissue motion is less severe outside the heart, and the microbubble enhancement is very well demonstrated on conventional Doppler. Mostly they are used for 'Doppler rescue', e.g. enabling transcranial Doppler to be performed reliably, differentiating between a tight internal carotid artery stenosis with trickle flow and a total occlusion, and enabling otherwise difficult portal vein and renal artery studies (plate 4). For the renal arteries, reducing the examination time is as important clinically as the reduction in failed studies [36]. Similar considerations underlie their use in peripheral arterial disease and they may also be useful for demonstrating deep vein thrombosis [37].

Since the effect of the microbubbles is to increase the signal intensity without altering blood flow, the intensity or loudness of the spectral signals is the aspect that changes. For colour Doppler, the equivalent is the increase in intensity in power Doppler, and generally this is the best mode to choose. Indeed it may be asked why anything is displayed in velocity colour Doppler: the very obvious increases in both the number of colour pixels and the frequency shifts that are observed are likely to be effects of the sensitivity filters; weak signals that were not Doppler-detected rise above this threshold after enhancement and are now registered. This phenomenon also explains the finding of an increased spectral broadening after enhancement: both higher maximum and lower minimum velocities are registered, corresponding with flow streams that were not detected before enhancement. This may necessitate a re-evaluation of criteria for disease detection, especially where peak velocity values are used, for example, to estimate the degree of a stenosis. Presumably ratios of, for example, the

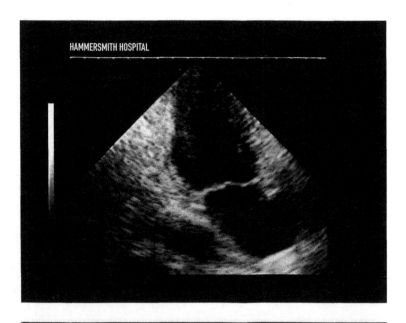

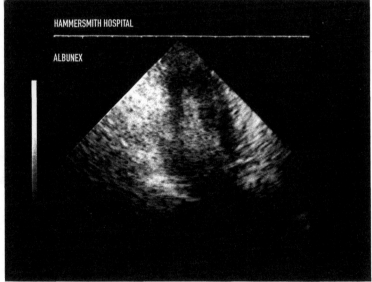

Plate 3.

Left ventricular enhancement. The normally echo-free heart chambers (top) are enhanced (opacified) following intravenous injection of a microbubble agent. In this patient, Albunex has improved delineation of the left ventricular endocardial border in a long-axis view (bottom). (Images courtesy of Dr Petros Nihoyannopoulos, Hammersmith Hospital, London.)

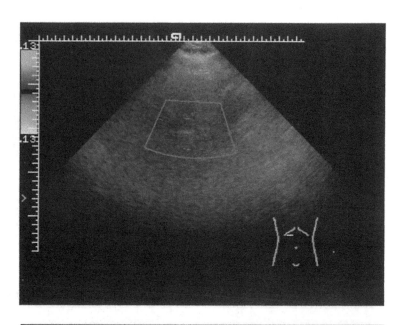

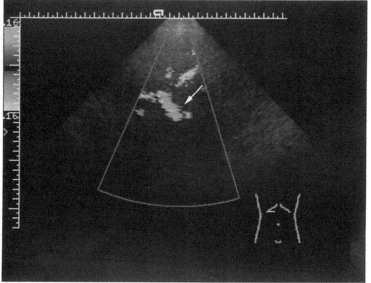

Plate 4.

Enhancement of Doppler signals from the portal vein. In this patient with cirrhosis, no Doppler signals could be obtained from the portal vein (top) because of the highly attenuating overlying liver. After enhancement with the microbubble agent Levovist, useful signals were obtained (arrowed, bottom), and the red colour coding indicated that the flow was in the opposite direction to normal, an important sign of portal hypertension.

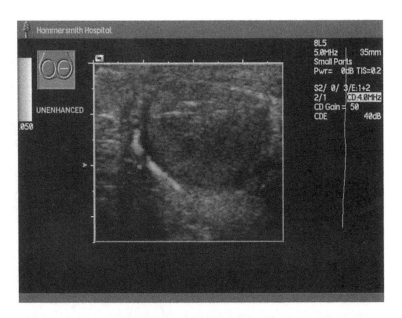

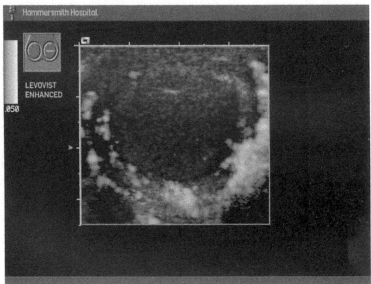

Plate 5.

Ischaemic testis. No colour signals could be obtained from the testis itself either before (top) or after (bottom) enhancement with Levovist in this man who presented with acute testicular pain, though there were abundant signals from the surrounding tissue layers. This suggested the diagnosis of testicular torsion that was confirmed at subsequent surgery.

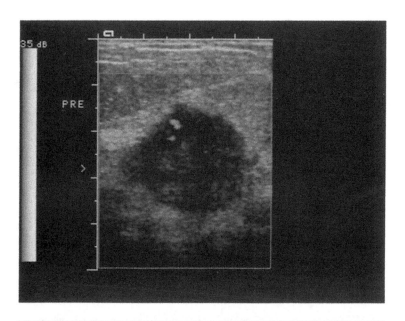

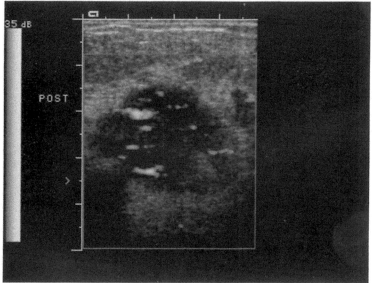

Plate 6.

Carcinoma of the breast. A small amount of vascularisation is demonstrated in this power Doppler scan (top) but much more of the neovascularisation is seen after enhancement with the perfluoro agent EchoGen (bottom).

peak velocity in the internal to the common carotid artery, will not suffer from this effect.

In the microcirculation, B-mode will be the preferred form of ultrasound use when suitably reflective agents become available because of its better resolution (both spatial and temporal) but, at present, Doppler is essential to detect the weak signals. Flow in vessels down to perhaps 100 μm in diameter can be demonstrated under ideal situations, the limit being set by the overwhelming clutter from bulk movement of the surrounding tissues, as with the myocardium. This allows at least arterioles and venules to be picked up and shows regions of infarction or ischaemia as colour voids (plate 5).

The improved display of the neovascularisation of tumours is an important clinical area: enhanced colour Doppler improves display of the tortuous and tangled pattern and can help differentiate benign from malignant lesions, for example in the breast and liver (plate 6). Since malignant neovascularisation is a predictor of tumour aggressivity, its direct and non-invasive display by enhanced Doppler might be useful, for example, in planning adjuvant chemotherapy. Similarly, changes during treatment may form useful indices of response or the development of chemo-resistance and have been used to determine whether sufficient interstitial treatment by alcoholisation or RF heating has been given [38]. With the development of specific blockers to angiogenesis factors, this application is expected to become increasingly important [39].

Tissue blood flow is also likely to be useful in inflammatory conditions such as for assessing the severity of arthritis and its response to treatment and similarly in inflammatory bowel disease. Evaluation of regions of reduced flow is also interesting: demonstrating myocardial ischaemia has been mentioned as one of the most important goals of microbubble research and similar considerations apply to the spleen and the kidneys, especially renal transplants. A hitherto unexplored area that could become very important is the placenta. Placental dysfunction is a major cause of fetal growth retardation and of pre-eclampsia; at present the diagnosis is clinical and therefore late. If early changes in the perfusion of the maternal vasculature of the placenta could be demonstrated by structural changes in the vasculature or by transit time abnormalities, management might become more effective. Since microbubbles are unlikely to cross into the fetus, safety is not a special concern.

12.2.5. *Quantification and functional studies*

The fact that microbubbles can be injected as a bolus opens up a whole new field for ultrasound by quantifying the Doppler signal strength to generate time–reflectivity curves. Videodensitometry has been used widely for grey scale quantification and has proved to be a useful tool for the pharmaceutical industry and in clinical echocardiography. Systems for measuring the signal intensity have been developed for all forms of Doppler [40] and, despite the nonlinearity of Doppler signal processing, have been shown to correlate very closely with relative microbubble concentration both *in vitro* [41, 42] and *in vivo* [43]. Thus the basis for measuring changes in B-mode and Doppler reflectivity has been established. Quantification systems are beginning to be incorporated into commercial scanners.

From these measurements, transit time curves of the wash-in and wash-out of the microbubbles can be generated, and important functional indices can be calculated from them, in the same way as with MRI and CT contrast agents. Because microbubbles act as blood pool markers, they are complementary to these agents which diffuse into the extracellular space. Both simple measurements such as the transit time, the peak value, the area under the curve and the fractional vascular volume, and complex features such as those requiring deconvolution, can be performed and promise to yield useful diagnostic criteria. For example, the transit seems to be longer and perhaps more complex in malignant tumours than in benign masses. There is every likelihood that physiological indices such as the mean transit time and true tissue perfusion can be calculated.

These indices could be presented as data for a region or point of interest or could be displayed as functional images [44]. Initially only two-dimensional calculations and representations are likely to be achieved but the possibility of extending this to three spatial dimensions is realistic in view of the fast frame rate that ultrasound allows. Since the faster wash-in phase of the transit needs only be sampled approximately once per second, sweeping the probe continuously through a volume would seem feasible. Thus eventually four-dimensional anatomico-functional datasets could be generated.

12.2.6. New uses: agents and techniques

A series of new agents is being developed, the goal being to improve the microbubbles' echogenicity and persistence. The increase in echogenicity allows B-mode to be used in the microvasculature, a desirable feature in view of the simpler equipment required, the better spatial and temporal resolution and the lower level of artefacts. The longer duration of enhancement will simplify some clinical applications where currently repeat injections are needed.

Some microbubbles have interesting new properties such as sonoscintillation (see below). Envisaged, but only beginning to be developed, are targeted agents such as Aerosomes to which ligands for activated platelets have been attached: this preparation sticks to fresh thrombus and is expected to be useful to highlight its presence [45]. Drug delivery, whereby an active pharmaceutical is attached to a microbubble and then released locally by disrupting the bubbles using a high power ultrasound beam, could eventually become the most important use of microbubbles. The microbubbles can be ruptured ultrasonically at the chosen site to release the drug in a high concentration while the remainder of the body receives only a small amount of dilute drug so that the therapeutic ratio is improved. Localised delivery of genetic material has been demonstrated using this approach [46] and attachment of thrombolytic drugs to the platelet-targeted Aerosomes could provide safer thrombolysis.

The development of new ways to insonate microbubbles in order to improve their reflectivity or duration include triggered or intermittent insonation to reduce bubble destruction. This approach has proved especially useful for myocardial imaging in B-mode [47]. Allowing an interval between sequential pulses not only minimises bubble destruction but also allows refreshment of the bubbles in the scanned plane because of inflow of fresh blood and this improves the echo intensity, sometimes to a surprising extent [48]. The rate of this inflow of microbubbles may itself contain functional information regarding tissue perfusion because the bubbles are carried by the microcirculation. The geometry of this flow, however, is likely to be complex since the ultrasonic tomogram only samples a slice of tissue and does not contain complete information on the flow direction. Nevertheless, average tissue flow values could form a useful measure of the microperfusion.

The bubble destruction that intermittent imaging seeks to minimise can itself be exploited in a different way: if a high intensity beam, deployed to destroy all the bubbles at one location in a vessel, is abruptly switched to a low intensity beam for imaging, a sharp flush of undamaged bubbles flows into the vessel, the so-called 'negative bolus' effect [49]. This is the equivalent of a direct arterial injection of the agent, a situation that is ideal for transit time measurements.

Second harmonic imaging is an intriguing and promising field which depends on the nonlinear oscillation of microbubbles during which they emit overtones (harmonics) at twice the insonating frequency [50]. Since tissue has only minor nonlinear behaviour, it emits lower levels of second harmonic signals. This apart, the harmonic signals are a microbubble signature so that in the harmonic mode, echoes from tissue, both stationary and moving, are suppressed, thus increasing the signal-to-noise/clutter ratio. The effect (but not the process) is analogous to digital subtraction X-ray angiography. In the harmonic mode, myocardial microbubbles can be detected on B-mode despite the strong echoes from cardiac muscle [51] and, in Doppler, harmonic signals from blood flow and the movements of solid tissue are suppressed. This allows smaller vessels with even slower flow to be detected and, in an experimental set-up, vessels down to 40 μm in diameter have been demonstrated in the kidneys [52].

Even in the harmonic mode, stationary bubbles in the periphery are hard to detect, but another nonlinear bubble mode is able to achieve this. Variously known as 'sonoscintigraphy', 'stimulated acoustic emission (SAE)' and 'loss of correlation (LOC) imaging', these unique microbubble signals seem to result from a bubble within a cohort being inactivated by the insonating beam. Since the colour Doppler processor operates by comparing sequential echo trains, inactivation of a bubble produces a difference between pairs of signals that the scanner circuitry interprets as a Doppler shift. The alternative explanation, that the scintillation represents the direct Doppler visualisation of the collapse of a microbubble, seems less likely. Sonoscintillation is recognised as a shimmering, multicoloured mosaic, better seen in velocity colour Doppler than in power Doppler. Production of the effect is critically dependent on the transmitted ultrasonic power and so falls off with depth and is more strongly elicited by low ultrasonic frequencies.

Sonovist is an excellent sonoscintillator and in the liver the effect persists for many hours because these robust microbubbles are phagocytosed by the Kupffer cells [24]. In animal studies, it reveals tumours as striking colour voids. The same effect can be produced with Levovist and, though it is much more transient, is sufficient to highlight focal lesions that are difficult to detect on grey scale imaging [53].

A general aspect of the effects of microbubbles is the fact that the improved signal-to-noise ratio that they produce could be exploited in a number of ways beyond merely viewing or measuring the boost to the echoes. For example, it should be possible to utilize some of the 20 dB extra information to improve spatial resolution or frame rate or to reduce motion artefacts. Novel ways to use the presence of the microbubbles include estimates of the frequency dependence of attenuation in the liver (based on the depth to which the scintillation effect can be achieved for a given transmit power) and exploitation of individual scintillation events to correct for aberrations in the ultrasonic field.

12.3. CONCLUSION

The development and introduction of microbubbles as safe and effective echo enhancers has extended and opened up new opportunities for clinical applications and for fundamental research. They herald a new era for ultrasound.

REFERENCES

[1] Cosgrove D (ed) 1996 Ultrasound Contrast Agents *Clin. Radiol.* **51** Suppl. 1, 1–58
[2] Goldberg B B (ed) 1997 *Ultrasound Contrast Agents* (London: Dunitz)
[3] Cosgrove D O (ed) 1999 Echo-enhancing agents *Contrast Agents in Radiology* ed P Dawson *et al* (Oxford: Isis) section 3, in press
[4] Harisinghani M G, Saini S, Schima W, McNicholas M and Mueller P R 1997 Simethicone coated cellulose as an oral contrast agent for ultrasound of the upper abdomen *Clin. Radiol.* **52** 224–6

[5] Ayida G, Harris P, Kennedy P S, Seif M, Barlow D and Chamberlain P 1997 Hysterosalpingo-contrast sonography (HyCoSy) using Echovist-200 in the outpatient investigation of infertility patients *Br. J. Radiol.* **69** 910–3

[6] Limberg B 1990 Diagnosis of large bowel tumours by colonic sonography *Lancet* **335** 144–6

[7] Limberg B 1992 Diagnosis and staging of colonic tumors by conventional abdominal sonography as compared with hydrocolonic sonography *N. Engl. J. Med.* **327** 65–9

[8] Walter D F, Govil S, William R R, Bhargava N and Chandy G 1993 Colonic sonography: preliminary observations *Clin. Radiol.* **47** 200–4

[9] Hernandez-Socorro C R, Guerra C, Hernandez-Romero J, Rey A, Lopez-Facal P and Alvarez-Santullano V 1995 Colorectal carcinomas: diagnosis and preoperative staging by hydrocolonic sonography *Surgery* **117** 609–15

[10] Nagita A, Amenoto K, Yoden A, Yamazaki T, Mino M and Miyoshi H 1994 Ultrasonographic diagnosis of juvenile colonic polyps *J. Pediatr.* **124** 535–40

[11] Limberg B and Osswald B 1994 Diagnosis and differential diagnosis of ulcerative colitis and Crohn's disease by hydrocolonic sonography *Am. J. Gastroenterol.* **89** 1051–7

[12] Rohrschneider W K and Troger J 1995 Hydrostatic reduction of intussusception under US guidance *Pediatr. Radiol.* **25** 520–34

[13] Gramiak R and Shah P M 1968 Echocardiography of the aortic root *Invest. Radiol.* **3** 356–66

[14] Staudacher T, Prey N, Sonntag W and Stoeter P 1990 The basis for ultrasonic phenomena during the injection of x-ray contrast media *Radiologe* **30** 124–9

[15] Meltzer R S, Tickner G, Salines T P and Popp R L 1980 The source of ultrasound contrast effect *J. Clin. Ultrasound* **8** 121–7

[16] Feinstein S B, Heidenreich P A, Dick C D, Schneider K, Rubenstein G A, Applebaum J, Brehm J L, Aronson S, Ellis J and Roizen M 1988 Albunex: a new intravascular ultrasound contrast agent: preliminary safety and efficacy results *Circulation* **78** Suppl. II 565

[17] Keller M W, Feinstein S B, Briller R A and Powsner S M 1986 Automated production and analysis of echo contrast agents *J. Ultrasound Med.* **5** 493–8

[18] Smith M D, Elion J L and McLure R R 1989 Left heart opacification with peripheral injection of a new saccharide echocontrast agent in dogs *J. Am. Coll. Cardiol.* **13** 1622–8

[19] Albrecht T, Urbank A, Cosgrove D O, Mahler M, Blomley M J K and Schlief R 1996 Prolongation and optimization of Doppler enhancement with continuous infusion of a US contrast agent *Radiology* **201**(P) 195

[20] Schlief R, Staks T, Mahler M, Rufer M, Fritzsch T and Seifert W 1990 Successful opacification of the left heart chambers on echocardiographic examinations after intravenous injection of a new saccharide-based contrast agent *Echocardiography* **7** 61–4

[21] Quay S C 1994 Ultrasound contrast agent development: phase shift colloids *J. Ultrasound Med.* **13** S9

[22] Schneider M, Arditi M, Barrau M B, Brochot J, Broillet A, Ventrone R and Yan F 1995 BR1: a new ultrasonographic contrast agent based on sulfur hexafluoride-filled microbubbles *Invest. Radiol.* **30** 451–7

[23] Unger E, Fritz T, Shen D-K, Lund P, Sahn D, Ramaswami V, Matsunaga T, Yellowhair D and Kulik B 1994 Gas filled lipid bilayers as imaging contrast agents *J. Liposome Res.* **4** 861–74

[24] Burns P N, Fritsch T and Weitschies W 1995 Pseudo Doppler shifts from stationary

tissue due to the stimulated emission of ultrasound from a new microsphere contrast agent *Radiology* P197

[25]   Dittrich H C, Bales G L and Kuvelas T 1995 Myocardial contrast echocardiography in experimental coronary artery occlusion with a new intravenously administered contrast agent *J. Am. Soc. Echocardiogr.* **8** 465–74

[26]   Aeschenbacher B C 1996 Experience with Imagent US, a new ultrasoundcontrast agent *1st Eur. Symp. on Contrast Imaging (Rotterdam)*

[27]   Christiansen C, Vebner A J, Muan B, Vik H, Haider T, Nicolaysen H and Skotland T 1994 Lack of an immune response to Albunex, a new ultrasound contrast agent based on air-filled albumin microspheres *Int. Arch. Allergy Immunol.* **104** 372–8

[28]   Bommer W J, Shah P M, Allen H, Meltzer R and Kisslo J 1984 The safety of contrast echocardiography: Report of the Committee on Contrast Echocardiography for the American Society of Echocardiography *J. Am. Coll. Cardiol.* **3** 6–13

[29]   Dalecki D, Raeman C H, Child S Z, Cox C, Francis C W, Meltzer R S and Carstensen E L 1997 Hemolysis in vivo from exposure to pulsed ultrasound *Ultrasound Med. Biol.* **23** 315–20

[30]   ter Haar G 1999 Interactions between the ultrasound beam and microbubbles *Contrast Agents in Radiology* ed P Dawson *et al* (Oxford: Isis) in press

[31]   Ophir J and Parker K J 1989 Contrast agents in diagnnostic ultrasound *Ultrasound Med. Biol.* **15** 319–33

[32]   de Jong N, Ten Cate F J, Lancee C T, Roelandt J R and Bom N 1991 Principles and recent developments in ultrasound contrast agents *Ultrasonics* **29** 324–30

[33]   Blomley M and Cosgrove D 1996 Contrast agents in ultrasound (editorial) *Br. J. Hosp. Med.* **55** 6–7 (discussion 7–8)

[34]   Nihoyannopoulos P 1999 Echocontrast in cardiology *Contrast Agents in Radiology* ed P Dawson *et al* (Oxford: Isis) in press

[35]   Denbow M L, Blomley M J K, Cosgrove D O and Fisk N M 1997 Ultrasound microbubble contrast angiography monochorionic twin fetuses *Lancet* **349** 773

[36]   Missouris C G, Allen C M, Balen F G, Buckenham T, Lees W R and MacGregor G A 1996 Non-invasive screening for renal artery stenosis with ultrasound contrast enhancement *J. Hypertens.* **14** 519–24

[37]   Langholz J, Heidrich H and Behrendt C 1989 Ultrasonic contrast medium in the diagnosis of deep venous thrombosis *Phlebologie* **89** 366–8

[38]   Solbiati L, Ierace T, Crespi L and Rizzatto G 1996 Three-dimensional power Doppler with an intravascular echo enhancement agent and second harmonic imaging in radio-frequency ablation of liver metastases *Radiology* **201**(P) 196

[39]   Harris A L 1997 Antiangiogenesis for cancer therapy *Lancet* **349** (Suppl. II) 13–5

[40]   Bell D S, Bamber J C and Eckersley R J 1995 Segmentation and analysis of colour Doppler images of tumour vasculature *Ultrasound Med. Biol.* **21** 635–47

[41]   Schwarz K Q, Bezante G P and Chen X 1995 When can Doppler be used in place of integrated backscatter as a measure of scattered ultrasound intensity? *Ultrasound Med. Biol.* **21** 231–42

[42]   Blomley M J, Cosgrove D O, Albrecht T, Powers J E and Urbanc A 1996 Quantification of microbubble concentration by using color power Doppler *Radiology* **201**(P) 159

[43]   Blomley M J, Jayaram V, Cosgrove D O, Patel N, Albrecht T and Llull J 1996 Doppler intensitometry with BR1 in humans *Radiology* **201**(P) 158

[44]   Blomley M J K 1997 Functional techniques with microbubbles *Advances in Echocontrast* 2nd edn, ed N Nanda, R Schleif and B B Goldberg (Dordrecht: Kluwer)

[45] Unger E C, McCreery T P, Shen D K, Wu G L, Sweitzer R H and Yellowhair D 1997 Thrombus specific Aerosomes a novel ultrasound contrast agent *J. Ultrasound Med.* **16** S36

[46] Unger E 1996 Imaging and drug delivery applications of Aerosomes *1st Annu. Ultrasound Contrast Media Research Meeting (San Diego, CA)*

[47] Porter T R, Xie F, Kricsfeld D and Armbruster R W 1996 Improved myocardial contrast with second harmonic transient ultrasound respon.e imaging in humans using intravenous perfluorocarbon-exposed sonicated dextrose albumin *J. Am. Coll. Cardiol.* **27** 1497–501

[48] Kamiyama K, Moriasu F, Kono Y, Mine Y, Nada T and Yamazaki N 1996 Investigation of the flash echo signal associated with a US contrast agent *Radiology* **201**(P) 158

[49] Ivey J A, Gardner E A, Fowlkes J B, Rubin J M and Carson P L 1995 Acoustic generation of intra-arterial contrast boluses *Ultrasound Med. Biol.* **21** 757–67

[50] Burns P N, Wilson S R, Muradali D, Powers J E and Fritzsch T T 1996 Microbubble destruction is the origin of harmonic signals from FS 069 *Radiology* **201**(P) 158

[51] Mulvagh S L, Foley D A, Aeschbacher B C, Klarich K K and Seward J B 1996 Second harmonic imaging of an intravenously administered echocardiographic contrast agent: visualization of coronary arteries and measurement of coronary blood flow *J. Am. Coll. Cardiol.* **27** 1519–25

[52] Burns P N 1996 Harmonic imaging with ultrasound contrast agents *Clin. Radiol.* **51** (Suppl. 1) 50–5

[53] Blomley M J K, Albrecht T, Cosgrove D O *et al* 1997 Stimulated acoustic emission in the liver with Levovist *Radiology* submitted

CHAPTER 13

SONOCHEMISTRY AND DRUG DELIVERY

Gareth J Price

INTRODUCTION

Along with animal navigation and sonar methods, the medical uses are probably the most widely recognised applications of ultrasound. However, there is growing interest in the use of high power ultrasound to cause chemical and physical changes in systems [1–3]. In addition to the more familiar medical applications of lithotripsy (Chapter 10) and focused surgery (Chapter 9), and cleaning tanks, new industrial techniques are being developed which range from the welding of plastics to the synthesis of complex pharmaceutical intermediates and from the crushing of metal ores to the extraction of plant flavourings. This chapter will review briefly some of these applications with the emphasis on a physical explanation of why the effects happen. In order to narrow the field to manageable proportions, in the main, applications will be taken from water based (aqueous) chemistry. For more complete discussion of aspects of sonochemistry, the reader is referred to references [1] to [3]. The chapter will conclude by reviewing some recent work in drug delivery which bridges the study of chemistry and the medical applications which are the main focus of this book.

Chemistry is about making things and understanding why reactions and processes act in the way they do. The 'buzzword' in chemistry at present is *control*. We need to control the structure of the molecules we make so that we control, for example their pharmaceutical action; we must control the way that the molecules organise themselves so that we control the bulk properties of our materials; we need to control the progress of a reaction in terms of rate, yield and product distribution so that safety and economic factors are kept under control. Finally, we need to control energy usage and waste products to give environmentally 'clean' chemistry. Ultrasound has allowed chemists to exert these types of control over a wide range of reactions.

Applications of sound, and particularly ultrasound, in chemistry have become known as *sonochemistry*. Most chemists divide ultrasound (somewhat arbitrarily) into two categories. The type used in medical diagnostics, sonar or non-destructive testing applications normally uses high frequencies in excess of 1 MHz so as to obtain good resolution. Relatively low powers (<1 W) are used in order not to change the material. This is commonly referred to as *diagnostic* ultrasound. While this can be used in chemistry, for instance in following the course of reactions or in studying conformational changes in compounds, diagnostic ultrasound has limited application in influencing chemical processes or reactions. For sonochemistry, *power* ultrasound is used, usually in the frequency range 20–500 kHz, and at powers of up to several hundred watts. Such high powers lead to the formation of microbubbles in a liquid and it is the growth and rapid collapse of these bubbles, or *cavitation*, which provides the source of energy for chemical reactions as outlined in the following section.

Two types of apparatus are commonly used for performing sonochemistry although variations of these exist. These are illustrated in figure 13.1. The simplest and most common is a cleaning bath of the type normally encountered in laboratories. A reaction vessel is simply immersed into the bath and the ultrasound conducted into the reaction through the walls of the vessel. Its economy and ready availability have led to its widespread use for synthetic chemistry although it has the drawbacks of poor reproducibility and temperature control. The major disadvantage is that of limited sound intensity since much of the ultrasonic energy is dissipated in the reaction vessel.

In general, better results are obtained using a horn or probe system. These couple the ultrasound from the transducer into the reaction using a metallic rod which also, depending on its shape, amplifies the vibration. Very high intensities are available but care must be taken to ensure adequate temperature control as considerable amounts of heat can be generated. Also, the horn must be inert to the reaction being irradiated. Glassware can be customised to accommodate the horn, that shown in figure 13.1 having a thermostat jacket and ports for admitting inert gases and removing samples during the reaction.

Other commercially available systems are less commonly used and have been reviewed by Mason [4]. In addition to laboratory apparatus, equipment for operating at much larger, commercially realistic, scales is being developed [5–7] often in flow systems which can process volumes up to 150–200 l min^{-1} depending on the required intensity. Units capable of delivering 10 kW of acoustic power are available and can be placed in series to achieve the desired effects. Hence, the equipment exists to exploit sonochemical effects on at least pilot-plant and possibly low-volume production scale.

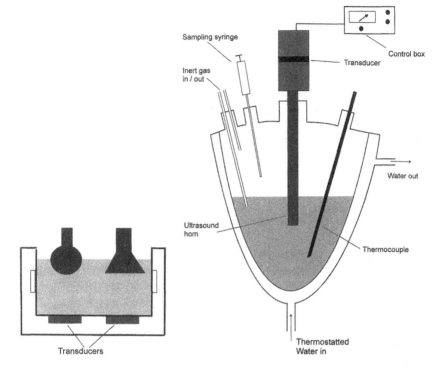

Figure 13.1. *Experimental apparatus for performing laboratory sono-chemistry. Left: ultrasonic cleaning bath. Right: 'horn' type apparatus.*

The generation of the ultrasound relies on the transducer. Often these are constructed from piezoelectric elements of the same type used for medical applications although for many industrial environments the greater robustness of magnetostrictive metals such as nickel or iron–cobalt alloy has been used to advantage.

13.1. CAVITATION AND ITS EFFECTS

In order to appreciate the potential advantages of using ultrasound, we need to understand the origin of sonochemical effects. It might be thought that the sound vibrations would couple into molecular bonds and so enhance their reactivity. However, the vibration frequencies of chemical bonds are in the gigahertz region and so at least 2–3 orders of magnitude higher than ultrasound frequencies. It is now accepted that virtually all sonochemical effects arise from cavitation.

A more complete description of cavitation appears elsewhere in this book (Chapter 11) and can also be found in the literature [8]. In order

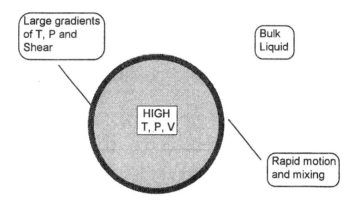

Figure 13.2. *Schematic representation of bubble effects during cavitation.*

to understand sonochemical events, a simplified model can be used. Most chemists essentially understand cavitation to be the nucleation, growth and implosive collapse of microscopic bubbles caused by passage of a sound wave through a liquid. There remains considerable doubt and discussion as to the precise effects of cavitation. On the assumption that the bubble collapse occurs adiabatically, Noltingk and Neppiras [9] showed that temperatures as high as 5000 K and pressures in the range of 30–40 MPa can be generated and there is now experimental evidence to support this so-called 'hot spot' theory [10, 11]. Other explanations in terms of the formation and rapid discharge of electric fields [12] or plasma discharges [13] have been suggested to explain some phenomena. What is certain though is that species exist in high energy, dissociated, vibrationally excited forms inside collapsing cavitation bubbles.

Three 'zones' in the reaction system can thus be identified as suggested by figure 13.2. The primary sonochemical activity such as the production of radicals and reactive intermediates takes place as a result of the harsh conditions inside the bubble. Secondly, the bulk liquid has no activity apart from reaction with sonochemically generated species which diffuse out of the bubble. Finally, the interfacial region around the bubbles has very large temperature, pressure and electric field gradients and also very large shear gradients caused by rapid movement of solvent molecules. This last factor has a special importance in situations where macromolecules are present in solution. There is an extra effect in heterogeneous systems where cavitation can occur in the region of an interface. In immiscible liquid mixtures, the interface is disrupted and efficient mixing and dispersion occurs. Bubble collapse near a solid surface is asymmetric and can result in jets of liquid impinging on the surface at high speed. Additionally, mass transport of reactants to and from the surface can be enhanced, speeding up the rate of chemical reactions. All of these effects have been used to advantage by chemists and some examples of the utility of each will be given.

The theme of this chapter is control over chemical reactions. Since virtually all the chemical effects can be traced back to cavitation, it is important to understand which factors affect the threshold and intensity of cavitation in a liquid. These factors have been described in detail [8, 14] but it is worthwhile summarising those of most importance to account for the behaviour described in the following sections.

Perhaps the least studied factor is the ultrasound frequency. It was thought that the amount of cavitation produced for the same intensity would fall as the frequency rose. However, there are some recent studies [15, 16] which show that there is an optimum value between 100 kHz and 1 MHz which depends on the system under investigation. The most important ultrasound parameter is the intensity which defines the amount of energy entering the reaction or process. As might intuitively be expected, the amount of cavitational activity rises with increasing intensity but there is an optimum value above which activity falls [17, 18].

Since a liquid is necessary for the transmission of sound, solvent properties such as density and viscosity which affect the mobility of the molecules are important, particularly in defining the cavitation threshold. However, of more influence on chemical activity is the solvent volatility. Volatile solvents will evaporate into the bubble as it grows and 'cushion' its collapse. This type of effect also explains the temperature dependence of sonochemical reactions. At lower temperatures, less vapour enters the bubbles so that collapse is more intense and chemical effects are enhanced. Thus, an acceleration at lower temperatures is often observed in contrast to the normal 'Arrhenius' behaviour where increase in rates is seen at higher temperatures.

Many chemical systems are heterogeneous and may contain two liquid phases or solid particles suspended in the liquid and this changes both the sound transmission and the cavitation thresholds. Thus, although complete characterisation of a sonochemical system is complicated by the large number of parameters that can affect cavitation [19], the role of most of them is at least qualitatively understood and their manipulation offers the possibility of controlling a reaction with considerable precision.

13.2.   WHAT CAN ULTRASOUND DO FOR CHEMISTS?

We will begin with a discussion of the usefulness in chemistry of what might be termed the 'physical' effects arising from cavitation. The first of these is the rapid motion of liquids caused by streaming and particularly the rapid motions around cavitation bubbles. Particles of solids suspended in the liquid are therefore set into motion and undergo inter-particle collisions. The shock waves caused by cavitation can also impinge on the particles, both factors leading to modification of the average particle size. In metallic powders, particles can either be fused together or fragmented depending on the physical properties such as hardness and melting point [20, 21]. With inorganic solids, two processes can be identified. Firstly, large (>10 μm) particles can be fragmented. Alternatively, loosely bound aggregates of small (<0.5–1 μm) particles can be broken up to give a better dispersion of the original material. Although the mechanisms may be different, this process has a medical analogue in the lithotripsy of solid deposits *in vivo*. A particular commercial application is in the production of suspensions of titanium dioxide, TiO_2, which is used as a filler and white pigment in many paints. The quality of the product depends on the quality of the dispersion. Stoffer and Fahim [22] studied some typical paint formulations containing TiO_2 and found that using ultrasound for the dispersal and mixing phases of the process led to smaller, more even particle sizes in shorter times than conventional methods, with consequent saving in energy consumption of up to 70%.

The efficient mixing and dispersion of liquids using ultrasound is a common commercial process, for example in the food industry. The large degree of motion induced by acoustic streaming efficiently disrupts the interface between the liquids while the very high shear forces around cavitation bubbles act to break up droplets of liquid and maintain a small and even distribution of droplet sizes. A good example of this is the production of organic polymers as emulsions dispersed in water, used to make paints and other surface coatings.

In addition to water and the organic components, stabilisers and dispersants are usually added to form a stable dispersion as well as initiators being needed to begin the reaction. Ultrasound was first applied to this type of reaction in the 1950s but recent work [23–25] has shown that ultrasound allows lower amounts (and in some cases none) of these additives to be used, improving the polymer properties, and that the reaction proceeds at a faster rate reducing the process time. This type of technology has also been applied on a larger, pilot-plant scale, for example, to prepare TiO_2 encapsulated in PVC where the coating was much more uniform when carried out in the presence of ultrasound. Scale-up studies were performed and the process carried out on a pilot plant to produce 200 kg of coated material [26].

There are a number of chemical reactions where powerful oxidising agents are needed in organic systems. However, these are often very toxic, dangerous and environmentally unfriendly. The use of aqueous agents would be beneficial but many do not mix with organic solvents. The rapid mixing of the reactants gives an obvious role for ultrasound here. For example, the oxidation of fatty acid esters is an important reaction and it has been shown to proceed smoothly using aqueous solutions of potassium permanganate [27]. Other significant phase transfer reactions have been reviewed by Luche [28].

A large number of chemical reactions are carried out using solid reagents or catalysts. The rates of reaction here are limited by the rate at which

Table 13.1. *The sonochemical enhancement of reactions using metallic lithium as a reagent (data taken from [29] and [30]).*

Reactant	Solid	Product	Conventional (%)	Sonochemical (%)
C_4H_9Br	Lithium	C_8H_{18}	No reaction[a]	55[b]
C_6H_5Cl	Lithium	$(C_6H_5)_2$	<5[a]	70[b]
C_6H_5Br	Lithium	$(C_6H_5)_2$	<5[a]	72[b]
Bromotoluene	LiAlH$_4$	Toluene	98[c]	98[d]
Iodotoluene	LiAlH$_4$	Toluene	92[c]	95[d]
Bromonaphthalene	LiAlH$_4$	Naphthalene	70[c]	99[d]
Iodonaphthalene	LiAlH$_4$	Naphthalene	72[c]	99[d]

[a] 12–24 h, 110°C
[b] 1 h, 25°C
[c] 6–24 h, 100°C
[d] 5–6 h, 25°C

reactants come to the surface and products are desorbed and also by the availability of a 'clean' surface. Active parts of the surface can become blocked, preventing the reaction. An everyday example is that of aluminium and magnesium which are highly reactive metals but they can be commonly used since they rapidly form strongly adhering oxide layers. A metal with similar properties much used in organic synthesis is lithium and this was involved in some of the earliest reported organic sonochemistry dating from the early 1970s. Table 13.1 shows some results from Han and Boudjouk [29, 30] which clearly demonstrate that reactions which do not take place under 'silent' conditions give respectable yields when performed under ultrasound from a cleaning bath. These examples are not very significant in themselves but represent classes of reaction which are commonly used when building up the structures of complex molecules. The results can be explained by the continual cleaning action removing the oxide layers and exposing fresh metal which can react further. Also, the reduction in particle size referred to earlier gives a higher surface area and hence more reaction sites. Many other examples of these effects have been published [31–34]. The effects are not confined to metallic surfaces and modifications to the surfaces of inorganic materials and polymers have been enhanced using sonochemical methods.

The final example of a physical effect is rather more subtle. When macromolecules (molecules consisting of long chains with backbones containing up to several thousand atoms) are present in solution near a cavitation bubble, they are caught in the movement and subjected to strong shear fields. Provided that the macromolecule is sufficiently long, it is stretched out and can break under the strain. This is, in fact, one of the earliest reported effects of power ultrasound in chemistry having been noted in the 1920s. It was first attributed to thixotropic effects but it was then realised

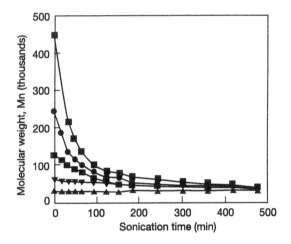

Figure 13.3. *Ultrasonic degradation of toluene solutions of polystyrenes with differing chain lengths. (The lines refer to differing starting molecular weights.)*

that the polymer chains were being broken and hence shortened. This led to a large amount of work over the succeeding decades to characterise the process in terms of the rate of bond cleavage for a wide range of polymers and the effect of the solution and ultrasound parameters. Further work [35, 36] has been carried out more recently with the advent of better apparatus for sonicating chemical reactions and, in particular, better methods of polymer analysis.

The basic effects [37] of irradiating a polymer solution with power ultrasound are shown in figure 13.3 using polystyrene in toluene as an example. The degradation proceeds more rapidly at higher molecular weights and approaches a limiting value below which no further degradation takes place, in this case ~30 000. Polymers with this, or lower values, are unaffected by ultrasound under these conditions. These effects appear to be universal in that they have been seen for a wide range of organic and inorganic polymers in organic solvents [35, 36] and for aqueous systems such as polyethylene oxide [38], cellulose [39], polypeptides, proteins [40] and DNA [41].

Most studies have shown that the degradation is relatively insensitive to the nature of the polymer. However, there is a wide range of behaviour depending on the physical properties of the solvent and of the ultrasound. In summary, the degradation proceeds faster and to lower molecular weights at lower temperatures, in more dilute solutions and in solvents with low volatility as would be expected from the discussion of the effect of these parameters on cavitation, above. Other factors which have been quantified are the ultrasound intensity and the nature of dissolved gases. There is no evidence that the extreme conditions of temperature found in cavitation

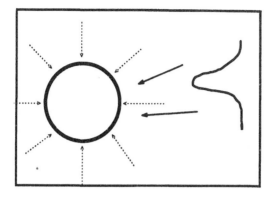

Figure 13.4. *Schematic diagram of ultrasonic polymer degradation.*

bubbles contribute to the degradation. A number of workers have shown that cleavage occurs preferentially near the middle of the chain [42, 43] and have found that the best fit to experimental data was given when the point of breakage was distributed in a Gaussian manner within ~15% of the centre of the chain. Furthermore, it has been clearly shown that neither a random model nor a model where breakage occurred exclusively at the chain centre fitted the experimental results as well as that derived from the Gaussian distribution.

Under conditions which suppress cavitation, no degradation has been found. The mechanism can briefly be best described as the polymer chain being caught in the rapid flow of solvent molecules caused by the collapse of cavitation bubbles and in the shock waves generated after the implosion of the bubbles. The chains are thus subjected to extremely large shear forces resulting in the stretching of the chain and eventual breakage of a bond in the chain as schematically shown in figure 13.4 (note that this is not drawn to scale; a polymer chain will be <1 μm in length and so fully extended compared with the 50–200 μm of the bubble). Some workers have also interpreted the effect in terms of frictional forces between the solvent molecules and polymer chains. The 'centre cleavage' model is also consistent with the stretching and breakage mechanism outlined above since the maximum stress will be generated in the middle of the chain. Degradation of polymer solutions in shear fields formed by extensional flow in narrow capillaries [44, 45], in other flow systems [46] as well as in high shear stirrers also results in preferential breakage at the chain centres.

Clearly, macromolecular degradation would be unwelcome *in vivo*. However, the physical properties of synthetic polymers depend markedly on the average chain length. The ultrasonic degradation is now understood with sufficient detail to make it commercially applicable in a number of areas as a method for controlling the chain length. The chain breakage leaves a very reactive radical chain end and this can be used to initiate further chemistry. By suitable manipulation of the experimental conditions, we can exert a great deal of control over the process, exploitation of which allows the modification of existing polymers into new materials.

All of the above examples involve heterogeneous systems where there is more than one phase present. When ultrasound is applied to a single liquid or solution, sonochemical effects rely on the generation of reactive intermediates. The extreme conditions at the centre of cavitation bubbles are sufficient to decompose all molecules which are volatile enough to enter the bubble in appreciable amounts and to form highly reactive intermediates including free radicals: very reactive species with unpaired electrons. The timescale of cavitational collapse is such that a significant proportion of these can escape from the bubble and react with other intermediates in the bubble or with compounds in solution.

In an early example, Suslick and co-workers [47] exposed long chain alkanes to high intensity ultrasound from a horn and showed that shorter chain compounds were formed in significant yields. The distribution of products suggested that the reaction was similar to those occurring in high temperature pyrolysis or 'cracking' processes such as those used to treat oil. Other, synthetically more useful examples were studied by Henglein and Fischer [48] who used the fragments formed when chloroform is exposed to ultrasound—a carbene—to introduce chlorinated groups into molecules.

A novel use of the cavitational effects was developed by Suslick *et al* [49] to prepare metal particles in unusual forms. Volatile, metal-containing compounds enter the bubbles and decompose to form metal atoms in the vapour. The lifetime of cavitational collapse is such that the temperature conditions experienced by these atoms changes from several thousand kelvin to room temperature in less than a microsecond giving huge rates of cooling. This means that as the metal atoms condense, they have no time to crystallise into regular structures and so form random aggregates of atoms. With metals such as iron, these aggregates have very different magnetic properties from iron particles produced by conventional means. The surface chemistry is also different so that the new metals have different catalytic actions.

One of the major uses (although it is the author's research area and so there may be an element of bias!) of free radicals in organic systems is to initiate polymerisation reactions to give, for example, polystyrene and PMMA. These are normally made by heating the monomers to around 120°C where radicals are formed by decomposition or by adding a compound—the initiator—which breaks down to form radicals at lower temperatures (~70–80°C). By trapping the radicals and measuring their concentration, we have shown that, by using ultrasound at 25°C, we can produce radicals at the same rate as using a conventional initiator at high temperature [50]. Carrying out the reaction at lower temperatures has some advantages in controlling the polymer structure.

Water is particularly susceptible to cavitation. Aqueous sonochemistry is dominated by a single reaction: the decomposition to hydrogen and hydroxyl radicals

$$H_2O \rightarrow H \bullet + OH \bullet$$

The hydrogen radicals are thought to rapidly combine to form hydrogen gas and are lost from the solution. However, the OH• remain and form hydrogen peroxide, H_2O_2, a powerful oxidising agent which can be used to perform much useful chemistry. There is considerable evidence for the presence of OH• from sonoluminescence [51], from electron spin resonance spectroscopies [52] and from hydrogen–deuterium isotope exchange experiments [53].

In common with the organic systems described above, the hydroxyl radicals can be used to initiate polymerisation to water soluble polymers [54]. However, some of the most interesting aqueous chemistry has involved the use of OH• and H_2O_2 as oxidising agents to destroy contaminants in water. In addition to the oxidation, volatile components can vaporize into the bubble and are literally pyrolysed to relatively harmless components. This has been an active area of research over the past few years and has now reached the point where it has been considered as a viable method of commercial waste treatment. Petrier and co-workers [55] have shown that a wide range of potential contaminants can be destroyed simply by application of high intensity ultrasound. For example, the pesticide atrazine as well as a wide range of other organic compounds have been converted totally to water, carbon dioxide and nitrates in a short period of treatment. In our work [56], we have studied aromatic compounds which can be considered as models for polycyclic aromatics (PAHs) and polychlorinated biphenyls (PCBs). Hoffmann *et al* [57] have modelled sonochemical degradation of a number of contaminants and demonstrated that the energy consumption is more efficient than high intensity ultraviolet light treatment.

Clearly, the production of large concentrations of OH• in medical applications of ultrasound would not be desirable. However, it has been detected in several systems. The techniques for detecting OH• mentioned above are rather specialised so that in our work we have used a more straightforward technique. Hydroxyl radicals react rapidly with terephthalate ions in solution to form a highly fluorescent species which can easily be monitored as shown in figure 13.5. We have used this method to extensively characterise OH• production in sonochemical systems but have also adapted it to *in vitro* studies of medical systems [58]. The apparatus in figure 13.6 was designed to allow the beam from an ultrasound source—a Therasonic 1032 physiotherapy unit which was operated at frequencies of 1.1 and 3.3 MHz— to pass through an acoustically transparent chamber contair.ing terephthalate.

Similar techniques were used by Miller and Thomas [59] who found it necessary to periodically rotate the chamber to achieve consistent results and Holland *et al* [60] who were able, using this method, to demonstrate cavitation activity in a 2.5 MHz diagnostic pulsed beam. We adopted the

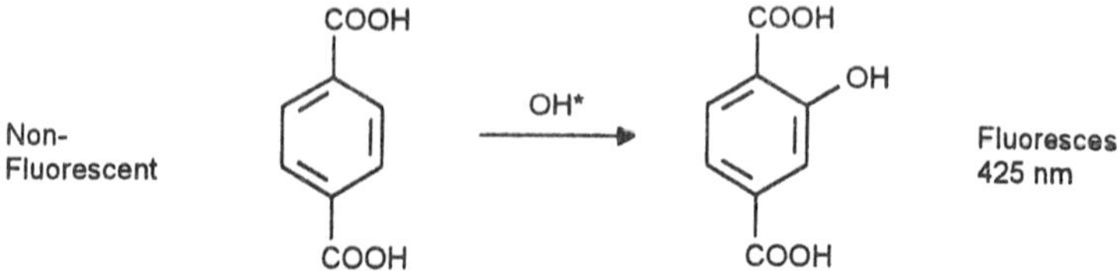

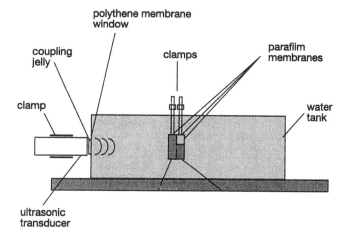

Figure 13.5. *Reaction between hydroxyl radicals and terephthalate ions in solution to form a highly fluorescent species.*

Figure 13.6. *Apparatus for monitoring hydroxyl production from medical ultrasound sources.*

technique [58] of 'seeding' the solution by adding a small quantity of solid particles to act as cavitation nuclei. Significant amounts of OH• were detected with the threshold for OH• production being 0.5 ± 0.15 W cm^{-2} at 1.1 MHz. It was noticeable that lower concentrations (by two orders of magnitude) were produced at the higher frequency.

This has been a very rapid survey of a limited range of chemical applications of ultrasound. It is impossible to do justice to a rapidly rising field of work in such a short review and, among active research topics which have not been described are detailed organic synthesis, electrochemistry, colloid science and areas of food science. The remainder of this paper is devoted to a discussion of potential benefits to therapeutic uses of ultrasound.

13.3. BIO-EFFECTS AND DRUG DELIVERY

A discussion of the various *in vivo* effects caused by ultrasound would be out of place here. It is sufficient to note that all of the effects of cavitation described above could potentially be seen *in vivo* if ultrasound is applied at the relevant combination of frequency and intensity. Possible bio-effects have been reviewed by Barnett *et al* [61] and by Miller and co-workers [62]. Clearly in many if not most medical situations (with the exception of lithotripsy) major cavitational events will have undesirable consequences. However, the remainder of this chapter is given over to a potentially more profitable application of cavitational ultrasound in therapy.

There has been considerable interest over recent years in the targeted delivery of pharmaceutical agents [63]. The aim is to ensure that a drug is present at appreciable concentrations only at the site of its action rather than throughout the body. Delivery is therefore via some type of implanted device and the aim is to develop a method of controlled release. In most cases, there is a threshold concentration below which the drug is ineffective but also a maximum concentration at which it becomes toxic. Of particular interest for this form of therapy are applications requiring long term administration such as contraception or insulin treatment of diabetes.

One form of delivery system is to implant a polymer impregnated with the active agent which is released by diffusion into surrounding tissue or body fluids [64]. However, while this gives steady concentrations of drug over long periods, the dose rate cannot be changed after implantation. One approach to overcome this deficiency was adopted by Kost *et al* [65, 66] who used ultrasound to promote the release of model compounds from a biodegradable polyanhydride matrix. The ultrasound (1 MHz, 1.7 W cm^{-2}) acted both by surface erosion of the polymer and also by accelerating the natural hydrolysis reaction. This approach has also been adopted more recently by Agrawal and co-workers [67], who used short exposures (2 min day^{-1} for 30 days) of 1 MHz, 1.5 W cm^{-2} ultrasound to promote the release of proteins from a matrix of polylactic and polyglycollic acids, both biodegradable in themselves. In an interesting twist to this technique, Kost *et al* [68] used ultrasound to rapidly remove all of the agent, in this case a contraceptive, from implanted microspheres over several days rather than the projected lifetime of several weeks. This enabled the therapy to be terminated without the need for surgical intervention.

The major method to which ultrasound has been applied has been transdermal delivery. This has the advantage that application can be made local to the desired site of action so that large concentrations of the active agent are not present elsewhere in the body. Also, the harsh conditions of the stomach and intestines are avoided so that compounds which cannot be administered orally can be used.

The outer layer of human skin—the stratum corneum—is a tough, relatively impenetrable barrier so that permeation of compounds into the bloodstream is too slow to provide therapeutic concentrations. The rate of permeation can be increased [69] by the use of chemical enhancers such as water, ethanol, N-methyl pyrollidone (NMP), polyethylene glycols or dimethyl sulphoxide (DMSO). These act by hydrating or swelling the cell and lipid structure of the skin to increase the rate of permeation. However, this approach is limited by the potential toxicity of large amounts of some of these compounds.

External modification of permeability has been achieved by the application of electric fields from electrodes placed on the skin (iontophoresis) but the widespread use of ultrasound for medical diagnosis and therapy led to its development as a technique for regulating transdermal permeation (sonophoresis or phonophoresis).

The technique was first reported some 30 years ago by Griffin *et al* [70, 71]. A large number of studies of this area has been undertaken since but there remains considerable doubt over the interpretation of the results. Many of these papers have simply reported the effect with no mechanistic information or interpretation provided. This review will summarise the findings of some studies where such interpretation was attempted. Papers in the former category are referenced in those given here.

There has been a considerable resurgence of interest over the past five years or so and in one of the earliest of these studies, Hofmann and Moll [72] included benzyl nicotinate in the coupling gel used on the skin and showed that the compound was both released from the gel at a faster rate and also permeated through skin faster than in the absence of the ultrasound. They therefore suggested that enhanced transport to the surface of the skin was the primary mechanism.

There are a number of potential effects of ultrasound which could be responsible for the observed results. For example, diffusion and permeation are accelerated by an increase in temperature which could be generated by the passage of sound. Secondly, streaming effects and the acoustic pressure could enhance transport through the gel, eliminating boundary and interfacial effects and literally forcing the drug through the skin. Finally, cavitation effects such as those noted above could disrupt the structure of the stratum corneum to form channels though which the permeants could rapidly move.

Machet and co-workers [73] found that the permeation rate of digoxin through mouse or human skin was accelerated slightly by the use of 1.1 or 3.3 MHz ultrasound at 1 W cm^{-2} but more than doubled when the intensity was increased to 3 W cm^{-2}. Significantly, they measured the temperature on the receiving side of the diffusion cell and found an increase of over 25°C for 10 min sonication at the higher intensity and also reported overt damage to the microstructure of the epidermis.

Yamashita *et al* [74] made an *in vitro* study of the permeation of prednisolone from a gel spread onto hairless mouse skin. The use of ultrasound (1 MHz) significantly enhanced the permeation and the enhancement was in proportion to the time over which it was applied. However, they also showed that any thermal effects on the skin were insufficient to account for the enhancement.

Table 13.2. *Combined effect of chemical enhancers and 1 MHz ultrasound on the transdermal permeation of corticostrone (data taken from tables 14.1 and 14.3 of [76]).*

Compound	Relative flux enhancement	
	No ultrasound	Ultrasound
Control (phosphate buffered saline at pH 7.4)	1	5
Polyethylene glycol 200	1.9	3.5
Isopropyl nyristate	4.5	16
Glycerol trioleate	0.8	1.1
50% ethanol	40	75
Linoleic acid in 50% ethanol	903	>13 000

Ueda *et al* [75] studied the effect of ultrasound (150 kHz, 0.11 W cm^{-2}) in conjunction with a series of common chemical enhancers on the rate of permeation of aminopyrine and reported a synergistic effect as did Johnson and co-workers [76]. In their work, illustrated in table 13.2, a range of chemical enhancers which operate by a variety of mechanisms could be combined with therapeutic ultrasound (1 MHz, 1.4 W cm^{-2}) to give enhanced perfusions in excess of those achieved using only one of the approaches.

Simonin [77] reviewed the literature which had been published prior to 1995 and critically re-evaluated the data. In some cases, the effect of ultrasound was not positive and depended critically on the nature of the drug, the gel used and the exact conditions. He estimated that, in a number of literature studies, the effect of heating would change the permeation by no more than around 20% and an additional increase of 10% might be expected from streaming and radiation pressure phenomena. He also suggested that, given the relative sizes of cavitation bubbles, the wavelength of the sound (~1.5 mm at 1 MHz) and the thickness of the stratum corneum, cavitation was unlikely to cause appreciable disruption to its structure. Speculation was therefore made that, although the sweat ducts may play an important role, the major factor in influencing the permeation was the disruption of the lipid layers and intracellular spaces.

These conclusions are somewhat at odds with those of Mitragoti *et al* [78] who agreed that cavitational effects were dominant but who used confocal microsopy to demonstrate that therapeutic ultrasound (1–3 MHz, 0–2 W cm^{-2}) did cause cavitation inside the stratum corneum and disrupted the structure to form channels through which the permeants could flow. A theoretical model was also developed to quantitatively explain the results. In an extension to this work, lower frequency, 20 kHz ultrasound was used rather than the 1–3 MHz 'therapeutic' frequencies used in the above work. The results here were dramatic. In most of the work at high frequencies, ultrasonic enhancements in permeation of up to 10–50 were the maximum noted. Use of 20 kHz gave enhancements of up to 3000. The authors used this as further evidence of their cavitation model since the effects of cavitation should be more pronounced at the lower frequency [79].

Meidan *et al* [80] and Byl [81] have also considered much of the literature evidence and reviewed many of the *in vitro* and *in vivo* studies. They suggest that much of the work is subjective since proper control experiments have not been performed or the precise ultrasound parameters not fully reported. However, of the proper documented trials they consider, the results confirm the suggestion that there is no single sonophoretic effect. In many cases, heating effects alone can explain the observed enhancement to permeability. However, there are other cases where the temperature changes cannot account for the observations. There does appear to be evidence in favour of the structural disruption resulting from cavitation damage.

There is a need for further work in this area and it seems to me that it is an area ripe for collaboration between the medical physicists, the analytical chemists and the sonotherapists.

REFERENCES

[1]   Suslick K S 1990 *Ultrasound: Its Chemical, Physical and Biological Effects* (New York: VCH)
[2]   Price G J (ed) 1992 *Current Trends in Sonochemistry (Royal Society of Chemistry Special Publication 116)* (Cambridge: Royal Society of Chemistry)
[3]   Margulis M A 1995 *Sonochemistry and Cavitation* (Amsterdam: Gordon and Breach)
[4]   Mason T J 1991 *Practical Sonochemistry* (Chichester: Ellis Horwood)
[5]   Martin P D 1992 Sonochemistry, a chemical engineer's view *Current Trends in Sonochemistry (Royal Society of Chemistry Special Publication 116)* (Cambridge: Royal Society of Chemistry) p 158
[6]   Berlan J and Mason T J 1992 Sonochemistry—from research laboratories to industrial plants *Ultrasonics* **30** 203
[7]   Mason T J and Cordemans E D 1996 Ultrasonic intensification of chemical processing and related operations—a review *Chem. Eng. Res. Design* **74** 511
[8]   Leighton T 1994 *The Acoustic Bubble* (London: Academic Press)
[9]   Noltingk B E and Neppiras E A 1951 *Proc. Phys. Soc.* B **64** 1032
[10] Suslick K S and Flint E B 1991 The temperature of cavitation *Science* **253** 1397
[11] Misik V and Riesz P 1996 EPR study of free radicals initiated by ultrasound in organic liquids *Ultrasonics Sonochem.* **3** 25
[12] Margulis M A 1994 The nature of sonochemical reactions and sonoluminescence *Adv. Sonochem.* **1** 13
[13] Lepoint T and Mullie F 1994 What exactly is cavitation chemistry? *Ultrasonics Sonochem.* **1** 13
[14] Hengein A, Herberger D and Guiterrez M 1992 Sonochemistry—some factors which determine the ability of a liquid to cavitate in an ultrasonic field *J. Phys. Chem.* **96** 1126

[15] Petrier C, Jeunet A, Luche J L and Reverdy G 1992 Unexpected frequency effects on the rate of oxidative processes induced by ultrasound *J. Am. Chem. Soc.* **114** 3148

[16] Entezari M E and Kruus P 1994 The effect of frequency on sonochemical reactions *Ultrasonics Sonochem.* **1** 75

[17] Sata N and Nakashina K 1943 *Bull. Chem. Soc. Japan* 18–20

[18] Price G J and Clifton A A 1996 Sonochemical acceleration of persulphate decomposition *Polymer* **37** 3971

[19] Young F R 1990 *Cavitation* (London: McGraw-Hill)

[20] Suslick K S, Casadonte D J, Green M L H and Thompson M E 1987 Effect of high intensity ultrasound on inorganic solids *Ultrasonics* **25** 56

[21] Doktycz S and Suslick K S 1990 Interparticle collisions driven by ultrasound *Science* **247** 1067

[22] Stoffer J O and Fahim M 1991 Ultrasonic dispersion of pigments in water based paints *J. Coating Technol.* **63** 61

[23] Hatate Y, Ikeura T, Shinonome M, Kondo K and Nakashio F 1981 Suspension polymerization of styrene under ultrasonic irradiation *J. Chem. Eng. Japan* **14** 38

[24] Stoffer J O, Sitton O C and Kim Y H 1992 The ultrasonic polymerization of acrylic monomers *A. C. S. Polym. Mater. Sci. Eng. Prepr.* **67** 242

[25] Biggs S and Greiser F 1995 Preparation of polystyrene latex with ultrasonic irradiation *Macromolecules* **28** 4877

[26] Lorimer J P, Mason T J, Kershaw D, Livsey I and Templeton-Knight R 1991 Effect of ultrasound on the encapsulation of titanium dioxide pigment *Coll. Polym. Sci.* **29** 392

[27] Misfik J and Kalluri P 1996 Ultrasound assisted oxidative cleavage of bonds in fatty acid esters *Lipids* **31** 1299

[28] Luche J L 1996 Sonochemical activation in organic synthesis *C. R. Acad. Sci. Paris* B **323** 337

[29] Han B H H and Boudjouk P 1981 Ultrasound promoted coupling of organic halides in the presence of lithium wire *Tet. Lett.* **22** 2757

[30] Han B H H and Boudjouk P 1981 Ultrasound promoted reduction using lithium aluminium hydride *Tet. Lett.* **22** 3813

[31] Ley S V and Low C M R 1989 *Ultrasound in Chemistry* (London: Springer)

[32] Compton R G, Ecklund J, Page S D, Mason T J and Walton D J 1996 Voltammetry in the presence of ultrasound *J. Appl. Electrochem.* **26** 775

[33] Birkin P R and Martinez S 1995 The effect of ultrasound on mass transport to a microelectrode *Chem. Commun.* 1807

[34] Madigan N A, Hagan C R S, Zhang H and Coury L A 1996 Effects of sonication on electrode surfaces and metal particles *Ultrasonics Sonochem.* **3** 249

[35] Price G J 1990 The use of ultrasound for the controlled degradation of polymer solutions *Adv. Sonochem.* **1** 231

[36] Basedow A M and Ebert K H 1977 Ultrasonic degradation of polymers in solution *Adv. Polym. Sci.* **22** 83

[37] Price G J and Smith P F 1993 Ultrasonic degradation of polymer solutions III *Eur. Polym. J.* **29** 419

[38] Chen K, Shen Y, Li H and Xu X 1995 Ultrasonic degradation of poly(ethylene oxide) solutions *Gaofenzi Tongxun* **6** 401

[39] Sato T and Nalepa D E 1977 Mechanical degradation of cellulose derivatives in aqueous solution using sonic irradiation *J. Coating Technol.* **49** 45

[40] Bradbury H and O'Shea J 1973 Denaturing and degradation of proteins *Aust. J. Biol. Sci.* **26** 583

[41] Davison P F and Freifelder D 1962 Studies on the sonic degradation of DNA *Biophys. J.* **2** 235

[42] Koda S, Mori H, Matsumoto K and Nomura H 1994 Ultrasonic degradation of water soluble polymers *Polymer* **35** 30

[43] Van der Hoff B M E and Gall C E 1977 Ultrasonic degradation of polymers *J. Macromol. Sci.* A **11** 1739

[44] Odell J A and Keller A 1986 Flow induced fracture of isolated linear macromolecules in solution *J. Polym. Sci. Polym. Phys.* **24** 1889

[45] Muller J, Odell J A, Nahr K A and Keller A 1990 Degradation of polymer solutions in extensional flows *Macromolecules* **23** 3090

[46] Nguyen T Q and Kausch H H 1992 Mechanochemical degradation in transient extensional flow *Adv. Polym. Sci.* **100** 73

[47] Suslick K S, Gawienowski J W, Chubert P F and Wang H H 1983 Alkane sonochemistry *J. Phys. Chem.* **87** 2299

[48] Henglein A and Fischer C H 1984 Sonolysis of chloroform *Ber. Bunsenges. Phys. Chem.* **88** 1196

[49] Suslick K S, Hyeon T W and Fang M M 1996 Nanostructured materials generated by high intensity ultrasound—sonochemical synthesis and catalytic studies *Chem. Mater.* **8** 2179

[50] Price G J, Norris D and West P J 1992 Polymerization of methyl methacrylate using ultrasound *Macromolecules* **25** 6447

[51] Didenco Y and Pugach S P 1994 Optical spectra of water sonoluminescence *Ultrasonics Sonochem.* **1** 19

[52] Riesz P, Kondo T and Krishna C M 1990 Free radical formation by ultrasound in aqueous solution—a spin trapping study *Free Radical Res. Commun.* **10** 27

[53] Fischer C H, Hart E J and Henglein A 1986 Ultrasonic irradiation of water in the presence of $^{18}O_2$—isotope exchange *J. Phys. Chem.* **90** 1954

[54] Lindsrom O and Lamm O 1951 Initiation of the polymerization of acrylic acid using ultrasound *J. Phys. Colloid. Chem.* **55** 1139

[55] Petrier C, David B and Laguian S 1996 Ultrasonic degradation of atrazine and pentachlorophenol in aqueous solution—preliminary results *Chemosphere* **32** 1709

[56] Price G J, Matthias P and Lenz E J 1994 The use of high power ultrasound for the destruction of aromatic compounds in solution *Proc. Safety and Environ. Protect.* B **72** 27

[57] Hoffmann M R, Hua I and Hochemer R 1996 Application of ultrasonic irradiation for the degradation of chemical contaminants in water *Ultrasonics Sonochem.* **3** 163

[58] Digby M, Duck F A, Lenz E J and Price G J 1995 Measurement of collapse cavitation in ultrasound fields *Br. J. Radiol.* **68** 1244

[59] Miller D L and Thomas R M 1993 A comparison of haemolytic and sonochemical activity of ultrasonic cavitation in a rotating tube *Ultrasound Med. Biol.* **19** 83

[60] Holland C K, Roy R A, Apfel R E and Crum L A 1992 In-vitro detection of cavitation induced by a diagnostic ultrasound system *IEEE Trans. Ultrasonics, Ferroelectr. Freq. Control* **39** 95

[61] Barnett S B, Ter Harr G R, Ziskin M C, Nyborg W L, Maeda K and Bang J 1994 Current status of research on biophysical effects of ultrasound *Ultrasound Med. Biol.* **20** 205

[62] Miller M W, Miller D L and Brayman A A 1996 A review of in-vitro bioeffects of inertial ultrasonic cavitation from a mechanistic perspective *Ultrasound Med. Biol.* **22** 1131

[63]   Johnson P and Lloyd-Jones J G 1987 *Drug Delivery Systems: Fundamentals and Techniques* (Chichester: Ellis Horwood)
[64]   Langer R 1990 Novel drug delivery systems *Chem. Br.* 232
[65]   Kost J, Leong K and Langer R 1979 Ultrasound enhanced polymer degradation and release of incorporated substances *Proc. Natl Acad. Sci.* **86** 7663
[66]   Liu S L, Kost J, D'Emanuele A and Langer R 1992 Experimental approach to elucidate the mechanism of ultrasound enhanced polymer erosion and release of incorporated substances *Macromolecules* **25** 123
[67]   Agrawal C M, Kennedy M E and Micallef D M 1994 The effects of ultrasound on biodegradable 50–50 copolymer of polylactic and polyglycollic acids *J. Biomed. Mater. Res.* **28** 851
[68]   Kost J, Liu S L, Gabelnick H and Langer R 1993 Ultrasound as a potential trigger to terminate the activity of contraceptive delivery implants *J. Control. Release* **30** 77
[69]   Walters K A and Hadgraft J 1993 *Pharmaceutical Skin Penetration Enhancement* (New York: Dekker)
[70]   Griffin J E and Touchstone J C 1963 Ultrasonic movement of cortisol into pig tissues *Am. J. Phys.* **42** 77
[71]   Griffin J E, Enternach J L, Price R E and Touchstone J C 1967 Patients treated with ultrasonic driven hydrocortisone and with ultrasound alone *Phys. Therapy* **47** 594
[72]   Hoffmann D and Moll F 1993 The effect of ultrasound on in-vitro liberation and in-vivo penetration of benzyl nicotinate *J. Control. Release* **27** 185
[73]   Machet L, Pinton J, Patat F, Arbeille B, Pourcellot L and Vaillant L 1996 In vitro phonophoresis of digoxin across hairless mice and human skin *Int. J. Pharm.* **133** 39
[74]   Yamashita K, Hirai Y and Tojo K 1996 Effect of ultrasound on rate of drug absorption through skin *J. Chem. Eng. Japan* **29** 812
[75]   Ueda H, Isshiki R, Ogihara M, Sugibayashi K and Morimoto Y 1996 Combined effect of ultrasound and chemical enhancers on the skin permeation of aminopyrine *Int. J. Pharm.* **143** 37
[76]   Johnson M E, Mitragoti S, Patel A, Blankschtein D and Langer R 1996 Synergistic effects of chemical enhancers and therapeutic ultrasound on transdermal drug delivery *J. Pharm. Sci.* **85** 670
[77]   Simonin J P 1995 On the mechanism of in-vitro and in-vivo phonophoresis *J. Control. Release* **33** 125
[78]   Mitragoti S, Edwards D A, Blankschtein D and Langer R 1995 A mechanistic study of ultrasonically enhanced transdermal drug delivery *J. Pharm. Sci.* **84** 697
[79]   Mitragoti S, Blankschtein D and Langer R 1996 Transdermal drug delivery using low frequency ultrasound *Pharm. Res.* **84** 697
[80]   Meidan V M, Walmsley A D and Irwin W J 1995 Phonophoresis—is it a reality? *Int. J. Pharm.* **118** 129
[81]   Byl N N 1995 The use of ultrasound as an enhancer for transcutaneous drug delivery *Phys. Therapy* **75** 539

PART 5

RESEARCH TOPICS IN MEDICAL ULTRASOUND

CHAPTER 14

IMAGING ELASTIC PROPERTIES OF TISSUE

James F Greenleaf, Richard L Ehman, Mostafa Fatemi and Raja Muthupillai

14.1. INTRODUCTION

Palpation is routinely used for the evaluation of mechanical properties of tissue in regions that are accessible to touch. Detecting pathology using the 'stiffness' of the tissue is more than 2000 years old. It is common for surgeons to find lesions during surgery that have been missed by advanced imaging methods. However, palpation is subjective and limited to individual experience and to the accessibility of the tissue region to touch. A means of non-invasively imaging elastic modulus (the ratio of applied stress to strain) anywhere in the body may be useful to distinguish tissues and pathologic processes based on mechanical properties such as elastic modulus (Sarvazyan *et al* 1994). Many approaches to imaging mechanical properties have been developed over the years (Krouskop *et al* 1987, Parker *et al* 1990, Lerner *et al* 1990, Bertrand *et al* 1989, Ophir *et al* 1991, Skovoroda *et al* 1994, O'Donnell *et al* 1994). Conventional imaging methods are typically used to measure mechanical strain of tissue in response to mechanical stress. Static, quasi-static or cyclic stresses have been applied. The resulting strains have been measured using ultrasound (Sarvazyan *et al* 1994, Krouskop *et al* 1987, Parker *et al* 1990, Lerner *et al* 1990, Bertrand *et al* 1989, Ophir *et al* 1991, Skovoroda *et al* 1994) or MRI (Lewa and de Certaines 1995a,b, Plewes *et al* 1994, Muthupillai *et al* 1995a,b, Fowlkes *et al* 1995) and the related elastic modulus has been computed from visco-elastic models of tissue mechanics.

14.1.1. *Exogenous transverse waves: imaging with MRE*

Recently a new MRI phase contrast technique has been reported in which transverse strain waves propagating in tissue are imaged (Muthupillai *et al* 1995a,b, 1996). Because the wavelengths of propagating waves are related to density and the shear modulus, this method promises to have good resolution and to be sensitive to the shear modulus. The wavelengths of low frequency transverse waves are of the order of millimetres. This paper reviews the theory of the method, presents some applications and discusses the implications of the method.

14.1.2. Stimulated acoustic emission: imaging with USAE

Other methods can be used to look at the global (bulk) mechanical parameters of objects. For example, in *Resonant Ultrasound Spectroscopy* (Maynard 1996), an ultrasound source and detector are used to measure the resonance frequencies of a sample with known size and mass. These values are used to calculate the bulk mechanical parameters, including the elastic constants, of the material. In this chapter, we present a method to interrogate, or image the object at high spatial resolution, and at the same time, to obtain the low frequency response of the object at each point. The imaging technique used here is based on the radiation force (Torr 1984, Westervelt 1951) of two ultrasound beams reported by the authors elsewhere (Fatemi and Greenleaf 1998). Briefly, two ultrasound beams of slightly differing frequency interfere on, or within, an object and the resulting dynamic radiation force produces an acoustic field that is dependent on mechanical properties of the object. The acoustic field is detected with a microphone or hydrophone.

14.2.   MAGNETIC RESONANCE ELASTOGRAPHY (MRE)

Because of the effect of motion of spins in MRI, tissue displacement can be measured with MRI. When the resulting images represent the elasticity of the object, the imaging method is termed MRE (magnetic resonance elastography).

14.2.1.   Theory

We will term the microscopic region of tissue that responds to the MRI signal a spin. The transverse magnetization phase $\Phi(\tau)$, of a spin moving in the presence of a varying magnetic field gradient $G(t)$, is given by

$$\Phi(\tau) = \gamma \int_0^\tau G(t) \cdot r(t) \, dt \qquad (14.1)$$

where $r(t)$ is the position vector of the moving spin, and γ is the gyromagnetic ratio which is characteristic of the spin under investigation.

In the case of a propagating transverse wave in the acoustic frequency range, the position vector can be described by

$$r(t) = r_0 + \xi_0 \exp(\mathrm{i}(k \cdot r - \omega t + \phi)) \tag{14.2}$$

where r_0 is the mean position of the spin, ω is the angular frequency of the mechanical excitation causing the strain wave, ϕ is the initial phase offset, k is the wave number of the strain wave, and ξ_0 is the peak displacement of the spin from its mean position.

If the gradient $G(t)$ is a square wave with magnitude alternating between $+|G|$ and $-|G|$ for N cycles with a period T equal to the strain wave period $2\pi/\omega$, then the resulting phase shift in the received signal is given by

$$\Phi(r, t) = \gamma \int_0^{\tau=NT} G(t) \cdot [r_0 + \xi_0 \exp(\mathrm{i}(k \cdot r - \omega t + \phi))] \, \mathrm{d}t$$
$$= \frac{2\gamma N T (G \cdot \xi)}{\pi} \sin(k \cdot r + \phi). \tag{14.3}$$

Equation (14.3) illustrates that the phase shift is related to the scalar product of the displacement and gradient vectors, thus projections of the motion can be obtained for each of the three coordinate axes if three sets of data are obtained with mutually orthogonal gradients. The phase shift is also proportional to the product NT which is the total on-time of the gradient. By varying ϕ, the phase offset between the gradients and the mechanical excitation, images for different points in time can be obtained, producing cine loops that depict the wave propagation through time.

14.2.2. Methods

Tests were performed in a series of experiments using a 1.5-T Signa imager†, to provide examples of applications of the theory.

Phantoms which simulated tissues with varying stiffness were made by mixing varying amounts of agarose‡ in distilled water at around 70°C.

A schematic of the experimental setup for generating strain waves is shown in figure 14.1. We constructed an electromagnetic electromechanical actuator, similar to a speaker mechanism, to produce transverse waves in the acoustic frequency range. A waveform generator fed a sinusoidal signal to the actuator coil through a stereo amplifier. The sinusoid was synchronized with the gradient cycles provided by the MR imager.

A phase contrast gradient echo sequence with cyclic gradient waveforms was used for imaging (figure 14.2). The number of trigger pulses was

---

† GE Medical Systems, Milwaukee, Wisconsin.
‡ Bacto Agar, Difco Laboratories, Detroit, Michigan, 1.0–3.0% w/w.

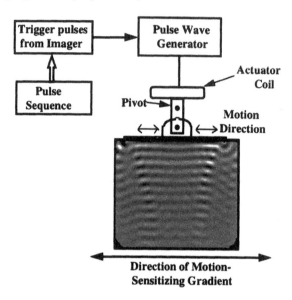

Figure 14.1. *Schematic for producing low frequency transverse waves in phantoms or tissues. The actuator coil is on a shaft that is hinged. When current is passed through the coil, the current causes the coil to align with the B field of the main magnet forcing a plate on the phantom to shift back and forth. This produces transverse waves in the phantom. The wave generator is triggered from the MR imager ensuring that the gradients within the imager are switched in synchrony with the excitation wave frequency. Reproduced with permission of Williams and Wilkins from Muthupillai et al (1996).*

adjustable and could be initiated prior to the imaging cycle, producing images of waves that had propagated into the phantom for various depths.

Each repetition of imaging pulses was done twice, alternating polarity of the motion-sensitising gradients (figure 14.2). This acquisition scheme reduced phase noise and doubled the sensitivity of the phase measurements. In the examples shown the gradients $G(t)$ were collinear with the displacement ξ_0 to measure transverse waves. Ranges of data acquisition parameters were: repetition time, 50–300 ms; echo time, 15–60 ms; acquisition matrix, 128–256; acquisition time, 20–120 s; flip angle, 10–60°. The frequency of mechanical excitation ranged from 100 to 1100 Hz and the number of gradient pulses (N) ranged from 2 to 30 cycles. The images are of phase displacement computed from equation (14.3), after correction for the balanced acquisition.

14.2.3.   MRE results

Transverse strain waves in an agarose gel phantom are shown in figure 14.3. The motion of the actuator was orthogonal to the plane of the figure and

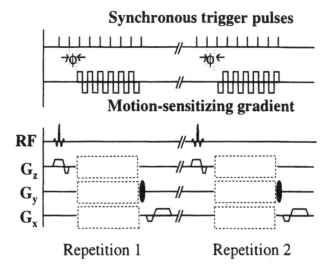

Figure 14.2. *Pulses to trigger the waveform generator, shown in figure 14.1, are produced in synchrony with the motion-sensitising gradients. The phase relationship between the gradients and the mechanical drive φ, can be altered. The motion-sensitising gradients can be applied along arbitrary axes allowing measurement of displacement along any axis. Two repetitions are used with opposite motion-sensitising gradients to subtract out systematic phase errors and to double sensitivity (NT in equation (14.3)) to synchronous motion. Reproduced with permission of Williams and Wilkins from Muthupillai et al (1996).*

collinear with the sensitising gradients. In one case, contact on the phantom was a point and in the other a plane. The results were spherical transverse waves and plane transverse waves, respectively. The images were produced from motion-sensitising gradients that were initiated after a near steady state condition was achieved by the actuator. The point source image illustrates the inverse radius relationship of amplitude and shows some faint reflections from the walls of the phantom. The plane source image illustrates edge waves caused by the edge of the 'aperture' and some reflections from the walls.

Figure 14.4 illustrates the propagation of a transverse pulse through a heterogeneous phantom consisting of two regions having different hardnesses. The harder gel (2% agar) is at the top and the softer gel (1.25%) is poured in the container at an angle. Refraction of wave propagation is shown in the area marked by the arrow.

The shear modulus for inhomogeneous material can be calculated as shown in figure 14.5. Two agarose gel cylinders of high and low stiffness were embedded in an agarose phantom of medium stiffness. A displacement image was obtained at 250 Hz using plane wave excitation. Figure 14.5(top) displays the displacement image of transverse waves propagating through the phantom and the two cylinders. The stiffer cylinder, on the left, displays

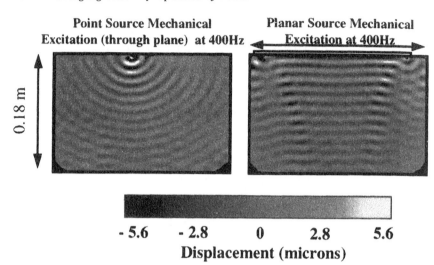

Figure 14.3. *Images of transverse waves produced at 400 Hz in agar phantom material using the actuator of figure 14.1. A point contact produced spherical transverse waves (left) and a planar contact produced plane waves (right). The number of gradient cycles used for the measurement was 15. Equation (14.3) was used for the calibration scale. Reproduced with permission of Williams and Wilkins from Muthupillai et al (1996).*

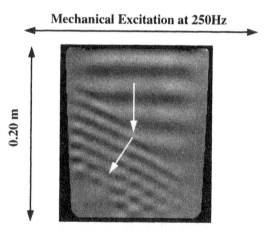

Figure 14.4. *Refraction of a propagating transverse wave is exhibited in this displacement image for mechanical excitation of 250 Hz. Six gradient-sensitising cycles were used. The white arrow depicts wave propagation direction for the agarose gel poured into the phantom container at an angle. Reproduced with permission of Williams and Wilkins from Muthupillai et al (1996).*

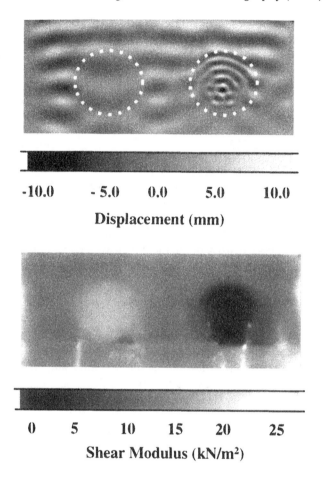

Figure 14.5. *Magnetic resonance elastography images of two cylinders of agarose with high and low stiffness embedded in a background material with average stiffness. Plane transverse waves were applied at the top of the phantom. The phase image taken at 250 Hz shows large wavelength in the hard cylinder and short wavelength in the soft cylinder (top). Computation of the local wavelength and the assumption of density of 1 g cm^{-3} results in a quantitative image of shear modulus (bottom). The indicated hardness of the cylinder on the left is about 22 kN m^{-2} compared to stiffness measurements of large samples of the same gel of 23.8 kN m^{-2}. Reprinted with permission from Muthupillai et al, Science, vol 269, 1854–7, copyright 1995 American Association for the Advancement of Science.*

larger wavelengths and the softer cylinder on the right, shorter wavelengths than the background material. This image was processed with a local wavelength estimation filter (Manduca *et al* 1996) to compute the local shear modulus assuming the density of the material to be constant at

1.0 g cm^{-3}. The resulting quantitative image of shear modulus is shown in figure 14.5(bottom).

14.3.   ULTRASOUND STIMULATED ACOUSTIC EMISSION (USAE)

14.3.1.   Theory of USAE

The general principle is that force is generated by a change in the energy density of an incident acoustic field. Let us consider a collimated ultrasound beam interacting with an object of arbitrary shape and boundary impedance that scatters and absorbs. The radiation force arising from this interaction has a component, F, in the beam direction. This component is proportional to the time-averaged energy density of the incident wave $\langle E \rangle$ and the projected area of the object, S, as shown in the following relationship (Westervelt 1951):

$$F = DS\langle E \rangle \tag{14.4}$$

where D is the drag coefficient, which is a function of the scattering and absorbing properties of the object (see Chapter 3).

Application of this force for object interrogation and imaging, as will be described later, ideally requires the stress field to be confined to a point, while its amplitude oscillates at selected frequencies. To generate a localised oscillatory field, two intersecting continuous wave (CW) focused ultrasound beams of different frequencies are used. It is only in the intersection region that the ultrasound field energy density is sinusoidally modulated, and hence the field can generate an oscillatory radiation force by interaction with the object. The ultrasound beams can be shaped in a variety of ways for this purpose. An interesting configuration is obtained when two coaxial, confocal transducers are used (figure 14.6). In this case, the elements of a two-element spherically focused annular array (consisting of a central disc and an outer ring) are excited by two CW signals at frequencies ω_1 and ω_2. Let us assume that the beams are propagating in the $+z$ direction with the joint focal point at $z = 0$. The resultant field on the $z = 0$ plane may be written as:

$$S(t) = g_1(x, y) \cos(\omega_1 t + \psi_1(x, y)) + g_2(x, y) \cos(\omega_2 t + \psi_2(x, y)) \tag{14.5}$$

where $g_1(x, y)$ and $g_2(x, y)$ are the beam profiles with ψ_1 and ψ_2 being the associated phase functions across the $z = 0$ plane. For focused beams $g_1(x, y)$ and $g_2(x, y)$ diminish quickly away from the origin. Let the difference frequency $\Delta\omega = \omega_2 - \omega_1 \ll \omega_1, \omega_2$, then it can be shown that the short time average of the acoustic energy density in the intersection region

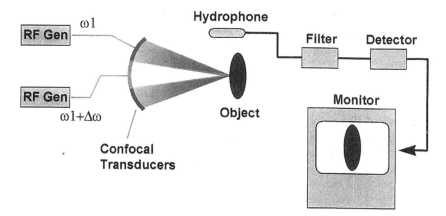

Figure 14.6. *Experimental system for ultrasonically stimulated acoustic emission. The system employs a confocal ultrasound annular array transducer, consisting of a centre disc and an outer ring. Transducer centre frequency is 3 MHz, outer diameter 22.5 mm and it is focused at 70 mm. The elements are driven by two CW sources, at frequencies equal to ω_1 and $\omega_2 = \omega_1 + \Delta\omega$, where these frequencies are very close to the centre frequency of the elements, and $\Delta\omega$ is much smaller than ($<1\%$) the centre frequency. The transducer and its beam pattern is illustrated on the left, showing that beams interact only at a small region around the joint focal point, where the amplitude of the field oscillates at the difference frequency $\Delta\omega$. The object is placed at the joint focal point vicinity, and interrogated point-by-point by scanning. The sound field resulting from object vibrations at each position is received by a hydrophone and recorded after digitisation. The recorded signal at one or more difference frequencies is used to form an image of the object. The experiments are conducted in a water tank.*

has slow variations at frequency $\Delta\omega$ about its long time average. Denoting this low frequency component by $E_1(t, x, y)$, we can write

$$E_1(t, x, y) = Ag_1(x, y)g_2(x, y)\cos[\Delta\omega t + \psi_2(x, y) - \psi_1(x, y)] \quad (14.6)$$

where A is a constant relating the field quantity to the energy density. Now, consider a planar target on the $z = 0$ plane. Referring to equation (14.4), and considering that the average energy density is position and time dependent in this case, the normal component of the time varying force on this target is found by the following integration:

$$F_1(t) = D \int \int_S E_1(t, x, y) \, dx \, dy$$
$$= CD \cos(\Delta\omega t + \alpha) \quad (14.7)$$

where C is a constant and α is a phase constant. S is the area over which $E_1(t, x, y)$ has significant values. This area, defined as the interaction region of the beams, can be very small for well focused beams. Hence, $F_1(t)$ can be thought of as an oscillating point force applied to the object at the origin.

Object vibrations result in an emission of acoustic energy which can be detected by a microphone (or hydrophone in water). The complex amplitude of this field, $\Phi(\Delta\omega)$, can be written as (ignoring α)

$$\Phi(\Delta\omega) = CDH(\Delta\omega)Q(\Delta\omega) \qquad (14.8)$$

where $Q(\Delta\omega)$ is a complex function representing the mechanical frequency response of the object at this point, and $H(\Delta\omega)$ represents the combined frequency response of the propagation medium and the microphone. $H(\Delta\omega)$ is assumed to be unchanged for any target point in the object. By keeping the source intensity constant, C in equation (14.8) will remain fixed at a given point in the object. Now, knowing $H(\Delta\omega)$, by changing $\Delta\omega$ and recording $\Phi(\Delta\omega)$, we can obtain $Q(\Delta\omega)$ for that point within a constant multiplier. Repeating the same procedure for other points of the object at a given plane (e.g. the $z = 0$ plane), we can collect a set of responses. These data can be mapped into a pictorial format. For instance, at a fixed frequency $\Delta\omega = \Delta\omega_0$, the value of $|\Phi(\Delta\omega)|$ at various points represents the spatial distribution of $D|Q(\Delta\omega_0)|$, which can be mapped into a monochromatic image, displaying object morphology. The spatial resolution of the resulting image is determined by the area S, where its size is of the order of the ultrasound power beamwidth at the focal point. To display frequency dependency information in addition to D distribution, we can use colour. For instance, the frequency response at each point can be divided into three frequency bands using three bandpass filters. The average amplitude in each band is mapped into a monochrome image. By assigning a different colour for each image, a multi-colour image (spectrograph) is formed by superimposing the three colour components. In this case, D influences the brightness, while the frequency variations of $|Q(\Delta\omega_0)|$ are represented by the hue of each image pixel. In practice, $|H(\Delta\omega)|$ may not be known, but since it is position independent, variations of image hue versus position qualitatively represent $|Q(\Delta\omega)|$ variations.

14.3.2.   USAE results

The experimental setup used to test the method is shown in figure 14.6. The experiment was conducted in a water tank because water is a good acoustic coupling medium while providing freedom of movement for the prototype scanner mechanism. A two-element confocal ultrasound transducer array was positioned such that the beams meet the object at their joint focal point. Sound produced by the object vibrations was detected by a submerged hydrophone placed within the water tank or on the object. In a system designed for *in vivo* imaging, one would use soft tissue as the coupling medium instead of water.

Basic parameters of an imaging system, such as the spatial resolution, can be evaluated using an isolated point as the elementary test object. The

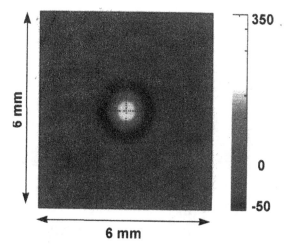

Figure 14.7. *Experimentally measured point spread function of the system. An isolated point target was simulated by a 380 μm diameter glass bead placed on a piece of thin latex sheet. The latex surface was scanned at 0.2 mm increments in either direction. The difference frequency was fixed at 7.3 kHz. The acoustic spectrogram of the bead is shown in grey scale. The latex sheet is almost transparent to the imaging system. The glass bead presents a high acoustic impedance discontinuity resulting in significant radiation force, and thus produces a significant acoustic field when exposed to the two beams and stands out in the image with high contrast. The image shows that the system resolution is about 500 μm, similar to that of the confocal transducer.*

resulting image is called the point spread function (PSF) of the system. For this purpose, we used a 380 μm diameter glass bead as a model for a point and placed it on a piece of thin latex sheet. The latex sheet produces only a small change in the incident energy, and is almost transparent to the ultrasonic beam. The entire object was placed in a water tank and the latex sheet surface was scanned in a raster format. The difference frequency, in this case, was fixed at 7.3 kHz. The resulting PSF is shown in figure 14.7. Fundamental image resolution, defined as the width of the PSF, was approximately 500 μm in either dimension. This result shows that the system is capable of detecting objects smaller than its resolution cell.

To show the capability of the method in displaying the frequency responses of an object versus position, a set of three tuning forks was chosen as the test object. The forks are made from identical material, and have identical finger cross-sections (lengths are different). Resonance frequencies in water are: 407 Hz (right), 809 Hz (middle) and 1709 Hz (left). The forks were scanned in a water tank using the system shown in figure 14.6. The scanning plane covers the front fingers at the bottom part of the forks. The digitised hydrophone signal was filtered by three overlapping bandpass filters each having −6 dB bandwidth of 500 Hz. The outputs of the filters with centre

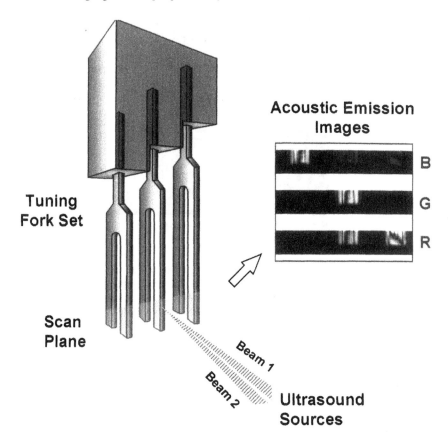

Figure 14.8. *Spectrogram of the tuning forks. The filtered components are shown in grey scale: (R) the red component, (G) the green component and (B) the blue component.*

frequencies 500, 1000 and 1500 kHz were used to produce the red, green and blue image components, respectively. The three components are shown in figure 14.8 as three grey scale images. To obtain a colour image these images can be coded in appropriate colours and superimposed (not shown).

Next we investigated the possibility of using the both the amplitude and phase of the acoustic emission signal to image the differences in the mechanical properties of biological tissues. For this, we used specimens of calcified and non-calcified human iliac arteries. The arteries were scanned in a plane perpendicular to the beam axis. The amplitude and phase of the acoustic emission field were recorded at each position. The difference frequency was fixed at 6 kHz. Resulting images are shown in figure 14.9. Calcifications are clearly identified as they exhibit distinctive amplitude and phase values when compared to the arterial walls.

X-ray:

Acoustic Emission:

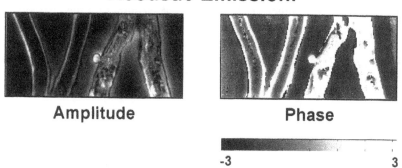

Amplitude   **Phase**

-3   3

Figure 14.9. *Acoustic spectrography of human iliac arteries. Top: X-ray image of normal (N) and calcified (C) excised human iliac arteries obtained from a 35-year-old woman and a 67-year-old man, respectively. Bright areas indicate calcifications. Bottom left: Acoustic spectrogram amplitude image at the difference frequency of 6.0 kHz. Calcification details appear distinctively bright, while the arterial walls are dim. Bottom right: Corresponding phase image, indicating that the calcified regions vibrate out of phase with respect to the soft tissues.*

14.4. CONCLUSIONS

Both MRE and USAE can obtain images of mechanical properties of tissues. Each has its place in imaging, perhaps in complementary applications.

14.4.1. MRE

Using synchronous motion-sensitising gradients, magnetic resonance elastography (MRE) is capable of producing either snapshots or cine loops of cyclic displacement caused by externally applied transverse waves in the acoustic range. The displacement can be measured in the three orthogonal directions providing a vector valued function ($\xi(r, t)$ in equation (14.3)). MRE can be used to study propagation effects such as refraction and diffraction and can provide quantitative measures of shear modulus.

The advantages of MRE are that it can provide images virtually anywhere in the body and in any orientation. For qualitative studies of wave propagation, such as refraction and diffraction (figures 14.3 and 14.4), MRE has the same resolution as a standard MR imaging system of about 0.5 mm. For shear modulus or wavelength measurements (figure 14.5), MRE is quantitative with a spatial resolution of the order of the wavelength of the transverse elastic waves and a contrast resolution of several per cent (Smith *et al* 1997) for stiffness or attenuation.

A disadvantage of MRE is that each two-dimensional displacement image requires separate repetitions of the mechanical excitation over a time span of about one minute to acquire data for the image. This acquisition time is multiplied by the number of phase offsets, ϕ, acquired for the cine loop and is still longer for measurements of three-dimensional displacement vectors.

MRE certainly provides a practical new experimental tool for measuring propagating transverse waves in tissues and tissue-like media. MRE may also provide a new clinical imaging tool, extending the classical palpation methods to anywhere in the body and to full three-dimensional quantitative levels.

14.4.2. USAE

Measurement of the PSF of USAE, shown in figure 14.7, indicates that the system resolution is about 0.5 mm, which is about the beam width of the ultrasound system. This is in agreement with the experimental images of the tuning forks. This system can be used as an imaging tool. However one must note that the PSF of the system is angle sensitive.

The method described in this chapter provides a combination of high resolution measurement of local and bulk physical parameters of the object. Applications of this method include non-destructive evaluation of materials and characterisation of biological tissues.

ACKNOWLEDGMENTS

The authors thank Elaine Quarve for secretarial assistance and Randy Kinnick, Ultrasound Laboratory, and T C Hulshizer, MRI Research Laboratory, for technical assistance.

REFERENCES

Bertrand M, Meunier J, Doucet M and Ferland G 1989 Ultrasonic biomechanical strain gauge based on speckle tracking *IEEE 1989 Ultrasonics Symp. Proc.* vol 2, pp 859–63
Fatemi M and Greenleaf J F 1998 Ultrasound-stimulated vibro-acoustic spectrography *Science* **280** 82–5
Fowlkes J B, Emelianov S Y, Pipe J P, Skovoroda S R, Carson P L, Adler R S and Sarvazyan A P 1995 Magnetic resonance imaging techniques for detection of elasticity variation *Med. Phys.* **22** 1771–8

Krouskop T A, Dougherty D R and Vinson F S 1987 A pulsed Doppler ultrasonic system for making noninvasive measurements of the mechanical properties of soft tissue *J. Rehabil. Res. Dev.* **24** 1–8

Lerner R M, Huang S R and Parker K J 1990 Sonoelasticity images derived from ultrasound signals in mechanically vibrated tissues *Ultrasound Med. Biol.* **16** 231–9

Lewa C J and de Certaines J D 1995a MR imaging of viscoelastic properties *J. Magn. Reson. Imaging* **5** 242–4

—— 1995b Stejskal Tanner method application to viscoelastic properties of NMR detection *Proc. SMR and ESMRBMB, 3rd Scientific Meeting (Nice)* p 690

Manduca A, Muthupillai R, Rossman P J, Greenleaf J F and Ehman R L 1996 Image processing for magnetic resonance elastography *Medical Imaging 1996: Image Processing, Proc. SPIE* **2710** 616–23

Maynard J 1996 Resonant ultrasound spectroscopy *Phys. Today* 26–31

Muthupillai R, Lomas D J, Rossman P J, Greenleaf J F, Manduca A and Ehman R L 1995a Magnetic resonance elastography by direct visualization of acoustic strain waves *Science* **269** 1854–7

Muthupillai R, Lomas D J, Rossman P J, Greenleaf J F, Manduca A, Riederer S J and Ehman R L 1995b Magnetic resonance imaging of acoustic strain waves *Proc. SMR and ESMRMB, 3rd Scientific Meeting (Nice)* p 189

Muthupillai R, Rossman P J, Lomas D J, Greenleaf J F, Riederer S J and Ehman R L 1996 Magnetic resonance imaging of transverse acoustic strain waves *Magn. Reson. Imaging* **36** 266–74

O'Donnell M, Skovoroda A R, Shapo B M and Emelianov S Y 1994 Internal displacement and strain imaging using ultrasonic speckle tracking *IEEE Trans. Ultrasonics Ferroelectr. Freq. Control* **41** 314–25

Ophir J, Cespedes I, Ponnekanti H, Yazdi Y and Li X 1991 Elastography: a quantitative method for imaging the elasticity of biological tissues *Ultrasonics Imaging* **13** 111–34

Parker K J, Huang S R, Musulin R A and Lerner R M 1990 Tissue response to mechanical vibrations for sonoelasticity imaging *Ultrasound Med. Biol.* **16** 241–6

Plewes D B, Betty I and Soutar I 1994 Visualizing tissue compliance with MRI *Proc. SMR, 2nd Ann. Meeting (San Francisco, CA)* p 410

Sarvazyan A P, Sovoroda A R, Emelianov S Y, Fowlkes J B, Pipe J G, Adler R S, Buxton R B and Carson P L 1994 Biophysical bases of elasticity imaging *Acoustical Imaging* vol 21, ed J P Jones (New York: Plenum)

Skovoroda A R, Emelianov S Y, Lubinski M A, Sarvazyan A P and O'Donnell M 1994 Theoretical analysis and verification of ultrasound displacement and strain imaging *IEEE Trans. Ultrasonics Ferroelectr. Freq. Control* **41** 302–13

Smith J A, Muthupillai R, Greenleaf J F, Rossman P J, Hulshizer T C and Ehman R L 1997 Characterization of biomaterials using magnetic resonance elastography *Review of Progress in Quantitative Nondestructive Evaluation* vol 16, ed D O Thompson and D E Chimenti (New York: Plenum) pp 1323–30

Torr G R 1984 The acoustic radiation force *Am. J. Phys.* **52** 402–8

Westervelt P J 1951 The theory of steady force caused by sound waves *J. Acoust. Soc. Am.* **23** 312–15

Yamakoshi Y, Sato J and Sato T 1990 Ultrasonic imaging of internal vibration of soft tissue under forced vibration *IEEE Trans. Ultrasonic Ferroelectr. Freq. Control* **37** 45–53

CHAPTER 15

THE SIGNAL-TO-NOISE RELATIONSHIP FOR INVESTIGATIVE ULTRASOUND

Christopher R Hill

Ultrasound offers the potential basis for a substantial set of clinical investigative techniques, only a few of which are at present in widespread use. At the same time, ultrasound is itself a subset of the full range of physical techniques that are available for investigation of particular clinical problems. This chapter analyses quality of investigative performance in this context, in terms of the relationship between tissue-informative signals and 'noise' (in the broad sense of the term), and discusses the derivation of a quantitative figure of merit for investigative performance. This approach has more general application than simply to ultrasound but indicates, in particular, that some hitherto underdeveloped ultrasound-based techniques could show considerable promise in the clinical investigation of tumours and other pathologies.

There is now of course a considerable armoury of physical techniques which are, or potentially might be, used for clinical investigation. Ultrasound provides the basis for one subset of these methods and, in spite of its widespread use, it is still far from being exploited to its full potential. Contemporary investigative ultrasound† is based almost entirely on two techniques that happened to be technically feasible some 30 or more years ago: backscatter amplitude imaging and backscatter Doppler processing. With modern technology, a whole range of additional approaches has become possible: speed of sound reconstruction imaging, and movement correlation processing, to name just two.

---

† Although 'diagnostic ultrasound' is conventional jargon, the clinical investigations to which ultrasound can contribute encompass much more than simply diagnosis. Hence the preferred use here of the term 'investigative'.

We should therefore be asking ourselves a number of questions, such as:

- What, technically, are we trying to do in investigative ultrasound?
- How well are we doing this job?
- How do we define and quantify 'doing well' in this context?
- Could we 'do better' using alternative, ultrasound-based methods?
- How does ultrasound performance in doing a particular job compare quantitatively with other investigative modalities?
- Thus, in general, how good is ultrasound?

Finding answers to such questions may take a long time but the purpose of this chapter is, first, to suggest that the questions are worthy of study and, second, to indicate some approaches that may help towards finding solutions.

Physical methods of medical investigation, whether based on X-rays, radioisotopes, NMR, ultrasound or any other means are all exercises to which the methods of communications science can usefully be applied (see, e.g. ICRU 1996). Whatever the particular method, the investigative process essentially entails retrieval of a 'signal' carrying information about some aspect of a patient's condition, and conveying to an 'observer' as much of that information as is not obscured by 'noise'. In simplistic terms, we are thus concerned with a 'signal-to-noise ratio' for the process. Before this can be helpful, however, we need to consider the nature both of the signals and of the noise, and we also need to think about the characteristics of the observer.

Particularly in the present context of investigative ultrasound, these are complex issues on which no general consensus can be found in the literature. They have however been discussed in some detail in two papers by the present author (Hill *et al* 1990, Hill *et al* 1991), only some of the content of which it will be possible to summarise here.

Clinical investigations generally require information signals that will help them answer two separate but interrelated classes of question: 'What is it?' and 'Where is it?' Thus the quality of performance of an investigative process needs to be characterised correspondingly. First, how accurately and precisely can a given anatomical structure be located in space (and/or, related to this, with what spatial resolution can two neighbouring structures be distinguished)? And, second, how accurately and precisely can the material/tissue of an anatomical region be identified (and/or, related to this, with what sensitivity can two such materials/tissues be distinguished)? A third, practically important consideration is the speed with which the process can provide these answers.

The situation is, however, complicated by the interrelated nature of these three aspects of performance. Spatial resolution is, for example, generally best for high contrast structures, while the sensitivity with which a tissue can be identified will generally increase with the quantity of tissue available for study (thus inversely with spatial resolution of the identification process) and also with the time available for the process.

In all this the concept of contrast resolution turns out to be central. Hill *et al* (1990) analyse this in terms that any individual tissue exhibits a value, P_i, of some physical property and that this gives rise, in some ultrasonic investigative process, to a signal value, S_i. One can then consider systematically the processes that give rise to uncertainty ('noise') in S_i and, correspondingly, in the value of P_i, which is what, ideally, the observer wishes to know.

Formally, if an investigative process is required to differentiate between two different targets/tissues on the basis of their physical property values P_i and P_j, the criterion for success is

$$\Delta = P_i - P_j > U \qquad (15.1)$$

where U is the numerical uncertainty (or 'noise') of the system and the scale of the inequality constitutes a figure of merit for the system. Uncertainty, in this context, arises from a number of different sources, all of which can affect the accuracy of an observation but only some of which may affect its precision†. These sources include the following:

• The relative property separation, P_i/P_j, as between two tissue types.
• The 'biological' variance, $s^2(P)_b$, in P corresponding to nominally identical tissue types. A related variance, $s^2(S)_b$, will occur for the corresponding signals.
• The technical precision, $s^2(S)_t$, with which the technique transforms the numerical property measure, P, into a signal, S.

The foregoing is treated in more detail in Hill *et al* (1990), where it is shown that a general expression for the figure of merit of an investigative system is

$$\Gamma = (dS/dP)P_iN^{1/2}/\sigma \qquad (15.2)$$

where σ is the effective standard deviation (noise level) of the system (i.e. the square root of the sum of the variances) and N is the number of uncorrelated measurements employed.

It is interesting to note here that this is essentially a more general form of the expression that has been derived independently for comparing the virtue of various pulse sequences in obtaining precise values of NMR relaxation times (Edelstein *et al* 1983). A special consideration arises, however, when this treatment is applied to investigative ultrasound. This relates to the third of the above factors, the *P–S* transformation precision, and arises because of the coherent nature of the ultrasound sources employed, which gives rise to a fundamentally important source of 'noise': coherent radiation speckle.

---

† 'Accuracy' is an index of the extent to which an experimentally measured value of a quantity approximates its true value; 'precision' is an index of the extent to which different measured values of some particular quantity approximate to each other.

At this point it will be helpful to digress briefly with a discussion of the various uses that we have been making of the term 'noise'. In its most general sense, the term 'noise' may connote any artefact that can be mistaken for a wanted signal and cannot be distinguished from it. Conventionally, however, and particularly in radio engineering, 'noise' has the restricted connotation of a random (and thus broad frequency-band) source of pseudo-signal. Coherent speckle, however, is an artefact that has a relatively narrow frequency spectrum set by the driving source (in our case the ultrasonic transducer), and may be more appropriately termed, as in radar parlance, 'clutter'.

For investigative ultrasound, coherent radiation speckle is doubly bad news. In the first place, its frequency spectrum peaks exactly with that of the system modulation transfer function (MTF), and thus specifically degrades perception of detail. On top of this, it may also interact negatively with the particular nature of the visual physiology of a human observer. This situation is discussed in Hill *et al* (1991) and illustrated here in figure 15.1, which is reproduced from that paper. The problem arises because human vision is maximally sensitive to image features exhibiting a particular spatial frequency, of the order of 2 cycles degree^{-1}. Presenting images to an observer at such a magnification that the spatial resolution of the ultrasonic imaging system appears in the presented image at this spatial frequency would thus seem, in one sense, to optimise overall machine observer performance. Unfortunately, by simultaneously maximising the perception of speckle, this arrangement is also a worst case for achievement of contrast discrimination.

Two approaches have been considered for mitigation of this problem. In the first place, attempts have been made to use partially incoherent systems but, without incurring unacceptable degradation in sensitivity and spatial resolution, this has proved to be of only very limited value. Secondly, however, the role of human observer performance suggests substitution of a more nearly ideal, machine observer and, in certain circumstances, this has been shown to yield significant improvement in overall performance (Wagner *et al* 1991).

Referring back to the discussion on equations (15.1) and (15.2), speckle noise contributes to the uncertainty in the $P–S$ transformation. A further contribution comes from electronic noise and the relationship between the two is illustrated in figure 15.2. This shows the condition—unusual elsewhere—of an S/N ratio constant with depth up to a boundary where electronic noise takes over from speckle; a situation that can be observed on suitably displayed scans. It therefore seems that we have to live with speckle, and thus accept, in terms of equation (15.2), the limitation that this imposes on attempts to minimise the value of σ. To achieve good contrast discrimination we therefore have to look for ways to obtain high values for the (dS/dP) factor. More specifically, for any given differentiation task,

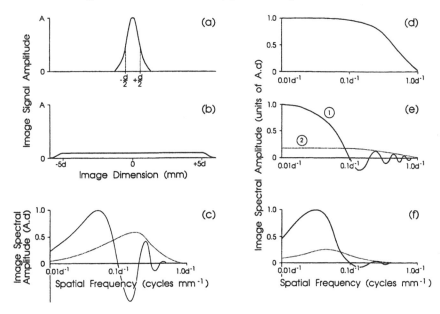

Figure 15.1. *Illustration of how the characteristics of human vision can influence the signal-to-noise ratio (SNR) of an investigative procedure that uses coherent radiation. (c) shows the form of the line spread function (lsf) for an imaging system in which the parameter 'd' represents the full width of the lsf at half maximum amplitude (fwhm). (d) is the corresponding modulation transfer function (MTF) of the system, which is identical in shape to the MTF of the system speckle noise. Curve (b), drawn here to have the same area as (a), is the spread function of a low-contrast object of linear dimension 10d (e.g. a tumour) and (e, 1) is its MTF. (e, 2) is the corresponding speckle noise spectrum, as in (d) but drawn here to have the same numerical area under the curve as (e, 1), taking into account the logarithmic scale. The difference in area between two such signal and noise spectra (the 'modulation transfer function area' or MTFA) is a well recognised measure of SNR. Introduction of the human observer into the calculation now entails appropriately multiplying the data of (e) by the spatial angular frequency sensitivity of the human eye–brain system, i.e. taking into account the effect of viewing distance: see e.g. Pearson (1975). (c) and (f) show the resulting spectra for viewing distances of 0.5 and 2 m respectively where, in each case, image luminance is of the order 500 cd m^{-2} and the parameter 'd' has the value 1 mm in object space (e.g. on a viewing screen). From such data SNR can be calculated as 10 log$_{10}$(A$_s$/A$_n$), where A$_s$ and A$_n$ are the areas under the signal and noise spectra respectively. In the above examples, for (f) SNR $\approx$ 10 dB while, for (c), SNR is negative: 'Now you see it, now you don't'.*

we need to look for a class of signal whose value will differ substantially between the two property conditions to be differentiated.

In one sense, this is not saying anything new. It is well known, for example, that the reason why ultrasound is generally better than CT for

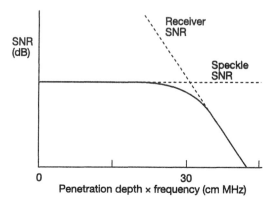

Figure 15.2. *Approximate form of the dependence of SNR of an echographic system on penetration depth, for the common situation where signal arises as contrast between different regions of a scattering medium, such as a soft tissue (from Hill et al 1991).*

Table 15.1. *Classes of signal and corresponding physical properties.*

Class of signal	Physical property
Backscatter echo amplitude	Bulk backscattering coefficient
Slope of echo amplitude with depth	Attenuation coefficient
Slope of echo amplitude with frequency (and/or features of image texture)	Characteristic spacing of tissue texture
Speed of sound	Bulk modulus/density
Nonlinearity parameter, B/A	Bulk modulus/density
Tissue movement correlation	Shear modulus

solid–cystic differentiation is that, while the X-ray attenuation coefficient in the two tissues may be very similar, amplitudes of ultrasonic backscatter received from them are very different. What this does suggest, however, is that, particularly when conventional ultrasonic backscatter imaging is unsatisfactory, we might look for different classes of signal for which the (dS/dP) value may be more favourable.

What are the possible classes of signal that we might consider? This is the subject of what has come to be called ultrasonic *tissue characterisation*, or, more descriptively, if less conventionally, *telehistology*. There has been considerable interest in this subject in recent years and what is remarkable in the present context is the large number of 'classes of signal' that have been considered and investigated. A number of these are listed here in table 15.1, together with the corresponding principal physical property that such a signal represents.

Some of the merits and demerits of the employment of such signals are

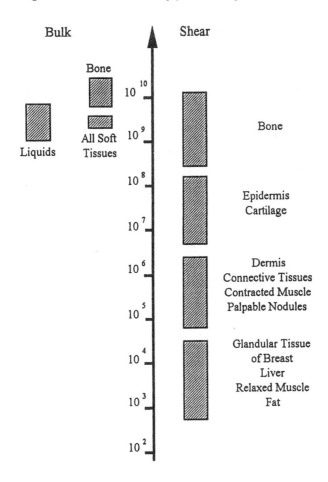

Figure 15.3. *Comparative spread of measured values of bulk and shear elastic moduli (pascal) for various mammalian tissues (adapted from Sarvazyan and Hill (1999)).*

discussed in Hill *et al* (1990). As an example, we can consider here the last on the list in table 15.1: investigation of tissue shear modulus via observation of movement correlation. All the more conventional sources of signal, as listed in this table, correspond in some way to tissue properties based on some combination of tissue density and bulk elastic modulus, and local variations of these quantities. While these can often be productive of usable signals, the actual variation of these quantities between different soft tissues is typically rather small. This is illustrated in figure 15.3 (taken from Sarvazyan and Hill (1999)). This figure contrasts graphically the ranges of variation that occur within various tissue groups for bulk and shear moduli, and demonstrates the relatively very small spread of bulk modulus values. It is the much higher numerical range of shear modulus values that underlies the promising results of the technique that has become known as 'elastography': essentially the quantitation, via ultrasonic imaging, of the rheological properties of tissues (Tristam *et al* 1988). An illustrative example of the potential value of this approach is in the investigation of some neoplasms, whose structure may lead them to show appreciable anomalies in shear modulus (and thus, if sufficiently superficial, may be manually palpable), while still yielding undetectable bulk mechanical, and even X-ray or NMR, contrast.

In conclusion, part of the answer to the question: 'How good is ultrasound?' is that it depends on how intelligently it is applied to the particular task in hand. There are systematic ways of determining a figure of merit for the application of a given technique to a given task, although such a quantitative approach has generally not been followed hitherto. At the same time, ultrasound provides, potentially, a very rich and varied set of approaches for obtaining informative 'signals' relating to human tissues. The reasons why many of these have not been substantially exploited hitherto may have more to do with the mechanics of commercial marketing than with their inherent potential as investigative tools. Thus we may eventually see some very interesting developments from ideas that are already current.

REFERENCES

Edelstein W A, Bottomley P A, Hart S R and Smith L S 1983 Signal, noise and contrast in nuclear magnetic resonance (NMR) imaging *J. Comput. Assist. Tomogr.* **7** 391–401

Hill C R, Bamber J C and Cosgrove D O 1990 Performance criteria for quantitative ultrasonology and image parameterisation *Clin. Phys. Physiol. Meas.* **11** (Suppl. A) 57–73

Hill C R, Bamber J C, Crawford D C, Lowe H J and Webb S 1991 What might echography learn from image science? *Ultrasound Med. Biol.* **17** 559–75

ICRU 1996 *Medical Imaging—the Assessment of Image Quality* ICRU Report No 54 (Bethesda, MD: International Commission on Radiation Units and Measurements)

Pearson D E 1975 *Transmission and Display of Pictorial Information* (London: Pentech)

Sarvazyan A and Hill C R 1999 Physical chemistry of the ultrasound–tissue interaction *Physical Principles of Medical Ultrasonics* ed C R Hill, J C Bamber and G R ter Haar (London: Wiley) 2nd edn, ch 7, in press

Tristam M, Barbosa D C, Cosgrove D O, Bamber J C and Hill C R 1988 Application of Fourier analysis to clinical study of patterns of tissue movement *Ultrasound Med. Biol.* **14** 695–707

Wagner R F, Insana M F, Brown D G, Garra B S and Jennings R J 1991 Texture discrimination: radiologist, machine and man *Vision, Coding and Efficiency* ed C Blakemore (Cambridge: Cambridge University Press)

CHAPTER 16

CHALLENGES IN THE ULTRASONIC MEASUREMENT OF BONE

John G Truscott and Roland Strelitzki

INTRODUCTION

In 1994 the UK's Advisory Group on Osteoporosis reported that the prevalence of low bone density is 22% in women over 50 years of age and results in 60 000 hip fractures a year in the UK. The most frequent osteoporotic fractures are of hip, spine and wrist and by the time they are 70 years old almost half of all women will have experienced such a fracture [1]. The cost to the UK National Health Service of such fractures is some £750 000 000 per year.

Current measurement techniques used to identify osteoporosis involve the use of ionising radiation (e.g. dual energy X-ray absorptiometry (DXA)), or single photon absorptiometry (SPA) to measure bone mineral density (BMD) or bone mineral content (BMC) in the lumbar spine and the hip. Over the last 5 years, however, a number of ultrasonic bone measurement devices have come onto the market. The calcaneus has been the measurement site of choice for several reasons. Vogel and co-workers brought out five key points in a recent review [2]:

- The calcaneus is more than 90% trabecular (see below) by volume.
- In women, age-related bone loss in the calcaneus and lumbar spine are similar, with both commencing before the menopause.
- Calcaneal BMD reflects spinal osteoporosis as effectively as spinal BMD.
- Fracture risk in both spinal and non-spinal sites may best be determined from calcaneal bone mineral content measurement.
- Although bone measurements at many sites are correlated, indicating the systemic nature of osteoporosis, it is not possible to predict a bone value at one site from a measure made elsewhere in the body, suggesting

Figure 16.1. *Photomicrograph of cortical bone. Concentric lamellae surround the central Haversian canal. Reproduced with permission of Lippincott-Raven from [16].*

that correlation of calcaneal ultrasound with other techniques (e.g. DXA of spine/femur) is of questionable value.

A recent European prospective study of hip fracture in 5662 elderly women [3] demonstrated ultrasonic measurement of the calcaneum to be as good a predictor of future hip fracture risk as DXA of the femoral neck. This work has since been confirmed by an American study of 6189 women aged over 65 [4]. Because of these factors, and the accessibility of the heel, the majority of ultrasound systems measure parameters in the calcaneus and we shall concentrate on that site in this chapter.

16.1. BONE

Bone may be classified into two basic types: cortical and trabecular. Cortical bone is hard and compact and is typically found in the shafts of long bones. Figure 16.1 shows a photomicrograph of this type of bone and its dense, lamellar nature can be seen. In the body some 80% of bone is of this type. The other 20% is trabecular bone which comprises an open network of struts and bars of bone with the interstices containing bone marrow. This network can be seen in figure 16.2 which is a scanning electron micrograph of healthy trabecular bone. Such bone is typically found at the head of long bones (e.g. the head of the femur), in the vertebrae and also in the calcaneus.

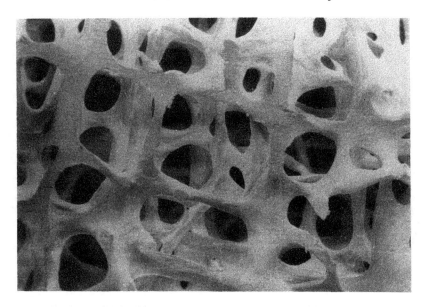

Figure 16.2. *Scanning electron micrograph of normal trabecular bone showing thick trabecular plates all connected. Courtesy of Alan Boyde.*

Bone is not an inert substance but an organ which undergoes continuous remodelling throughout life. Cells known as osteoclasts resorb bone and are followed by osteoblasts which form and deposit bone matrix which is then mineralised. Any imbalance in this process can lead to pathological bone being created. The processes of major interest to us in osteoporosis are due either to over-resorbtion or to inadequate formation of bone. Bone remodelling in health and osteoporosis are shown in figure 16.3. As trabecular bone has a structure which gives it a large surface area per unit volume it is metabolically very active, typically having an eight times higher turnover rate than cortical bone, making sites containing this type of bone ideal sites for observing bone loss. The effect of osteoporosis on trabecular bone is dramatically illustrated in the scanning electron micrograph in figure 16.4 where marked thinning and disconnection of the trabeculae can be noted. The strength of the bone is severely undermined by such changes and can lead to fractures being caused by very small impacts. In fact some fractures in very osteoporotic bone appear to occur spontaneously. As mentioned in the introduction, the calcaneus is a site of high trabecular content making it ideal for the measurement of bone condition.

16.2. ULTRASONIC MEASUREMENTS SUITABLE FOR BONE

Current commercial bone measurement systems concentrate on two measurements of the ultrasonic wave: attenuation due to the calcaneus and

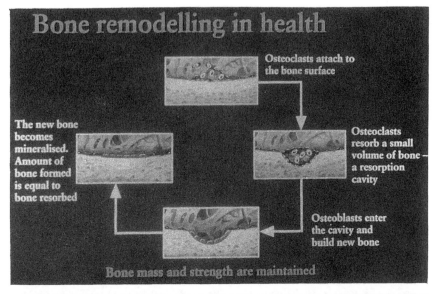

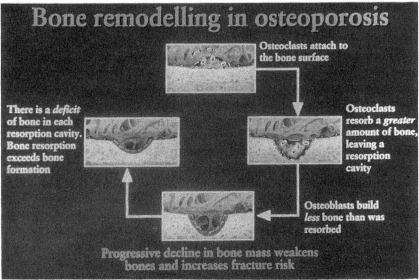

Figure 16.3. *Bone remodelling in health and in osteoporosis. Courtesy of Merck, Sharpe and Dohme.*

the speed of sound along the acoustic path. Because of the density of bone and its highly attenuative properties the only mode suitable for making such measurements is by transmission (insertion technique). This rules out such frequently used methods as 'sing around' and 'pulse echo overlap' (see

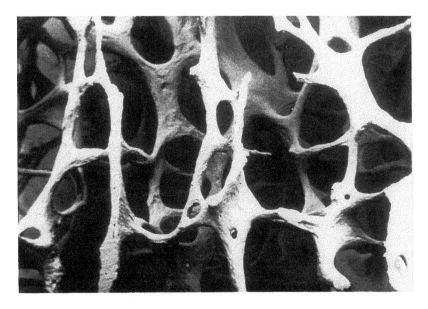

Figure 16.4. *Scanning electron micrograph of osteoporotic trabecular bone showing marked thinning and disconnection of trabeculae. Courtesy of Alan Boyde.*

Chapter 4). Measurements are often made in a water bath to aid transducer-to-subject coupling. Dry systems also exist and these assume the same insertion technique as wet systems using a virtual water path to allow for a comparison trace to assess changes in the signal when the heel is placed in the beam.

16.2.1. Speed of sound (SOS)

Using a water bath with two fixed, co-axial transducers, one acting as a transmitter and one as a receiver, it is possible to measure the change of the time-of-flight of an acoustic pulse caused by the insertion of the heel into the acoustic path. Because the sound travels faster in bone than in water, the acoustic pulse will arrive earlier by a time Δt, which, if the speed of sound in water is known, may be used to calculate the speed of sound in the calcaneus. This measurement is shown in figure 16.5 from which the velocity in the subject (c_s) can be calculated as

$$c_s = \frac{d_s c_w}{d_s - c_w \Delta t} \tag{16.1}$$

where c_w is the velocity in water and d_s the heel width.

In current water bath systems the manufacturers' assumptions about heel width are not known. In systems which use surface contact the distance

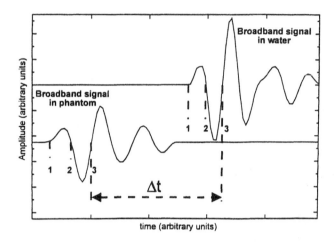

Figure 16.5. *Schematic diagram of the principle used for measurements of the velocity with a broadband signal. The numbers represent the zero-crossing points of the broadband signal in water and bone.*

between the transducers is measured and assumed to be the heel width. Contact systems assume an insertion technique and use an effective bone-free signal which is based on the time of passage of the signal through a soft-tissue-equivalent material of known transmission speed (e.g. water). The assumption of water as a soft tissue equivalent or of homogeneity of soft tissue and fat coverage of the heel bone may lead to inaccuracy in the measurement of speed of sound. For example in people with a high fat content in the soft tissue of the heel the measured speed of sound may be reduced independent of bone condition.

16.2.2. Attenuation

Many commercial systems confine attenuation measurements to a technique known as broadband ultrasonic attenuation (BUA), a measurement first developed by Langton *et al* in 1984 [5]. The following description of the measurement of BUA is again based on a system using water as a coupling medium between transducers and heel. Two transducers are located co-axially in a waterbath, one acting as a transmitter and the other as a receiver. The heel is positioned between the two as shown in figure 16.6. A broadband frequency-rich pulse is generated at the transmitter (typically having frequencies in the band 0.2–1 MHz), passed through the heel and collected at the receiver. Similarly a pulse is transmitted through the water bath with the heel removed. The two received signals are analysed between 200 kHz and 1 MHz to establish the frequency related amplitude of the received signals. Two such signals are shown in figure 16.7. The difference

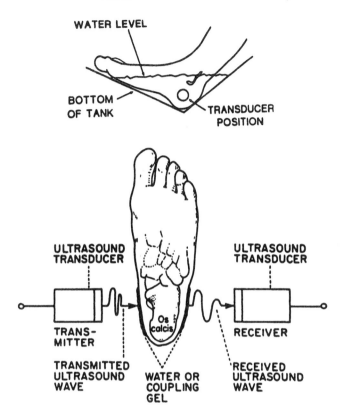

Figure 16.6. *The position of the foot in the water bath shown (top) medio-laterally and (bottom) in plantar view (i.e. from the sole of the foot). Courtesy of Christian Langton.*

between the two represents the attenuation due to the heel bone and when expressed as a logarithmic ratio gives us a diagnostic tool. A graph of frequency dependent attenuation is shown in figure 16.8 and the slope of this graph defines BUA which is expressed in dB MHz^{-1}. Under the assumption that the relationship between attenuation coefficient and frequency is linear, of which more later, BUA may be defined in terms of the log ratio of amplitudes of the Fourier transforms of the transmitted signals:

$$\text{BUA} = \frac{\mathrm{d}}{\mathrm{d}f}\left[20\log_{10}\left[\frac{A_\text{W}(f)}{A_\text{H}(f)}\right]\right] \tag{16.2}$$

where f is the frequency (200 kHz to 1 MHz), $A_\text{W}(f)$ is the amplitude of the Fourier transform without the heel present and $A_\text{H}(f)$ is the amplitude of the Fourier transform with the heel in the beam.

Figure 16.9 shows two graphs: one for a young healthy female and the other for a female with a fractured neck of femur. The higher value of BUA

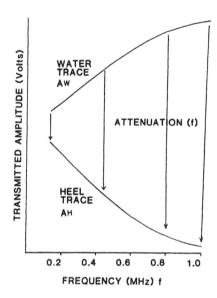

Figure 16.7. *Water and heel amplitude traces. Courtesy of Christian Langton.*

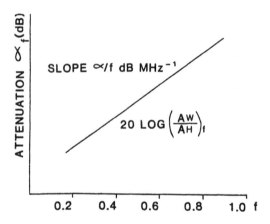

Figure 16.8. *Attenuation trace for the two curves in figure 16.7. Courtesy of Christian Langton.*

is assumed to be indicative of bone of better 'quality'. Some noise and nonlinearity is obvious in the higher frequency parts of the upper graph and for this reason the region between 200 and 600 kHz has been adopted for clinical use by most manufacturers.

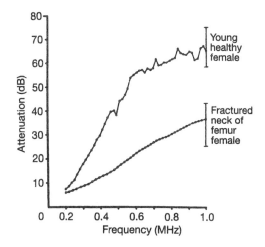

Figure 16.9. *Comparative traces for a typical young female adult and typical fractured neck of femur patient. Courtesy of Christian Langton.*

16.2.3. Problems

A number of factors lead to variability in the measurement of BUA in bone. For example, transducer coupling, phase cancellation and the fitting of a linear model are all sources of variation. Most important, however, is the structural and directional architecture of the bone which has an effect on the signals, which will be considered in the next section. As the measurement has been shown to be a good predictor of future hip fracture [3, 4] this variability has been tolerated.

Speed of sound measurement is inherently a more accurate and precise measurement to make. It is, however, not without problems and we shall explore these and some potential ways of understanding them later in this chapter.

16.3. EFFECT OF STRUCTURE ON BROADBAND ULTRASONIC ATTENUATION

In order to assess the effects of structure on measured BUA we measured a range of phantoms with porosities between 46.5% and 82.4%. There were two sets of these phantoms with uniform pore sizes of 0.6 mm and 1.3 mm respectively. These phantoms were manufactured using a technique described by Clarke *et al* [6]. Liquid epoxy (CW1302, Ciba Geigy, Duxford, Cambridge, UK) and hardener were degassed at 3 Torr in a vacuum chamber and then mixed with cubic granules of gelatine. The sieve size for the production of these granules was either 0.8 mm or 1.5 mm and thus pore sizes in the two phantom sets differed by a factor of about two. The gelatine

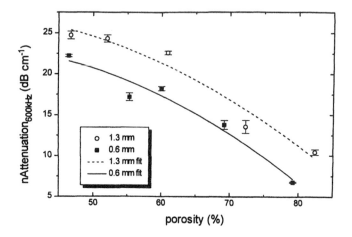

Figure 16.10. *Size normalised attenuation coefficient as a function of porosity for two different pore sizes.*

and epoxy were mixed to give various porosities and left to cure for four weeks. The phantom blocks were 40 mm long by 50 mm wide by 15 mm thick.

These porosities and pore sizes are well within those found *in vivo* [7]. The attenuation coefficient (dB cm^{-1}) of each phantom was measured at 600 kHz. The data and fitted curves are shown in figure 16.10. Attenuation coefficients normalised for phantom size (nBUA, dB MHz^{-1} cm^{-1}) were also measured in the range 200–600 kHz and these data and their fitted curves are shown in figure 16.11. Two striking factors emerge from these data:

- both attenuation coefficient and BUA reduce with increasing porosity;
- for a given porosity, a larger pore size produces an increase in both BUA and attenuation coefficient.

As porosity and density are inversely related, the measurement of BUA can be seen to reflect density. However, the changes in pore size demonstrate a change in BUA or attenuation coefficient with the internal structure of the material. This could go some way to explaining the different values of BUA found for similar density values in studies of excised bone [8–11]. As structure has such a marked effect upon BUA at any density value it is not surprising that attempts to correlate BUA with DXA measures of bone mineral density (which of course has no structural component in its measurement) have been inconclusive whereas those measuring some aspects of strength, albeit in terms of future fracture risk [3, 4], have shown BUA to be a useful parameter.

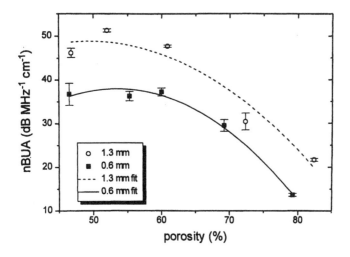

Figure 16.11. *Size normalised BUA as a function of porosity for two different pore sizes.*

16.4.   PROBLEMS IN THE MEASUREMENT OF SPEED OF SOUND

16.4.1.   Time domain (zero-crossing point measurement)

As stated above, the majority of commercial systems use the pulse insertion technique, either explicitly or implicitly, to make measurements of speed of sound. This involves the selection of a reference point on each of the two waveforms between which Δt can be measured. Figure 16.12 shows a broadband signal pair and five reference points which may be considered. Zero-crossing points have been selected to mirror the criteria used in commercial instruments. A broadband signal has also been used again, to be equivalent to commercial systems in which BUA is measured, allowing both measurements to be made from a single acquisition. The measured time difference between the reference points is changing from point to point and this is shown in figure 16.13. This figure also indicates the speed measured using a 750 kHz single frequency toneburst (centre frequency of the transducers). This single frequency measurement of speed is close to the speed of sound measurement associated with the use of the third zero-crossing points. Some systems use the first zero-crossing and others the third which could lead to differences in measured speed of sound of as much as 100 m s^{-1} in the same sample. This effect may be explained if we remember that BUA is such that the attenuation dependence on frequency is approximately linear. This would tend to attenuate higher frequencies more than lower frequencies producing a shift in the frequency content of the attenuated pulse towards the lower end of the spectrum and leading to an increase in spacing between zero-crossing points. Measured

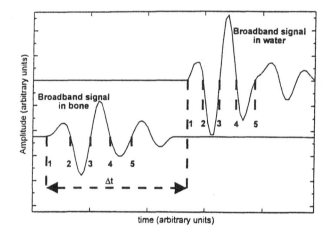

Figure 16.12. *Schematic diagram of the principle used for measurements of the velocity with a broadband signal. The numbers (1–5) represent the zero-crossing points of the broadband signal measured with water and bone.*

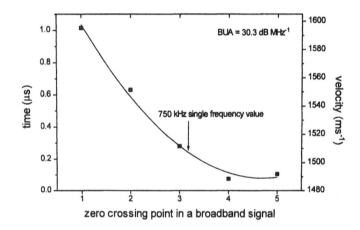

Figure 16.13. *Measured times and calculated velocities with a broadband signal on its zero-crossing points on the os calcis.*

BUA in a range of bone samples is compared to the time between second and third zero-crossing points, for that sample, in figure 16.14. An obvious linear proportionality can be seen. This prompted us to model the situation on a computer. We digitised a broadband pulse after passage through water, applied freqency-dependent attenuation at 20, 40 and 60 dB MHz^{-1} in the Fourier domain, and inverse transformed to give the waveforms shown in figure 16.15. For clarity, only the water, 20 dB MHz^{-1} and 60 dB MHz^{-1}, simulated waveforms are shown. No account was taken of phase in this

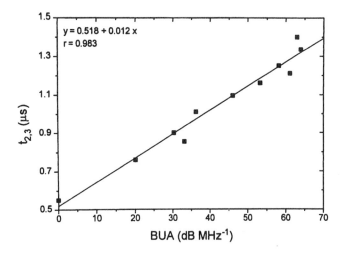

Figure 16.14. *Time between second and third zero-crossing point in the broadband signal.*

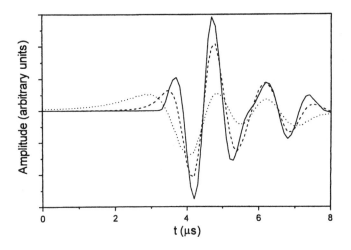

Figure 16.15. *Simulated BUA on a scan taken through water (solid line) for 20 dB MHz^{-1} (dashed line) and 60 dB MHz^{-1} (dotted line).*

analysis. The time differences for each zero-crossing point can be calculated and these are plotted in figure 16.16 (cf. figure 16.13) which is similar to the *in vivo* case. The time differences differ least among themselves around the third zero-crossing point and, remembering the 750 kHz toneburst measurement, one is tempted to conclude that this may represent some form of 'true' value.

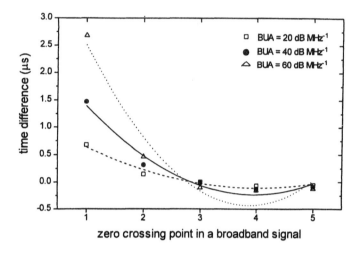

Figure 16.16. *Time differences taken from a broadband signal (water) to the simulated signals of different BUA (fitting curves: 20 dB MHz^{-1}, dashed; 40 dB MHz^{-1}, solid; and 60 dB MHz^{-1}, dotted) on the zero-crossing points.*

16.4.2. *Frequency domain measurements*

Because of the problems identified with zero-crossing methods it may be worthwhile examining some other techniques which are carried out in the frequency domain rather than the time domain. We are, in some ways, driven to these by the fact that there are practical difficulties associated with single frequency measurements in commercial systems. We have considered three measurements made in frequency space: phase velocity, group velocity and mean pulse velocity. Except for group velocity, a value of Δt needs to be obtained in order to calculate velocity from equation (16.1). Group velocity is obtained from phase velocity data. All measurements utilised broadband pulses and did not restrict themselves to the range 200–600 kHz currently in use in clinical practice.

16.4.2.1. Phase velocity. Measurements were carried out on 10 os calces taken from human cadavers. Details of age and sex were unknown. The samples were stored in 10% buffered formaldehyde solution at 4°C and were subsequently degassed at 3 torr in distilled water for four hours prior to measurements being made. For each bone sample used in this investigation digitised waveforms, at 40 ns resolution, were obtained with and without the sample present in a waterbath. Both waves were subjected to Fourier transformation to produce phase values similar to those shown in figure 16.17(top) which were then 'unwrapped' to give a continuous angular scale shown in figure 16.17(bottom). The time difference (Δt) at

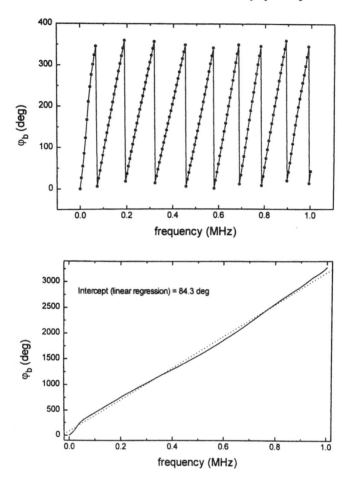

Figure 16.17. *(Top) phase of the signal and (bottom) unwrapped phase of the signal (solid line) and its linear regression (dotted line); with a bone inserted in a water tank.*

each frequency (f) could then be calculated from:

$$\Delta t(f) = \frac{\varphi_w(f) - \varphi_b(f)}{360 f} \tag{16.3}$$

where φ_w and φ_b are the phases through water and bone respectively at frequency f. These values of $\Delta t(f)$ can be used in equation (16.1) to calculate the phase velocity (c_p) which is frequency dependent. This velocity is plotted as a function of frequency for two bones in figure 16.18. A curious feature of these curves is the negative change in velocity with frequency (velocity dispersion dc_p/df) which can be seen in all the samples tested

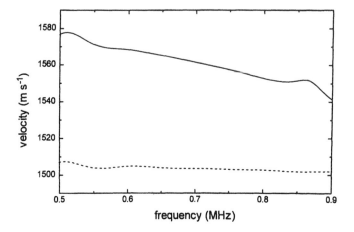

Figure 16.18. *Velocity as a function of frequency for two bones of different density (bone 6 (solid line) and bone 7 (dotted line) from table 16.1).*

Table 16.1. *Data for 10 samples of calcaneum: c_p phase velocity (700 kHz); dc_p/df slope of phase velocity versus frequency (velocity dispersion); c_g group velocity at 700 kHz; c_{cc} mean pulse velocity; $c_p(f_{centre})$ phase velocity at transducer centre frequency (750 kHz); nAtten attenuation coefficient at 700 kHz normalised for sample size; nBUA BUA for the range 600–800 kHz normalised for size.*

Sample	c_p m s^{-1}	dc_p/df m s^{-1} MHz^{-1}	c_g m s^{-1}	c_{cc} m s^{-1}	$c_p(f_{centre})$ m s^{-1}	nAtten dB cm^{-1}	nBUA dB MHz^{-1} cm^{-1}
1	1544	−35.3	1520	1547	1543	16.0	34.0
2	1511	−4.1	1508	1511	1511	7.0	12.5
3	1528	−39.2	1501	1528	1525	12.6	23.3
4	1552	−63.4	1509	1553	1549	16.1	26.5
5	1508	−12.4	1499	1507	1507	6.5	12.1
6	1561	−79.6	1507	1562	1557	16.4	39.9
7	1503	−11.0	1496	1504	1503	5.7	11.9
8	1511	−6.6	1507	1510	1511	7.8	15.9
9	1517	−12.0	1508	1515	1515	7.9	14.0
10	1540	−51.9	1504	1542	1537	14.4	29.2

(table 16.1). Such negative dispersion is rarely found in nature but has been reported in the lung [12] which is also highly attenuating and porous. The predominant finding in the lung was of slight positive dispersion but much of the lower frequency data showed a pronounced downward trend in group velocity values with increasing frequency. More recently Nicholson [13], in measurements on 30 os calces, obtained data which displayed negative dispersion although the authors did not comment on this phenomenon. Another study involving skull bone [14], however, demonstrates positive dispersion but this could be because both density and structure are so different in the skull from the calcaneum.

16.4.2.2. Group velocity. The group velocity (c_g) may be calculated from the phase velocity and its dispersion using the equation

$$c_p = \frac{c_g}{1 - (f/c_p)(dc_p/df)}. \tag{16.4}$$

As velocity dispersion is negative in all cases this always leads to group velocity being lower than phase velocity (see table 16.1).

16.4.2.3. Mean pulse velocity. The mean pulse velocity (c_{cc}) is obtained, indirectly, from a cross-correlation in Fourier space of the waves digitised for the calculation of phase velocity. The cross-correlation function, $s(\Delta t)$, as a function of time is calculated from the following inverse Fourier transform (IFT):

$$s(\Delta t) = \text{IFT}[Z_b(f)^* Z_w(f)] \tag{16.5}$$

where $Z_b(f)^*$ is the complex conjugate of the Fourier transformed signal in bone and $Z_w(f)$ is the Fourier transformed signal in water. The peak value of this function is taken to be Δt and used to calculate a velocity from equation (16.1). The results for each bone are given in table 16.1 as is the phase velocity calculated at the centre frequency of the transducers.

16.5. DISCUSSION

The measurement of speed of sound in the frequency domain may be more robust than measurement in the time domain. In the time domain variations of up to 7% can occur due solely to selection of a zero-crossing point for measurement of Δt. This occurs as a result of frequency dependent attenuation and velocity dispersion. Using frequency domain measurements of c_p, c_g and c_{cc}, correlation was carried out with the corrected attenuation and BUA values given in table 16.1. For c_p and c_{cc}, the value of r^2 was greater than 90% for both attenuation terms, whereas for c_g these values were less than 30%. The phase velocity correlation would be expected from the well known Kramers–Kronig equation [15], but negative dispersion in the os calcis would not be predicted by this model. Since the group velocity is calculated from the dispersion of the phase velocity this may go some way towards explaining the lack of correlation.

A computer model was used to investigate the differences between c_p and c_{cc}. This model used a Gaussian broadband pulse centred on 750 kHz (the centre frequency of the transducers) as shown in figure 16.19. In the frequency domain, frequency dependent attenuation was introduced as a weight on amplitude, followed by velocity dispersion as a weight on phase. In each case transformation to the time domain was carried out. The results

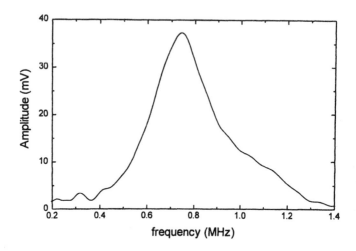

Figure 16.19.   *Received pulse spectrum after passage through water only.*

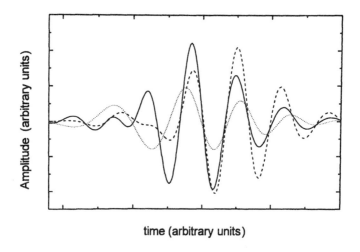

time (arbitrary units)

Figure 16.20.   *Simulated 750 kHz pulse (solid line), with $\mathrm{d}c_p/\mathrm{d}f = -84$ m s^{-1} MHz^{-1} (dashed line) and with $\mathrm{d}c_p/\mathrm{d}f = -84$ m s^{-1} MHz^{-1} and BUA = 80 dB MHz^{-1} (dotted line).*

are shown in figure 16.20. The mean pulse velocity and phase velocity are similar at the transducer centre frequency. Deviations between these velocities may be explained in terms of BUA and $\mathrm{d}c_p/\mathrm{d}f$ (dispersion) as follows:

- BUA $\neq 0$ and $\mathrm{d}c_p/\mathrm{d}f > 0 \Rightarrow\Rightarrow c_{cc} < c_p(f_{centre})$;
- BUA $\neq 0$ and $\mathrm{d}c_p/\mathrm{d}f < 0 \Rightarrow\Rightarrow c_{cc} > c_p(f_{centre})$;

but the deviations are always less than 2% and as such frequency-domain methods seem relatively robust compared to time domain methods.

More work is needed in the investigation of the relationship between these velocity measures and bone properties of clinical relevance. In particular the relationship between bone strength and the additional velocity parameters should be investigated as should the influence of pore size and other structural features on both velocity and attenuation. Although the relationship between BUA and bone structure has been accepted we feel that similar relationships between speed of sound and structural parameters may have been overlooked in the measurement of bone velocity using commercially available systems and could well repay further investigation.

ACKNOWLEDGMENT

RS acknowledges the financial support of the German Academic Exchange Transfer.

REFERENCES

[1] Barlow D H (ed) 1994 *Report of the Advisory Committee on Osteoporosis* (London: Department of Health)

[2] Vogel J M, Wasnich R D and Ross P D 1988 The clinical relevance of calcaneus bone mineral measurement: a review *Bone Mineral* **5** 35–58

[3] Hans D, Dargent-Molina P, Schott A M *et al* 1996 Ultrasonographic heel measurements to predict hip fracture in elderly women: The EPIDOS prospective study *Lancet* **348** 511–4

[4] Bauer D C, Gluer C C, Cauley J A *et al* 1997 Broadband ultrasound attenuation predicts fractures strongly and independently of densitometry in older women—a prospective study *Arch. Internal Med.* **157** 629–34

[5] Langton C M, Palmer S B and Porter R W 1984 The measurement of broadband ultrasonic attenuation in cancellous bone *Eng. Med.* **13** 89–91

[6] Clarke A J, Evans J A, Truscott J G *et al* 1994 A phantom for the quantitative ultrasound of trabecular bone *Phys. Med. Biol.* **39** 1677–87

[7] Aaron J E, Johnson D R, Kanis J A *et al* 1992 An advanced method for the analysis of bone structure *Comput. Biomed. Res.* **25** 1–16

[8] Tavakoli M B and Evans J A 1992 The effect of bone structure on ultrasonic attenuation and velocity *Ultrasonics* **30** 389–95

[9] Serpe L and Rho J Y 1996 The non-linear transition period of broadband ultrasound attenuation as bone density varies *J. Biomech.* **29** 963–6

[10] Han S, Rho J, Medige J and Ziv I 1996 Ultrasound velocity and broadband attenuation over a wide range of bone mineral density *Osteoporosis Int.* **6** 291–6

[11] Hodgskinson R, Njeh C F, Whitehead M A and Langton C M 1996 The non-linear relationship between BUA and porosity in cancellous bone *Phys. Med. Biol.* **41** 2411–20

[12] Pedersen P C and Ozcan H S 1986 Ultrasound properties of lung tissue and their measurements *Ultrasound Med. Biol.* **12** 483–99
[13] Nicholson P H F 1995 personal communication
[14] Barger J E 1979 Attenuation and dispersion of ultrasound in cancellous bone *Ultrasonic Tissue Characterisation* 2 ed M Linzer (Washington, DC: US Govt Printing Office) special publication **525** pp 197–201
[15] O'Donnell M, Jaynes E T and Miller J G 1978 General relationship between ultrasonic attenuation and dispersion *J. Acoust. Soc. Am.* **63** 1935–7
[16] Dempster D W 1992 Bone remodelling *Disorders of Bone and Mineral Metabolism* ed F L Coe and M J Favus (New York: Raven)

INDEX

Milton Keynes UK
Ingram Content Group UK Ltd.
UKHW021630071024
449327UK00020BA/1267